METHODS IN MOLECULAR BIOLOGY

Series Editor
John M. Walker
School of Life and Medical Sciences
University of Hertfordshire
Hatfield, Hertfordshire, AL10 9AB, UK

For further volumes:
http://www.springer.com/series/7651

High Content Screening

A Powerful Approach to Systems Cell Biology and Phenotypic Drug Discovery

Second Edition

Edited by

Paul A. Johnston

Department of Pharmaceutical Sciences, School of Pharmacy, University of Pittsburgh, Pittsburgh, PA, USA

Oscar J. Trask

Bahama Bio, LLC, Bahama, NC, USA;
Perkin Elmer Inc., Waltham, MA, USA

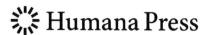

Editors
Paul A. Johnston
Department of Pharmaceutical Sciences
School of Pharmacy
University of Pittsburgh
Pittsburgh, PA, USA

Oscar J. Trask
Bahama Bio, LLC
Bahama, NC, USA

Perkin Elmer Inc.
Waltham, MA, USA

ISSN 1064-3745   ISSN 1940-6029   (electronic)
Methods in Molecular Biology
ISBN 978-1-4939-8461-9   ISBN 978-1-4939-7357-6   (eBook)
DOI 10.1007/978-1-4939-7357-6

This Humana Press imprint is published by Springer Nature
The registered company is Springer Science+Business Media LLC
The registered company address is: 233 Spring Street, New York, NY 10013, U.S.A.

Preface

The year 2017 marks the 20-year anniversary of the commercial launch of automated imaging platforms that were integrated with software to control both the acquisition and analysis of digital images of cells arrayed in multi-well microtiter plates and other substrates. Reagent development converged with the automation of cell handling procedures, cell staining methods, fluorescent microscopy, computer-assisted image analysis, and informatics technologies to enable the interrogation of complex cellular biology processes at throughputs and capacities compatible with drug discovery, developmental biology, and toxicology. In stark contrast to most high-throughput screening assays that have univariate readouts, automated cellular imaging platforms create multivariate data that extracts multiparametric quantitative data and information from digital images at the pixel level in an unbiased manner, a process that has been variously branded as high content screening (HCS), high content analysis (HCA), high content imaging (HCI), or image cytometry (IC). For example, in addition to specific target readouts such as protein expression levels, post-translational modifications (phosphorylation, acetylation, methylation, etc.), subcellular localization, and/or translocation events, HCI assays also provide data on literally hundreds or thousands of features that can be used to define and/or discriminate among cellular phenotypes. In recent years, there has been a resurgence in the implementation of phenotypic drug discovery strategies, and the ability of HCI assays to acquire multiplexed fluorescent images and extract quantitative multiparameter data at the individual cell and/or population levels has provided an ideal platform for phenotypic screening. Moreover, the field has expanded into many disciplines including environmental toxicology, developmental biology, and even preclinical studies that utilize a wide range of model systems such as histological preparations, live cells, and whole organisms.

Since its introduction in the late 1990s the field of HCS/HCA/HCI has continued to evolve and expand culminating with the founding of a grassroots organization called the Society for Biomolecular Imaging and Informatics (SBI^2) in 2012 (http://www.sbi2.org/) [1]. SBI^2 is an international community of leaders, scientists, and students promoting technological advancement, discovery, and education to quantitatively interrogate biological models to provide high context information at the cellular level. SBI^2 is dedicated to increasing the understanding, utility, and rigor of quantitative imaging or high content analysis [2]. In addition to SBI^{2}s annual scientific conference and exhibitor shows in the fall each year, the society has organized numerous regional meeting, developed and produced beginner and advanced level educational courses at these venues, and organized focused colloquia and special interest group discussions on selected HCS/HCA/HCI topics.

The HCS/HCA/HCI community continues to innovate and produce enhancements to reagents, instrumentation platforms with novel imaging modalities, improved image analysis algorithms, multivariate analysis and informatics tools, and the development and implementation of HCI assays for more complex and physiologically relevant cellular models. This volume is organized into four parts that highlight some of the recent trends in HCS/HCA/HCI. In Part I, the chapters describe methods that utilize reagents and kits that have been developed to measure cells and a variety of different cell phenotypes. In Part II, the chapters describe critical quality control measures including procedures and criteria for microplate selection, how to identify imaging artifacts in HCS data, and how to minimize the off-target effects of RNAi's. Also in Part II you will find overviews of the data handling issues associated with HCS/HCA/HCI, the benefits and challenges of live cell imaging assays, and a chapter describing methods to implement 1536-well ultra-high-throughput HCS assays. The emphasis in Part III is on methods for harnessing translocation events, signaling pathways, single cell analysis, and multidimensional data for phenotypic analysis. Part IV focuses on methods to implement more complex phenotypic models utilizing primary neurons, neural stem/progenitor cells, induced-pluripotent stem cells, multicellular tumor spheroids, endothelial/mesenchymal stem cell cocultures, and zebrafish embryos.

Pittsburgh, PA, USA   *Paul A. Johnston*
Bahama, NC, USA   *Oscar J. Trask*
Waltham, MA, USA

References

1. Trask OJ, Johnston PA (2014) Introduction and welcome to the Society of Biomolecular Imaging and Informatics (SBI^2). Assay Drug Dev Technol 12:369–374
2. Trask OJ, Johnston PA (2015) Standardization of high content imaging and informatics. Assay Drug Dev Technol 13:341–346

Contents

Contributors

ROBERT J. AGGELER • *Thermo Fisher Scientific, Eugene, OR, USA*

MARTYN BANKS • *Discovery and Optimization, Bristol-Myers Squibb, Wallingford, CT, USA*

MARC BICKLE • *High Throughput Technology Development Studio (HT-TDS), Max Planck Institute of Molecular Cell Biology and Genetics, Dresden, Germany*

JOHN L. BIXBY • *Miami Project to Cure Paralysis, University of Miami Miller School of Medicine, Miami, FL, USA*

MURRAY BLACKMORE • *Department of Biomedical Sciences, Marquette University, Milwaukee, WI, USA*

MARK-ANTHONY BRAY • *Novartis Institutes for BioMedical Research, Cambridge, MA, USA; Imaging Platform, Broad Institute of MIT and Harvard, Cambridge, MA, USA*

SEAN BURKE • *De Novo Software, Glendale, CA, USA*

DANIEL P. CAMARCO • *Department of Pharmaceutical Sciences, School of Pharmacy, University of Pittsburgh, Pittsburgh, PA, USA*

ANNE E. CARPENTER • *Imaging Platform, Broad Institute of MIT and Harvard, Cambridge, MA, USA*

KEVIN M. CHAMBERS • *Thermo Fisher Scientific, Eugene, OR, USA*

SUDHAKAR CHINTHARLAPALLI • *Discovery Oncology, Lilly Research Laboratories/Lilly Corporate Center, Eli Lilly and Company, Indianapolis, IN, USA*

SCOTT CLARKE • *Thermo Fisher Scientific, Eugene, OR, USA*

DAVID A. CLOSE • *Department of Pharmaceutical Sciences, University of Pittsburgh, Pittsburgh, PA, USA*

NORMAND CLOUTIER • *High Content Biology for Discovery IT, Bristol-Myers Squibb, Princeton, NJ, USA*

EVAN F. CROMWELL • *Protein Fluidics, Inc., Hayward, CA, USA*

NICK J. DOLMAN • *Thermo Fisher Scientific, Pittsburgh, PA, USA*

ROY EDWARD • *Biostatus Ltd, Leicestershire, UK*

LISA ELKIN • *Global Data Strategies and Solutions, Bristol-Myers Squibb, Wallingford, CT, USA*

M. EMMENLAUER • *University of Basel and SyBIT, Basel, Switzerland*

MILAN ESNER • *High Throughput Technology Development Studio (HT-TDS), Max Planck Institute of Molecular Cell Biology and Genetics, Dresden, Germany; Department of Histology and Embryology, Faculty of Medicine, Masaryk University, Brno, Czech Republic*

BEVERLY FALCON • *Discovery Oncology, Lilly Research Laboratories/Lilly Corporate Center, Eli Lilly and Company, Indianapolis, IN, USA*

VERENA FETZ • *Molecular and Cellular Oncology/ENT, University Medical Center, Mainz, Germany; Department of Chemical Biology, Helmholtz Institute of Infection Research, Braunschweig, Germany; Department of Experimental Toxicology and ZEBET, German Federal Institute for Risk Assessment (BfR), Berlin, Germany*

KAYA GHOSH • *De Novo Software, Glendale, CA, USA*

JENNIFER R. GRANDIS • *University of Pittsburgh Cancer Institute, Pittsburgh, PA, USA; Department of Otolaryngology, University of Pittsburgh, Pittsburgh, PA, USA; Department of Otolaryngology - Head and Neck Surgery, University of California, San Francisco, San Francisco, CA, USA*

STEVEN A. HANEY • *Cancer Biology and the Tumor Microenvironment, Discovery Oncology, Lilly Research Laboratories/Lilly Corporate Center, Eli Lilly and Company, Indianapolis, IN, USA*

JOSHUA A. HARRILL • *Center for Toxicology and Environmental Health, LLC, Little Rock, AR, USA*

YANHUA HU • *Bristol-Myers Squibb, Hopewell, NJ, USA*

YUN HUA • *Department of Pharmaceutical Sciences, School of Pharmacy, University of Pittsburgh, Pittsburgh, PA, USA*

MICHAEL S. JANES • *Thermo Fisher Scientific, Eugene, OR, USA*

PAUL A. JOHNSTON • *Department of Pharmaceutical Sciences, School of Pharmacy, University of Pittsburgh, Pittsburgh, PA, USA*

SHIRLEY K. KNAUER • *Institute for Molecular Biology, Centre for Medical Biotechnology (ZMB), Mainz Screening Center UG & Co. KG, University of Duisburg-Essen, Essen, Germany*

STANTON J. KOCHANEK • *Department of Pharmaceutical Sciences, University of Pittsburgh, Pittsburgh, PA, USA*

K. KOZAK • *Carl Gustav Carus University Hospital, Clinic for Neurology, Medical Faculty, Technical University Dresden, Dresden, Germany; Fraunhofer IWS, Dresden, Germany; Wroclaw University of Economics, Wroclaw, Poland*

JOHN S. LAZO • *Department of Pharmacology, University of Virginia, Charlottesville, VA, USA; Department of Chemistry, University of Virginia, Charlottesville, VA, USA*

VANCE P. LEMMON • *Miami Project to Cure Paralysis, University of Miami Miller School of Medicine, Miami, FL, USA*

O. LEVEN • *Screener Business Unit, Genedata AG, Basel, Switzerland*

BHASKAR S. MANDAVILLI • *Thermo Fisher Scientific, Eugene, OR, USA*

FELIX MEYENHOFER • *High Throughput Technology Development Studio (HT-TDS), Max Planck Institute of Molecular Cell Biology and Genetics, Dresden, Germany; Département de Médecine, Faculté des Sciences, University of Fribourg, Fribourg, Switzerland*

DARIO MOTTI • *Miami Project to Cure Paralysis, University of Miami Miller School of Medicine, Miami, FL, USA*

DEBRA NICKISCHER • *Primary Pharmacology Group, Pfizer Inc., Groton, CT, USA; Pfizer Inc., New Haven, CT, USA*

DAVID NOVO • *De Novo Software, Glendale, CA, USA*

JONATHAN O'CONNELL • *Biomolecular Screening, Forma Scientific, Watertown, MA, USA*

B. RINN • *Scientific IT Services, ETH Zürich, Zurich, Switzerland*

BRENT A. SAMSON • *Thermo Fisher Scientific, Pittsburgh, PA, USA*

MANUSH SAYDMOHAMMED • *Department of Developmental Biology, School of Medicine, University of Pittsburgh, Pittsburgh, PA, USA*

MALABIKA SEN • *Department of Otolaryngology, University of Pittsburgh, Pittsburgh, PA, USA*

FENG SHAN • *Department of Pharmaceutical Sciences, University of Pittsburgh, Pittsburgh, PA, USA*

Tong Ying Shun • *University of Pittsburgh Drug Discovery Institute, Pittsburgh, PA, USA*

Oksana Sirenko • *Molecular Devices, LLC, Sunnyvale, CA, USA*

Roland H. Stauber • *Molecular and Cellular Oncology/ENT, University Medical Center, Mainz, Germany*

Christopher J. Strock • *Cyprotex US, Watertown, MA, USA*

Michelle Swearingen • *Discovery Oncology, Lilly Research Laboratories/Lilly Corporate Center, Eli Lilly and Company, Indianapolis, IN, USA*

Oscar J. Trask • *Bahama Bio, LLC, Bahama, NC, USA; Perkin Elmer Inc., Waltham, MA, USA*

Michael Tsang • *Department of Developmental Biology, School of Medicine, University of Pittsburgh, Pittsburgh, PA, USA*

Mark Uhlik • *Translational Oncology, Biothera Pharmaceuticals, Eagan, MN, USA*

Judith Wardwell-Swanson • *InSphero, Inc., Brunswick, ME, USA*

Andrea Weston • *Primary Pharmacology Group, Pfizer Inc., Groton, CT, USA*

Michelle Yan • *Thermo Fisher Scientific, Eugene, OR, USA*

The original version of this chapter was revised. An erratum to this chapter can be found at DOI 10.1007/978-1-4939-7357-6_23

Part I

Reagents and Kits for Measuring Cells and their Phenotypes

Chapter 1

Applications and Caveats on the Utilization of DNA-Specific Probes in Cell-Based Assays

Roy Edward

Abstract

To perform cell-based assays using fluorescence as the readout there is a fundamental need to identify individual cellular objects. In the majority of cases this requires the addition of a DNA dye or so-called nuclear counterstain and these have become integral to assay design. End-point assays can use live or fixed cells and thus it is beneficial if such reagents are cell membrane-permeant.

Further, membrane-permeant DNA dyes can open new opportunities in dynamic real time assays with caveats according to the impact of their interaction with the chromatin in live cells. As cell-based assays offer information on the in vitro toxicity of treatments, cell viability has become a basic readout and cell membrane-impermeant fluorescent DNA-specific dyes can provide this information.

In the case of both nuclear counterstaining and viability reporting, it is beneficial if the DNA dyes employed are suitably spectrally separated to permit multi-color experimental design. Methods will be described for these two important assay readouts.

Key words Anthraquinones, DNA dyes, Fluorescent, Cell-based assays, High content screening, In-cell western, In vitro toxicology, siRNA, Cell health, Viability

---

1 Introduction

The established use of image-based, cellular assays demands reliable solutions for cellular compartment segmentation to track critical events—for example those reported by GFP fusion proteins within cell cycle control pathways, signaling pathway activation and protein translocation, and those associated with drug-induced toxicity such as mitochondrial membrane depolarization, plasma membrane permeabilization, and reactive oxygen species generation. In fluorescence-based end-point assay formats there is a general need to identify individual cells for automated image analysis. This can be achieved with the aid of a robust and readily identified constitutively expressed component of the cell (cytoskeletal or chromatin protein, for example) or more usually by the addition of a DNA-specific dye, a so-called nuclear counterstain. The choice

Paul A. Johnston and Oscar J. Trask (eds.), *High Content Screening: A Powerful Approach to Systems Cell Biology and Phenotypic Drug Discovery*, Methods in Molecular Biology, vol. 1683, DOI 10.1007/978-1-4939-7357-6_1,
© Springer Science+Business Media LLC 2018

of the latter is primarily based on its spectral fit with the other fluorescent signals being recorded and whether there is a requirement for cell permeability (i.e., live, unfixed cells) or not (i.e., fixed cells). These nuclear counterstains, many discovered as bi-products of drug discovery and used as bioprobe research tools, have widely varying properties: absorbance/emission profiles and maxima, water solubility, cell membrane permeability, brightness, fluorescence enhancement on binding, photostability, specificity, affinity, and toxicity. The combination of features offered by these bioprobe molecules becomes more critical as the variations and complexities of assay formats increase reflecting: an increasing number of parameters, the move to multiplexed treatments, the demand for live-cell end-point or increasingly real-time readouts, coping with dynamic live-cell assays, the introduction of coculture and 3-D systems including tumor spheroids and tissue histoids in attempts to re-iterate more closely critical aspects of in vivo cell behavior and microenvironment.

To explore the tradeoffs involved in selecting a suitable DNA bioprobe, a class of fluorescent DNA-targeting anthraquinone molecules is used as exemplars: namely, the live cell-permeant dyes DRAQ5 and CyTRAK Orange and live cell-impermeant DRAQ7 (Figs. 1 and 2). As red/far-red emitting dyes, they provide convenient fluorescent emission signatures that are spectrally separated from the majority of commonly used reporter proteins (e.g., eGFP, YFP, mRFP), and a wide range of fluorescent tags such as AlexaFluor 488, Fluorescein and Cy2, and fluorescent functional probes used to report cell health status or demark organelles. In

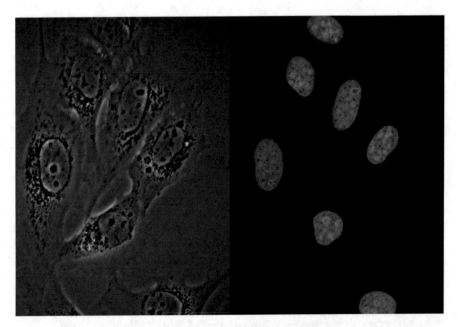

Fig. 1 Transmission (*L*) and fluorescence (*R*) microscopy of U-2 OS cells stained with DRAQ5™

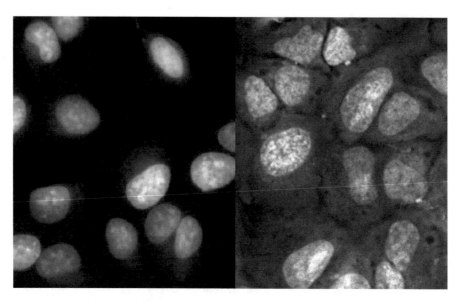

Fig. 2 U-2 OS cells (live—*left* and fixed with formaldehyde—*right*) stained with CyTRAK Orange™ to show segmentation on the basis of a differential intensity between nucleus and cytoplasm

addition, they are not excited by UV wavelengths thus avoiding the issue of pharmacophore UV-autofluorescence, and the potential photodamage from UV light sources in live-cell imaging applications. Conversely, their preferential red excitation reduces interference by biological sample autofluorescence. High water solubility and high affinity DNA binding properties provide a convenient method for the stoichiometric labeling of cell nuclei in live cells without the aid of DMSO that can equally be applied to fixed cells. Powerfully, they permit the simultaneous and differential labeling of both nuclear and cytoplasmic compartments in live and fixed cells to clearly render the precise location of cell boundaries that may be beneficial for quantitative expression measurements, cell-cell interactions, and most recently compound in vitro toxicology testing. In one case, DRAQ7™, the core structure, has been chemically modified to render it impermeable to the intact-cell-membranes of living cells. This far-red viability dye can be more widely combined with other fluorescent reporters to reveal temporally separated events and exhibits negligible cytotoxicity thereby enabling long-term, real-time experiments.

Advances in the sensitivity and spatial resolution of light microscopy have permitted the development of cell-based assays where any change is observed as the altered intensity, re-distribution or translocation of a fluorescently tagged protein species (GFP mutants, most commonly with eGFP). Examples of these include Cyclins, Rac-1, β-arrestin, MAP Kinase, NF-κB, and transcription factors such as FKHR. Separate developments in automation and image analysis software tools have further allowed these assays to be performed at increased throughputs so they can be contemplated for primary target and increasingly phenotypic screening of compound and siRNA libraries that may contain 10^5-10^6 members. A number of significant challenges arise from the combination of these features to allow development of robust assays that can be transferred to routine use. Given the normal and often wide distribution of cellular responses to a stimulus evident even in commonly used cell culture lines (such as CHO, HeLa, U-2 OS, SU-DHL-4, or PC-12), it is statistically beneficial to identify a representative number of individual cellular objects of interest in a field of view. This permits automated object identification, determination of well-to-well variations, anomalies such as plate "edge" effects, and generalized cytotoxicity. Proteins may have promiscuous expression in more than one compartment of the cell (e.g., between the cytoplasm and nucleus) that may make the establishment of baseline or threshold expression more difficult. In this case the use of a nuclear counterstain can aid the segmentation of nucleus from cytoplasm allowing the measurement of cytoplasmic (or, for example, mitochondrial) signals to the exclusion of nuclear area and vice versa. Further, such a fluorescence signal needs to be spectrally separated from that of the protein or functional "reporter" by excitation and/or emission, and the requirements for this will depend on the configuration of the imaging platform employed. For instance, although the semi-permeant DNA dye DAPI has an emission signature that somewhat overlaps with GFP or the commonly used AlexaFluor 488, FITC, or Cy2 fluorophores; it is excited by UV wavelength light while that of eGFP is from the blue region of the spectrum (typically 488 nm). Thus, sequential excitation permits generation of the desired and subsequently merged image but in most instances this unfortunately may require two scans of the cells and double image acquisition times. More concerning, however, is the recent description of the photoconversion of DAPI (and live cell-permeant dyes Hoechst 33342 and Vybrant DyeCycle Violet) [1, 2] resulting in spectral overlap with GFP upon any subsequently excitation. Unwanted FRET-like quenching of Hoechst 33342 upon increasing concentrations of a commonly used reference anti-cancer compound Doxorubicin, due to the close proximity of the emission maximum of Hoechst (458 nm) and excitation maximum of Doxorubicin (473 nm), was resolved by the use of the far-red counterstain DRAQ5, due to its spectral separation [3]. Alternatively, eGFP and the non-permeant nucleic acid propidium iodide (PI) can be co-excited with light at 488 nm while their emission signals can be separately detected by selection of appropriate filters. In reality, however, PI is not suited as a counterstain as the degree of RNA staining is such that the nucleus is not distinguishable in interphase cells, even following cumbersome permeabilization and RNase treatment [4]. An ideal counterstain would preferably limit the possibility of

FRET-like excitation of the reporter protein's fluorescent tag such that it would not quench, interfere with, or compete with the tag's signal. Importantly, the spectral properties of the assay and the counterstain should permit transferability of a method from a development platform to a high-throughput platform.

The advantage of the far-red fluorescent probe DRAQ5™ [5] is it has high specificity for dsDNA, strongly binding at $(dA\text{-}dT)_2$ sequences providing nuclear-specific staining while not being detectable on mitochondrial DNA [6]; its peak absorbance (excitation) is at 646 nm and emission at 697 nm (DNA-bound) yet it is detectable into the far-red region separating it both by excitation and/or emission from the majority of visible range fluorophores and most protein reporter tags, including CFP, GFP, YFP and RFP, as well as the antibody labels FITC, PE, PE-Texas Red and analogues allowing rapid, parallel image acquisition for up to four parameters; there is no evidence that DRAQ5™ (photo-) chemically interferes with the signal from another fluorophore by quenching or FRET-like mechanisms; and DRAQ5™ has been commonly used in cell-based assays and multi-color flow cytometry and is widely applicable to fluorescence microscopy [7] and HCS platforms [8] since most are equipped with a HeNe laser (635 nm) and a far-red detection channel (nominally associated with the Cy5 fluorescent label). Additionally, with a red-excited live cell-permeant counterstain the excitation wavelength light penetrates deeper and with less scatter into tissue sections than for the previously described UV excitation (for dyes such as Hoechst 33342 and DAPI) while the emitted light moreover, further red-shifted, similarly benefits from this effect [9]. When a suitable nuclear counterstain is utilized, image analysis software can then be applied to segment the nucleus based on optimized staining. Most algorithms available today "dilate" radially from the nuclear staining segmentation provide an assumed cytoplasmic zone around the nucleus. However, such an approach often requires a conservative perinuclear zone to be selected within the cytoplasm that can be measured for an expected change resulting from perturbation, stimulus, or insult. At the very least this can affect the statistical value of the measurement or may under-report the actual expression level within the region of the cytoplasm, and at worst may render the cellular data-point invalid if the area chosen for measurement overlaps a large subcompartment such as a mitochondrion or where the cytoplasmic compartment is significantly heterogeneous in shape, size, or volume across the cell population. This is less of an issue where there is expression of a cytoplasmic signal that can be used to delineate the compartment but this requires either the use of a detection threshold or other parameters that might limit the practical dynamic range between the reference or basal levels. This is clearly unwelcome where the stimulus applied to the cells causes significant changes in cytoplasmic distribution or "texture," for example via different endocytotic pathways [10]. One possible solution is the combination of probes to demark the two compartments [11], but the challenge is finding two probes that have similar spectral properties (avoiding occlusion of too much of the available spectrum), require little titration between different cell types, and can be used in either live or fixed cells. Further, the addition of another molecule into the analytical complex raises the risk of chemical interference, quenching and limits cross-assay and cross-platform compatibility. It would be advantageous, then, to have a nuclear counterstain that operates with a single spectral profile, separated from the common visible fluorescent reporters and antibody-/FISH probe-tags, yet which has a useful and differential staining intensity for and between the nucleus and the cytoplasm. This would perhaps be achieved where the DNA binding dye also exhibits a low level of promiscuity in binding to dsRNA. Excitation by a red wavelength (e.g., 633 or 647 nm) is preferable as this limits the background signal contribution due to biofluorescence and avoids local heating that might be encountered when detecting weak signals with a UV laser, typically with a much higher energy output. Likewise, a nuclear counterstain with high chemical and photo-stability (low photobleaching) would be advantageous. Again, DRAQ5™ meets these needs. There is a demonstrable and appropriate level of cytoplasmic staining to allow segmentation of that compartment in addition to the much brighter staining nucleus. It is a red-excited dye, well separated from the visible range fluors. DRAQ5™ shows excellent chemical and photo-stability in aqueous solution at ambient temperatures and very low photo-bleaching compared to several other commonly used fluorescent molecules [4]. In fact, more than double the number of laser scans were possible on DRAQ5™ than for GFP-Histone H2B expressed in a HeLa cell line. This becomes an important factor where the reporter signal is already weak, requiring extended exposures, or in FRAP experiments.

Cell-based high content screening experiments can be broadly separated into two types: with either fixed or dynamic endpoints. Where the change is temporally short-lived or unstable it may be beneficial to fix cells to allow imaging to take place independent of the end-point. However, where the reporter molecule leaches from one compartment to another [12] or even out of the cell there is then a need for a live-cell membrane-permeant counterstain. Such a requirement would also hold for dynamic end-point assays such as those used for a Rac-1 translocation assay system [13] where perhaps a wide natural variation in the unstimulated control setting might need to be accounted for. Equally, it might be that a cell-based assay moves from a dynamic end-point to a fixed end-point when implemented in high-throughput screening. Thus, for the ease of assay development the ideal agent would perform equally well on both fixed and live-cell preparations with minimum optimization and function in normal physiological buffers, culture media and, if possible, in pre-fixation, post-fixation, and "in-fixative" modes. It is essential that a dye used on live cells labels in a temporally stable manner since the need for a timed addition of that component to an assay procedure would be an unwelcomed complication, particularly when reading a 1536-well microtiter plate. In essence, this would require a dye probe with high avidity/affinity for the target DNA such that it does not bleed into other compartments rendering it "invisible" to ABCG2 pumps most often associated with drug clearance from cells (especially the case for multidrug resistant phenotypes in primary cancer cells or tumor cell lines) unlike the UV excitable, live-cell DNA dye Hoechst 33342 which is often used as a reference for efflux of other drugs [14] notwithstanding its unhelpful emission overlap with GFP/FITC. At a practical level, an agent that is supplied ready-to-use (i.e., water-soluble) and which does not require a washing step would be beneficial in automation and the streamlining of experimental design. Last, the staining should be at a satisfactory level in a matter of minutes.

DRAQ5™ is highly lipophilic and readily crosses cell and nuclear membranes in intact live cells and tissues to bind to DNA [15]. It has also been widely used on fixed tissues and cells, prepared with a variety of fixatives as well as prior to fixation and, usefully, in the commonly used fixative formaldehyde [16] thereby removing a pipetting and washing step, especially valuable in large compound or RNAi knockdown screening campaigns. Alternatively, DRAQ5 has been added to commonly used glycerol-based fluorescence microscopy mountants [17]. DRAQ5 exhibits excellent temporal binding stability in live cells over many hours [4, 18] and is not noticeably effluxed by cells that have either overexpression of ABCG2 [14] or demonstrable (multi-)drug resistance. Cell staining (live or fixed) with DRAQ5™ is effected in a few minutes rapidly reaching a stable equilibrium that so favors DNA that washing is not required since the low concentration of dye in the surrounding medium is typically below the threshold of detection. Despite being lipophilic, DRAQ5™ is readily soluble in an aqueous solution, is commercially supplied thus, and, importantly, is compatible with buffers such as PBS and cell culture media. Since DRAQ5™ labeling is stoichiometric to DNA [4], it is possible to take further quantitative advantage of the segmentation of the signal in the nucleus to report DNA content (cell cycle position) in any individual cell. Doing this for every cell permits cell cycle analysis for the whole field in view. This can be important in the study of anti-cancer agents, inflammatory responses or where the cell population response to a stimulus is heterogeneous. This has been demonstrated on a plate-based cytometer [19] and on a high content imaging platform [20]. It is preferable if staining is possible in both fixed and live cells as this will aid flexibility of the protocol.

This stoichiometric labeling by a DNA probe can also be applied to so-called wellular analysis. The total DNA signal from a field of diploid cells is directly proportional to the number of cells and can be used to normalize the labeled antibody signal, thereby taking into account well-to-well variations in cell numbers. DRAQ5™ has been robustly applied for "in-cell western" experiments [21, 22] showing the excellent linearity of DRAQ5™ dye signal to cell number. Interestingly, in studies on the impact of baculovirus transduction on chromatin distribution in HepG2 cells DRAQ5™ showed chromatin dispersion due to viral transduction, confirmed by monitoring histones [23] again supporting DRAQ5™ as a reliable reporter of DNA/chromatin in living eukaryotic cells.

A further challenge for cell-based assays is to show the full extent of the cytoplasmic envelope while retaining the ability to differentially segment the cytoplasm and nucleus. This would be very useful, for example, where one of the translocation termini was the cytoplasmic membrane. To meet this requirement, CyTRAK Orange™ was developed [24] as a derivative of DRAQ5™. It has significantly more promiscuous labeling of the cytoplasm than DRAQ5™ permitting segmentation to the cytoplasmic membrane. Like DRAQ5™, CyTRAK Orange™ is membrane-permeant but loses the ability to report DNA content in intact cells. Its spectral profile (Exλ_{max} 520/Emλ_{max} 610) allows co-excitation with GFP and detection in separate channels [25]. This makes it an ideal counterstain for early live-cell epifluorescent microscopes as these often do not allow far-red detection. CyTRAK Orange™ is not red-excited and therefore can be an alternative to DRAQ5™ when the far-red channel is required for another parameter (e.g., an APC-Cy7 tagged antibody, Mitotracker Deep Red FM, or a Bodipy-coupled GPCR ligand). CyTRAK Orange™ has been used in a live-cell end-point, four-color assay combined with AlexaFluor 350, GFP, and AlexaFluor 647 [26]. Both DRAQ5™ and CyTRAK Orange™ can also be combined with "nonspecific" cell labeling dyes. For instance, CyTRAK Orange™ has been combined with Whole Cell Stain Green (unpublished) to segment nuclei and cytoplasm. More recently, DRAQ5™ has been combined with nonyl acridine orange (NAO) to give patterning of cells for a machine learning approach to permit operator-independent inter-pretation of the effect of a compound or RNAi treatment on cells [27] as one approach to phenotypic screening.

High content screening allows early warning of drug toxicity and, even at its simplest, a reduction in the number of counterstain-labeled nuclei in a field relative to negative controls can infer a compound having catastrophic effects on cell viability and/or adherence. In a method to identify AKT-signaling pathway inhibi-tors [18] DRAQ5™ was used to report both relative cell numbers between a mixture of isogenic cells genetically engineered to be sensitive/insensitive to specific inhibition and unwanted generalized cytotoxicity which results in decrease of both subpopulations. Similarly, toxic compounds have been screened out based on reduced absolute cell numbers (due to apoptosis or other mechanisms leading to cell death) but also monitored for nuclear condensation by increased nuclear DRAQ5™ fluorescence intensity, inferring different modes of toxicity [28]. Likewise, it is possible to detect micronuclei [29] or nuclear fragmentation. A nuclear counterstain with a secondary cytoplasmic signal as described above offers further morphometric indications of cellular toxicity as one can now measure changes in shape/area of nuclear and cytoplasmic envelopes. Such an approach has been used to evaluate the cytotoxicity of anti-viral compounds against host/target cells in an HIV infection inhibitor assay [30] and in a HCV replication inhibitor assay [31]. In these, a combination of features such as degree of change of nuclear roundness, cell area, and cell roundness were found to be significant. The measurement of each of these parameters relied solely on the staining properties of DRAQ5™. Toxicity studies have been further advanced by a four-color, live-cell assay combining three functional probes: monochlorbimane, detecting glutathione (GSH); H_2DCFDA—reactive oxygen species (ROS); and TMRM—mitochondrial membrane potential; with DRAQ5™—cell events, enumeration, nuclear condensation, and lipid accumulation [32]. On primary human hepatocytes the data accurately classified idiosyncratic hepatotoxicants.

As mentioned above, one of the most obvious features of apoptosis or of a compound's toxicity is the loss of membrane permeability, leading to cell death. This can most easily be detected by the addition of a membrane-impermeant DNA dye. Commonly, PI (Em max 617 nm) or TOTO-3 (Em max 660 nm) have been used for this purpose. DRAQ7™ is a spectrally improved membrane-impermeant DNA probe based on the DRAQ5™ chromophore [33–35]. DRAQ7™ has identical spectral properties to DRAQ5™ avoiding overlap with most visible-range fluors and RNA binding is similarly very weak (undetectable by flow cytometry). However, it has been modified such that it does not cross the membrane of viable cells but rapidly enters "leaky" cells and labels nuclear DNA. DRAQ7™, therefore, can be used as an indicator of cell membrane permeabilization, apoptosis, necrosis, and dead cells. As such, the advantages of DRAQ7™ over agents such as PI and TOTO-3 are that it may provide additional information. Unlike PI, DRAQ7™ is highly DNA-specific avoiding false positives due to cytoplasmic RNA staining of viable cells [15, 36]. Some capability for formaldehyde stabilization of DRAQ7™ staining has been shown to permit batch analysis overnight (G. Griffiths, Imagen Biotech, personal communication), while DRAQ7™ can be combined with cell-permeant DNA-binding dyes such as Hoechst 33342 [33] and CyTRAK Orange™ (unpublished data). Far-red

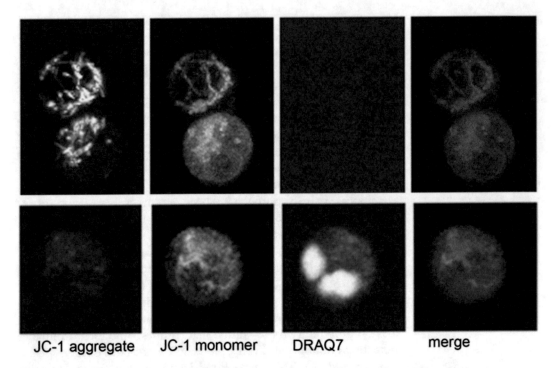

JC-1 aggregate   JC-1 monomer   DRAQ7   merge

Fig. 3 Cells co-labeled with mitochondrial membrane potential probe JC-1 and DRAQ7™ showing from *top*—healthy cell, cell with impaired mitochondrial health, and cell with compromised membrane and failed mitochondria

emission allows combinations with cell health probes, many of which fluoresce in the orange/red region of the spectrum and are otherwise occluded by PI. Events such as loss of mitochondrial-membrane potential (MMP) detected with the probe JC-1 can thus be temporally related to membrane-leakiness with DRAQ7™ [37] (Fig. 3). It can be applied as a viability reporter in RNAi knock-down assays, in mitotic index assays, and in studies into cell health in response to insults. Uniquely, it can be used as a reporter of cell damage in dynamic, real-time viability/cytotoxicity assays [34] in which it has been partnered with the MMP probe TMRM. The suitability of DRAQ7™ for this application was confirmed by the absence of DNA-probe interactions in intact cells as determined by highly sensitive bioassays for DNA damage response as indicated by H2AX and ATM kinase phosphorylation [34]. In addition to these assays, there is a continued move toward live-cell imaging (including time-lapse), coculture, and 3D (such as organoids, micro-tissues, tumor spheroids) with the goal of more truly reflecting in vivo responses. Beyond counterstaining for end-point analysis, one needs to consider carefully the use of cell-permeant DNA intercalating dyes in dynamic assays since these will likely have an impact on cell proliferation, resulting in cell cycle arrest (Hoechst 33342 [38], DRAQ5 [4]) or energy-dependent drug clearance or impaired motility [39] which may result from UV-induced photo-toxicity. Paradoxically, there are examples where cell-permeant DNA dyes appear not to have a noticeable impact on cell motility [39] or cell function (CyTRAK Orange [40]). The need for testing of suitability to dynamic cell analysis will hold true for any cell-permeant agent including functional and compartment-specific dyes. As described earlier, for the monitoring of viability in dynamic assays, DRAQ7 has been demonstrated to have ideal characteristics.

2   Materials

Phosphate-buffered saline (PBS) solution, culture or assay medium as required.

1%, 4%, 8% formaldehyde.

Nuclear counterstain: DRAQ5 or CyTRAK Orange (5 mM), DRAQ7 (0.3 mM).

95% Ethanol, Triton X-100, Tween-20, Blocking solution, Primary and labeled secondary antibodies.

Annexin V/pSIVA™, JC-1, TMRM, H₂DC-FDA, mono-chlorbimane, FLICA® caspase substrates, Whole Cell Stain™ Green.

Microscopy mountant, e.g., Vectashield™, CyGEL™, Fluoro-mount-G™, ProLong Gold™.

3   Methods

3.1   Nuc:Cyto Counterstaining in Live Cells

1. Prepare cells for staining with the counterstain (*see* **Notes 1–3**).

2. Add counterstain (DRAQ5™/CyTRAK Orange™) directly at a final concentration of 5 μM (1:1000) in PBS, culture, or assay medium (*see* **Note 4**). This will be as an overlay for adherent cells/tissue sections, added to the chamber/well liquid directly in fresh medium following a wash step (*see* **Notes 5** and **6**).

3. Gently mix and then incubate for 10–30 min at room temperature (*see* **Notes 7** and **8**).

4. Cells can be analyzed directly without further treatment or washing, preferably within 2 h.

(*See* **Notes 9–12**).

3.2   Nuc:Cyto Counterstaining in Fixed Cells After Applying Other External Stains and Antibodies

(*See* **Notes 1–3** and **13**).

1. Remove culture medium from the cells.

2. Apply 4% formaldehyde fixative and incubate for 30 min at room temperature. Protect from light if the cells are expressing a fluorescent protein tag such as GFP.

3. Permeabilization may be performed at this stage as required.

4. Aspirate, wash, and stain cells with fluorescent antibodies and/or functional dyes. Incubate for the appropriate time and temperature in the dark.

5. Aspirate the antibody solution and wash the cells with PBS, and aspirate.

Separate counterstaining and mounting:

6. Apply counterstain (DRAQ5™/CyTRAK Orange™) directly at a final concentration of 5 μM in PBS (*see* **Note 7**). Incubate for 20–30 min at room temperature, protecting from light (*see* **Note 8**).

7. Aspirate liquid from the well/slide. Do not wash. Overlay with mountant.

Combined counterstaining and mounting:

6. Premix an aliquot of microscopy mountant (e.g., Fluoromount-G™, ProLong Gold™) with counterstain at 5 μM.

7. Apply admixed mountant/counterstain to the sample. Apply coverslip and incubate for 20–30 min. Store in the dark until imaging performed.

3.3 Nuc:Cyto Counterstaining Fixed Cells Expressing a Fluorescent Protein Reporter

1. Prepare a 10 μM solution of counterstain (DRAQ5/CyTRAK Orange; i.e., 1:500).

2. Thoroughly mix equal volumes of 8% Formaldehyde solution and 10 μM counterstain.

3. Remove culture medium from the cells.

4. Apply the fix and stain solution to the sample and incubate for 30 min at room temperature (*see* **Note 5**). Protect from light.

5. Aspirate and wash the cells with PBS.

6. Overlay with mountant and coverslip if required. Store in the dark until imaging is performed.

3.4 DRAQ5™ DNA Content Measurement by Imaging (with DRAQ5™)

(*See* **Note 14**).

1. After all other treatments or staining steps are complete aspirate and wash cells (live or fixed).

2. Stain cells with DRAQ5 at a final concentration of 10–20 μM. Incubate for 30 min at 37 °C. Protect from light.

3. Do not wash cells, simply aspirate excess liquid. Apply mountant as required. If cells are live (or otherwise unfixed) measure within 2 h of staining (*see* **Notes 15** and **16**).

3.5 Cell Enumeration in "In-Cell Westerns" (After 21)

1. Seed cells at 5–6000 cells per 96-well microtiter plate and treat with inhibitors, etc.

2. Cells are fixed by the addition of 0.5 volumes of 12% formaldehyde for a final concentration of 4%. Incubate for 1 h at room temperature.

3. Wash the cells with PBS. Repeat twice. Aspirate.

4. Permeabilize the cells with PBS/0.1% Triton X-100. Repeat 2×. Aspirate.

5. Apply the blocking solution and incubate for 2 h at room temperature. Aspirate.

6. Apply the unlabeled primary antibody, in blocking solution. Incubate for 2 h. Aspirate.

7. Wash with PBS/0.1% Tween-20. Repeat twice. Aspirate.

8. Dilute labeled secondary antibody as required with DRAQ5™ to 5 µM in PBS/0.5% Tween-20. Incubate for 1 h at room temperature.

9. Wash with PBS/0.1% Tween-20. Repeat twice. Aspirate.

10. Incubate in PBS and read wells.

The antibody signal for each well is corrected according to the corresponding DRAQ5 signal.

3.6 Identification of Membrane-Compromised Cells

(*Viability testing in real-time, dynamic cell-based assays and high content screening*) (*see* **Notes 17** and **18**).

1. Add DRAQ7™ directly to the cell culture medium in each well: diluted to a final concentration of 1–3 µM (i.e., 1:300–1:100). Incubate at 37 °C.

2. Image the wells without further treatment. With suspension cells—remove cell aliquots as required and analyze for far-red (>665 nm) fluorescing cells relative to negative/positive controls by flow cytometry (*see* **Notes 12** and **13**).

3. To further enumerate the *total* cell population after the last live-cell image:

   (a) Stain the cells with CyTRAK Orange. Prepare a solution of live cell-permeant counterstain CyTRAK Orange at 10 µM. Image at the appropriate wavelength (*see* **Notes 12–14**) and merge the DRAQ7 and CyTRAK Orange images, or:

   (b) Stain the cells with DRAQ5. Prepare a solution of live cell-permeant counterstain DRAQ5 at 5 µM/10–20 µM for DNA content analysis. Image at the appropriate wavelength (*see* **Notes 12–14**), using false color to properly visualize differential staining of the DRAQ7 and DRAQ5 images and to merge for overlap or colocalization using the appropriate acquisition and image analysis algorithm.

4 Notes

1. For suspension cells, resuspend cells in appropriate buffer such as PBS at a concentration of $<4 \times 10^5$/ml in a test tube. Cells will need to be tested for the ability to settle onto the slide surface or may require cytospin preparation by centrifugation. For adherent cells estimate the number of cells based on confluence level or tissue section dimensions.

2. Typical counterstain volume requirements: slides: 200 µl per coverslip; microtiter plates: 96-well—100 µl; 384-well—30 µl; 1536-well—10 µl.

3. It is important to consider the combinations of fluorochromes and instrument filters for the experiment.

4. For DNA content measurement with DRAQ5, it may be necessary to increase the staining concentration of DRAQ5 to 10–20 µM, to the higher end of this range in live cells. If possible, measure the DRAQ5 signal above 700 nm for the best G1 peak c.v. Segmentation of the nuclear compartment will provide the total DRAQ5 signal equating to the DNA content for each cell, which can then be plotted for all cells measured in that well to give a cell cycle DNA profile.

5. DRAQ5™ is usually added as the last stain in a labeling procedure since no washing is required to remove excess dye.

6. For time-lapsed assays in live cells (e.g., studying translocation of a GFP tagged protein) the counterstain may be added to the assay medium for the duration of the assay (typically 0.5–1 h) at 2–5 µM prior to any agonist/antagonist addition.

7. Counterstaining is accelerated in mammalian cells at 37 °C and may be reduced to 1–3 min.

8. Protect from the light during incubation periods if other (immuno-) fluorescent stains have been applied to the cells prior to the counterstaining, which may otherwise suffer photo-bleaching.

9. For imaging, DRAQ5™ has been successfully excited by wavelengths from 561 nm, and up to 647 nm (Exλ_{max} 646 nm). Typically, however, for cell imaging, excitation is performed with either 633 or 647 nm wavelengths. Using flow cytometry, DRAQ5 can be detected using 488 nm excitation. DRAQ7™ can also be excited following the settings for DRAQ5™. CyTRAK Orange™ may be excited by a wavelength range from 488 nm up to 540 nm (Exλ_{max} 520 nm). 488 nm excitation offers the most simple and optimal wavelength and additionally it is available on most imaging instruments. The key benefit of CyTRAK Orange™ is that is can be co-excited with eGFP or dyes with equivalent spectra to FITC, and the

emission profile enables it to be robustly separated from either of these fluorophores and it is not excited by red light (e.g., at 635 or 647 nm).

10. DRAQ5™ emission peaks at 697 nm (intercalated to dsDNA) and can be usefully detected from 670 nm, i.e., beyond a dichroic filter used to cut off the red excitation light. Suitable filters include "Cy5," 695LP, 715LP, or 780 LP. DRAQ5™ has no spectral emission overlap with FITC/GFP or R-PE and many other fluorescing proteins allowing image acquisition in one scan. DRAQ7™ can be detected following the same settings as DRAQ5™. CyTRAK Orange™ emission starts at 580 nm (Emλ_{max} 615 nm intercalated to dsDNA). See emission spectra for CyTRAK Orange™. Suitable filters include 590LP; or a 630/60 nm band-pass or 615/50 nm band-pass for imaging with GFP or FITC fluorophores.

11. To avoid confusion it may be advisable to false color DRAQ5/DRAQ7 images in red and CyTRAK Orange in orange, consistent with their respective peak emission wavelengths; or other suitable representative colors that demonstrate the contrast labeling of the fluorophores.

12. It is important to understand the topology of the adhered cells, especially with confocal imaging systems, if the nuclei and cytoplasm of cells are not in the same focal plane which may impact the automation of the acquisition and analysis.

13. Where fixation is required or preferred then the choice of fixation and permeabilization should be carefully evaluated since solvent-based methods favor DNA analysis while cross-linking methods generally favor proteins. In this context DRAQ5™ shows excellent stoichiometry with DNA content in both live and fixed cells.

14. It is generally preferable to treat cells for DNA measurement with ethanol fixation as this aids accessibility of DNA for the intercalating dye by releasing DNA binding proteins but may be detrimental to protein detection and must be investigated.

15. If possible, measure the DRAQ5 signal above 700 nm for the best G1 peak c.v.

16. Segmentation of the nuclear compartment will provide the total DRAQ5 signal equating to the DNA content for each cell, which can then be plotted for all cells measured in that well to give a cell cycle DNA profile.

17. DRAQ7™ has been shown not to have any effect on the proliferation rate of cells in long-term culture assays. DRAQ7™ can then be used to report, in real time, the cytotoxic or apoptotic effect of a specific treatment on cells, e.g.,

pharmacological agent, RNAi, virus, antibody-dependent complement-mediated killing, in vitro toxicity.

18. DRAQ5 can be used as a 100% positive control for DRAQ7 as they share the identical DNA target and spectral properties.

Acknowledgments

The author would like to thank Professor Laurence H Patterson, Institute of Cancer Therapeutics, University of Bradford; and Professor Paul J Smith and Professor Rachel J Errington, Institute of Cancer and Genetics, School of Medicine, Cardiff University; and Mr. Stefan Ogrodzinski, Biostatus Ltd., Shepshed, UK for their essential assistance in the preparation of this article.

References

1. Jez M, Bas T, Veber M et al (2013) The hazards of DAPI photoconversion: effects of dye, mounting media and fixative, and how to minimize the problem. Histochem Cell Biol 139:195–204

2. Zurek-Biesiada D, Kedracka-Krok S, Dobrucki JW (2013) UV-activated conversion of Hoechst 33258, DAPI, and Vybrant DyeCycle fluorescent dyes into blue-excited, green-emitting protonated forms. Cytometry A 83:441–451

3. Antczak C, Takagi T, Ramirez CN et al (2009) Live-cell imaging of caspase activation for high-content screening. J Biomol Screen 14:956–969

4. Martin RM, Leonhardt H, Cardoso MC (2005) DNA labeling in living cells. Cytometry A 67:45–52

5. Smith PJ, Blunt N, Wiltshire M et al (2000) Characteristics of a novel deep red/infrared fluorescent cell-permeant DNA probe, DRAQ5, in intact human cells analyzed by flow cytometry, confocal and multiphoton microscopy. Cytometry 40:280–291

6. van Zandvoort MA, de Grauw CJ, Gerritsen HC et al (2002) Discrimination of DNA and RNA in cells by a vital fluorescent probe: lifetime imaging of SYTO13 in healthy and apoptotic cells. Cytometry 47:226–235

7. Visconti RP, Ebihara Y, LaRue AC et al (2006) An in vivo analysis of hematopoietic stem cell potential: hematopoietic origin of cardiac valve interstitial cells. Circ Res 98:690–696

8. Loechel F, Bjorn S, Linde V et al (2007) High content translocation assays for pathway profiling. Methods Mol Biol (Clifton, NJ) 356:401–414

9. Helmchen F, Denk W (2005) Deep tissue two-photon microscopy. Nat Methods 2:932–940

10. Pelkmans L, Fava E, Grabner H et al (2005) Genome-wide analysis of human kinases in clathrin- and caveolae/raft-mediated endocytosis. Nature 436:78–86

11. Cogger VC, Arias IM, Warren A et al (2008) The response of fenestrations, actin, and caveolin-1 to vascular endothelial growth factor in SK Hep1 cells. Am J Physiol Gastrointest Liver Physiol 295:G137–g145

12. Foley KF, De Frutos S, Laskovski KE et al (2005) Culture conditions influence uptake and intracellular localization of the membrane permeable cGMP-dependent protein kinase inhibitor DT-2. Front Biosci 10:1302–1312

13. Amersham Biosciences (2003) GFP-Rac1 Assay. User manual, section 5.2.7. 25–8007-27UM, Rev A, 2003

14. García-Escarp M, Martínez MV, Barquinero J et al (2003) Characterization of the functional activity of the ABCG2 transporter for the development of new expression vectors with in vivo selectable potential. Paper presented at the II Reunión de la Sociedad Española de Terapia Génica, Barcelona, Spain

15. Rieger AM, Hall BE, Luong le T et al (2010) Conventional apoptosis assays using propidium iodide generate a significant number of false positives that prevent accurate assessment of cell death. J Immunol Methods 358:81–92

16. Haasen D, Wolff M, Valler MJ et al (2006) Comparison of G-protein coupled receptor

desensitization-related beta-arrestin redistribution using confocal and non-confocal imaging. Comb Chem High Throughput Screen 9:37–47

17. Reznichenko L, Amit T, Youdim MB et al (2005) Green tea polyphenol (−)-epigallocatechin-3-gallate induces neurorescue of long-term serum-deprived PC12 cells and promotes neurite outgrowth. J Neurochem 93:1157–1167

18. Rosado A, Zanella F, Garcia B et al (2008) A dual-color fluorescence-based platform to identify selective inhibitors of Akt signaling. PLoS One 3:e1823

19. Payne S, Wylie P, Edward R, Goulter A (2007) Use of far-red emitting DNA dye DRAQ5 for cell cycle analysis with microplate cytometry. In: Drug Discovery and Technology Conference, Washington DC, USA

20. May K, Preckel H, Schaaf S, Mumtsidu E (2008) Image-based quantification of cyclin B1 and DNA content during cell cycle using the Opera HCS Platform. In: SBS/ELRIG Drug Discovery Conference, Bournemouth, UK

21. Hannoush RN (2008) Kinetics of Wnt-driven beta-catenin stabilization revealed by quantitative and temporal imaging. PLoS One 3:e3498

22. Richardson C (2008) Development of a mechanistic cell-based assay for the identification of JAK2 inhibitors. In: SBS/ELRIG Drug Discovery Conference, Bournemouth, UK

23. Laakkonen JP, Kaikkonen MU, Ronkainen PH et al (2008) Baculovirus-mediated immediate-early gene expression and nuclear reorganization in human cells. Cell Microbiol 10:667–681

24. Errington RJ, Patterson LH, Edward R, Smith PJ (2006) CyTRAK™ probes: novel nuclear and cytoplasm discriminators compatible with GFP-based HCS and HTS assays. In: Society for Biomolecular Sciences Annual Conference, Seattle

25. Maiuri L, Luciani A, Giardino I et al (2008) Tissue transglutaminase activation modulates inflammation in cystic fibrosis via PPARgamma down-regulation. J Immunol 180:7697–7705

26. Sawada J, Shioda R, Matsuno K, Asai A (2009) Phenotype-based screening of chemical library for mitotic inhibitors. In: High Content Analysis Conference, San Francisco, USA

27. Andrews D (2011) Automated classification of images generated in high content screening of human cells. MipTec Congress, Basel

28. Simonen M, Ibig-Rehm Y, Hofmann G et al (2008) High-content assay to study protein prenylation. J Biomol Screen 13:456–467

29. Grieshaber SS, Grieshaber NA, Miller N et al (2006) Chlamydia trachomatis causes centrosomal defects resulting in chromosomal segregation abnormalities. Traffic (Copenhagen, Denmark) 7:940–949

30. Gustin E, Van Loock M, Van Acker K, Krausz E (2009) Cell health in drug discovery: morphological markers. Paper presented at the High Content Analysis Conference, San Francisco, USA

31. Berke JM, Fenistein D, Pauwels F et al (2010) Development of a high-content screening assay to identify compounds interfering with the formation of the hepatitis C virus replication complex. J Virol Methods 165:268–276

32. Xu JJ, Henstock PV, Dunn MC et al (2008) Cellular imaging predictions of clinical drug-induced liver injury. Toxicol Sci 105:97–105

33. Smith PJ, Wiltshire M, Chappell SC et al (2013) Kinetic analysis of intracellular Hoechst 33342–DNA interactions by flow cytometry: misinterpretation of side population status? Cytometry A 83:161–169

34. Akagi J, Kordon M, Zhao H et al (2013) Real-time cell viability assays using a new anthracycline derivative DRAQ7®. Cytometry A 83A:227–234

35. Edward R (2011) DRAQ7™ - a unique far-red viability dye provides new combinations for cell health & in vitro tox assays. Paper presented at the Society for Biomolecular Sciences Annual Conference, Orlando, USA

36. Zhao H, Oczos J, Janowski P et al (2010) Rationale for the real-time and dynamic cell death assays using propidium iodide. Cytometry A 77:399–405

37. Smith PJ, Falconer RA, Errington RJ (2013) Micro-community cytometry: sensing changes in cell health and glycoconjugate expression by imaging and flow cytometry. J Microsc 251:113–122

38. Siemann DW, Keng PC (1986) Cell cycle specific toxicity of the Hoechst 33342 stain in untreated or irradiated murine tumor cells. Cancer Res 46:3556–3559

39. Lacoste J, Young K, Brown CM (2013) Live-cell migration and adhesion turnover assays. Methods Mol Biol (Clifton, NJ) 931:61–84

40. Mierke CT (2011) Cancer cells regulate biomechanical properties of human microvascular endothelial cells. J Biol Chem 286:40025–40037

Chapter 2

General Staining and Segmentation Procedures for High Content Imaging and Analysis

Kevin M. Chambers, Bhaskar S. Mandavilli, Nick J. Dolman, and Michael S. Janes

Abstract

Automated quantitative fluorescence microscopy, also known as high content imaging (HCI), is a rapidly growing analytical approach in cell biology. Because automated image analysis relies heavily on robust demarcation of cells and subcellular regions, reliable methods for labeling cells is a critical component of the HCI workflow. Labeling of cells for image segmentation is typically performed with fluorescent probes that bind DNA for nuclear-based cell demarcation or with those which react with proteins for image analysis based on whole cell staining. These reagents, along with instrument and software settings, play an important role in the successful segmentation of cells in a population for automated and quantitative image analysis. In this chapter, we describe standard procedures for labeling and image segmentation in both live and fixed cell samples. The chapter will also provide troubleshooting guidelines for some of the common problems associated with these aspects of HCI.

Key words High content screening, High content imaging, CellMask, Segmentation, Nuclear segmentation, CellTracker

1 Introduction

HCI combines the spatial resolution provided by fluorescence microscopy with algorithm-based quantitation of a variety of parameters, providing multiplex data relating to many aspects of cellular function, structure, and toxicity across populations of cells. Rapid developments in instrumentation, software, and reagents over the last decade have enabled HCI to emerge as a powerful platform for cell-based interrogation of biological processes with higher throughput and increased statistical significance [1–3]. In order to generate reliable data by HCI, accurate identification and segmentation of each cell in a population is required. Segmentation by HCI typically employs the use of fluorescent probes which bind nucleic acids for nuclear staining, protein-reactive dyes for whole

Paul A. Johnston and Oscar J. Trask (eds.), *High Content Screening: A Powerful Approach to Systems Cell Biology and Phenotypic Drug Discovery*, Methods in Molecular Biology, vol. 1683, DOI 10.1007/978-1-4939-7357-6_2,
© Springer Science+Business Media LLC 2018

Table 1
Whole cell and nuclear stains for segmentation

Segmentation tool	Excitation/emission approximate maxima (nm)	Target
HCS NuclearMask™ Blue stain	350/461	Nuclear
HCS NuclearMask™ Red stain	622/645	Nuclear
HCS NuclearMask™ Deep Red stain	638/686	Nuclear
Hoechst 33,342 stain	350/461	Nuclear
HCS CellMask™ Blue stain	346/442	Whole cell
HCS CellMask™ Green stain	493/516	Whole cell
HCS CellMask™ Orange stain	556/572	Whole cell
HCS CellMask™ Red stain	588/612	Whole cell
HCS CellMask™ Deep Red stain	650/655	Whole cell
CellTracker™ Blue CMAC stain	353/466	Whole cell
CellTracker™ Blue CMF$_2$HC stain	371/464	Whole cell
CellTracker™ Blue CMHC stain	372/470	Whole cell
CellTracker™ Violet BMQC stain	415/516	Whole cell
CellTracker™ Green CMFDA stain	492/517	Whole cell
CellTracker™ Green BODIPY stain	522/529	Whole cell
CellTracker™ Orange CMTMR stain	541/565	Whole cell
CellTracker™ Orange CMRA stain	548/576	Whole cell
CellTracker™ Red CMTPX stain	577/602	Whole cell
CellTracker™ Deep Red stain	630/660	Whole cell
CellMask™ Orange Plasma membrane stain	556/572	Plasma membrane
CellMask™ Deep Red Plasma membrane stain	650/655	Plasma membrane

cell staining, and to some extent, plasma membrane probes to delineate the cellular border (*see* Table 1 for a list of segmentation stains and dyes and their excitation and emission wavelengths). In addition to the fluorescence characteristics of segmentation tools and determination of compatible hardware configurations, selection criteria should account for the ability of such probes to label live cell or fixed cell samples and whether or not targets of interest are localized to the nuclear region [4]. In this chapter, we provide detailed protocols for labeling live or fixed cells for segmentation from whole cells or nuclei along with representative images of

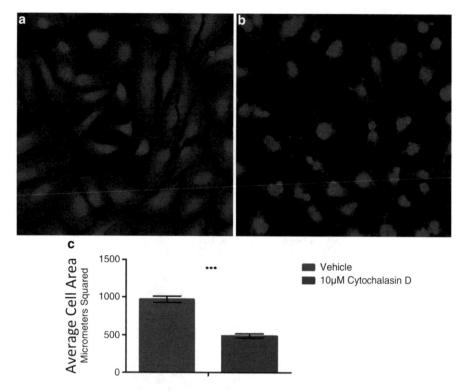

Fig. 1 HeLa cells labeled with HCS CellMask™ Blue stain and treated with vehicle (**a**) or 10 μM cytochalasin D for 3 h (**b**). Automated imaging and analysis of these two populations showed that the mean area of the cells decreased by half with 10 μM cytochalasin D treatment (**c**). $P \leq 0.0001$

proper segmentation and the quantitative data of drug-dependent change in morphology that can be measured with these segmentation tools (Figs. 1 and 2).

2   Segmentation Tools for Automated High Content Imaging and Analysis

Image segmentation is a critical step within automated image analysis, and reagents designed to facilitate this process represent integral tools in the development of HCI-based assays [1]. Image segmentation is a process that divides an image into one or more regions that represent defined objects of interest as cells or within cells and allows further determination and quantitation of their features in a spatial context. In order to obtain quantitative data from cell populations, the spatial distribution of the object and regions of interest within the object must be defined. Subsequently, the objects defined by segmentation tools and well-established image analysis algorithms may be utilized to measure changes in the number, fluorescence emission intensities, shape, texture, or motion of these objects within or across defined regions [4, 5]. Fluorescent dyes that stain nuclei, the cytoplasm or the plasma

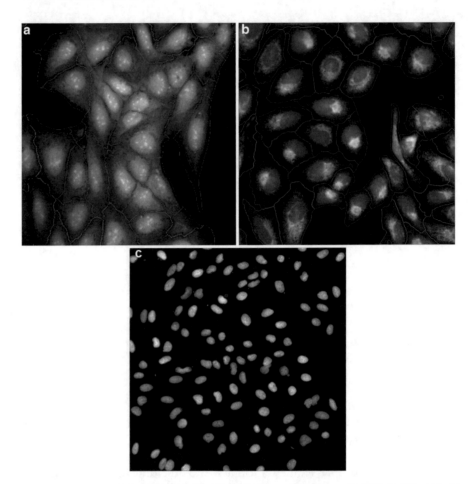

Fig. 2 Image segmentation of U-2 OS cells based on whole cell labeling with HCS CellMask™ Blue stain (**a**), CellTracker™ Deep Red stain (**b**), or based on nuclear labeling with Hoechst 33342 stain (**c**). Accepted objects are outlined in *blue* and rejected objects that cannot be completely quantified are in *orange*

membrane are very useful labeling tools that, in combination with segmentation algorithms, enable robust HCI-based assays (Table 1). In addition to demarcation of objects and regions of interest, stains used for image segmentation may also provide information about cell health including cellular and nuclear morphology. Here we describe example protocols that utilize HCS CellMask™ stains [2] and CellTracker™ probes to label the nuclei and cytoplasm of fixed and live cells, respectively. The whole cell staining that results from these probes enable robust segmentation for automated analysis of changes in cellular size and morphology across populations (Fig. 1). In addition, we provide an example protocol for cell segmentation based upon labeling of nuclei in live or fixed cells with the common, blue-fluorescent nucleic acid stain, Hoechst 33342 [2].

3 Fixed Cell Staining with HCS CellMask™ Stains

3.1 Materials

1. HCS CellMask™ Stain (cytoplasmic/nuclear stain; 2 Component Kit: Component A = HCS CellMask Stain, Component B = DMSO).

2. HeLa (Fig. 1) or U-2 OS (Fig. 2) cells.

3. Fetal bovine serum (FBS).

4. MEM (HeLa cells) or McCoy's 5A (U-2 OS cells) media.

5. 96-well tissue culture microplate (*see* **Note 1**).

6. 16% formaldehyde (*see* **Note 2**).

7. 100% Triton® X-100.

8. Laminar flow hood.

9. Cell culture incubator.

10. Phosphate buffered saline (PBS).

11. High content imaging instrument with appropriate objectives and filter sets for the dye being used (Table 2) (*see* **Note 3**).

3.2 Reagent Preparation

1. Prepare complete growth medium by supplementing MEM (HeLa cells) or McCoy's 5A (U-2 OS cells) with 10% FBS.

2. Prepare a 10 mg/mL HCS CellMask™ stock solution by dissolving the entire contents of the HCS CellMask™ vial (Component A) in 25 μL of DMSO (Component B) (*see* **Note 4**).

3. Prepare the fixative by diluting 16% aqueous formaldehyde solution in PBS to obtain a 4% formaldehyde solution.

4. Prepare the permeabilization solution by adding 10 μL of Triton® X-100 to 10 mL of PBS.

3.3 Methods

1. Seed U-2 OS or HeLa cells into a 96-well plate at a density of 3000 cells/well in 100 μL of McCoy's 5A (U-2 OS cells) or MEM (HeLa cells) supplemented with 10% FBS and incubate overnight under normal cell culture conditions (*see* **Note 1**).

2. Remove the medium and add 4% formaldehyde solution (prepared in Subheading 3.2, **step 3**) to each well and incubate for 15 min at room temperature (*see* **Note 5**).

3. Remove fixative and wash the fixed cells 2–3 times with PBS.

4. Add the 0.1% Triton® X-100 solution (prepared in Subheading 3.2, **step 4**) to each well and incubate for 15 min at room temperature (*see* **Note 6**).

5. Remove permeabilization solution and wash wells 2–3 times with PBS.

Table 2
HCS studio configuration for image segmentation using thermo fisher HCI platforms (see Note 7)

Parameter	HCS CellMask™ Blue	CellTracker™ Deep Red	Hoechst 33342
Bioapplication	Cytoskeletal rearrangement	Cytoskeletal rearrangement	General intensity measurement tool
Objective	40×	40×	20×
Exposure time	Fixed	Fixed	Fixed
Acquisition type	Widefield	Widefield	Widefield
Filter	386-23_ BGRFRN_BGRFRN	650-13_ BGRFRN_BGRFRN	BGRFR_386_23
Acquisition mode	1104 × 1104 (2 × 2 binning)	1104 × 1104 (2 × 2 binning)	1104 × 1104 (2 × 2 binning)
Autoexposure setting	25% peak target	25% peak target	25% peak target
Image preprocessing			
Channel	1	1	1
Object type	Bright	Bright	Bright
Background removal	Off	Off	Off
Primary object identification			
Smoothing	On	On	Off
Method	Uniform	Uniform	
Value	3	3	
Thresholding	On	On	On
Method	Fixed	Fixed	Fixed
Value	400	200	900
Segmentation	On	On	Off
Method	Intensity	Intensity	
Value	520	520	
Object cleanup	Off	On	On

6. Prepare a 1× HCS CellMask™ staining solution by adding 2 μL of the HCS CellMask™ stock solution (prepared in Subheading 3.2, **step 2**) to 10 mL PBS.

7. Add 100 μL of HCS CellMask™ staining solution (prepared in Subheading 3.3, **step 6**) to each well and incubate for 30 min at room temperature.

8. Wash each well 2–3 times in PBS to remove excess stain before plate sealing and imaging.

9. Optimize the exposure conditions to equal less than saturation limits of detector or camera for positive control, and then scan samples on any high content imaging instrument. The protocol used here was developed on a Thermo Scientific CellInsight CX7 platform with the Cytoskeletal Rearrangement Assay BioApplication and are available in Table 2 (Fig. 2a) (*see* **Note 7**).

4   Live Cell Labeling with CellTracker™ Deep Red Dye

4.1   Materials

1. CellTracker™ Deep Red Dye (Cell permeant dye that becomes impermeant inside the cell; 20 × 15 μg Kit, 1 component).

2. U-2 OS cells.

3. Fetal bovine serum (FBS).

4. McCoy's 5A Medium.

5. 96-well tissue culture plate (*see* **Note 1**).

6. Dimethylsulfoxide (DMSO).

7. Cell culture incubator.

8. Phosphate buffered saline (PBS).

9. High content imaging instrument with objectives and filter set appropriate for detecting CellTracker™ Deep Red stain (Table 2) (*see* **Note 3**).

4.2   Reagent Preparation

1. Prepare complete growth medium by supplementing McCoy's 5A (U-2 OS cells) with 10% FBS.

2. Prepare a 1 mM CellTracker™ Deep Red stock solution (1000×) by dissolving the entire contents of one vial of CellTracker™ Deep Red dye (Component A) in 20 μL of DMSO (*see* **Note 11**).

4.3   Methods

1. Seed U-2 OS cells into a 96-well plate at a density of 2500 cells/well in 100 μL of complete growth medium and incubate overnight under normal cell culture conditions (*see* **Note 1**).

2. Prepare a 1× CellTracker™ Deep Red working solution by adding 10 μL of the CellTracker™ Deep Red stock solution (prepared in Subheading 4.2, **step 2**) to 10 mL unsupplemented MEM.

3. Add 100 μL of CellTracker™ Deep Red working solution (prepared in Subheading 5.3, **step 2**) to each well and incubate for 30 min at room temperature.

4. Remove CellTracker™ Deep Red working solution from wells.

5. Add 100 μL of complete growth media to each well of the 96-well plate.

6. Optimize the exposure conditions and scan samples on any high content imaging instrument. The protocol used here was developed on a Thermo Scientific CellInsight CX7 platform with the Cytoskeletal Rearrangement Assay BioApplication (Fig. 2b) (Table 2) (*see* **Note 7**).

5   Live or Fixed Cell Labeling of the Nuclei with Hoechst 33342

5.1   Materials

1. Hoechst 33342 stain, trihydrochloride, trihydrate—10 mg/mL aqueous solution (nucleic acid dye).

2. U-2 OS cells.

3. Fetal bovine serum (FBS).

4. McCoy's 5A Medium.

5. 96-well tissue culture microplate (*see* **Note 1**).

6. 16% formaldehyde (*see* **Note 12**).

7. 100% Triton® X-100.

8. Laminar flow hood.

9. Cell culture incubator.

10. Phosphate buffered saline (PBS).

11. High content imaging instrument with objectives and filter sets appropriate for Hoechst 33342 (Table 2) (*see* **Note 3**).

5.2   Reagent Preparation

1. Prepare complete growth medium by supplementing McCoy's 5A (U-2 OS cells) with 10% FBS.

2. Prepare the fixative by diluting 16% aqueous formaldehyde solution in PBS to obtain a 4% formaldehyde solution (fixed cell labeling only).

3. Prepare the permeabilization solution by adding 10 μL of Triton® X-100 to 10 mL of PBS (fixed cell labeling only).

5.3   Methods

1. Seed U-2 OS cells into a 96-well plate at a density of 2500 cells/well in 100 μL of complete growth medium and incubate overnight under normal cell culture conditions (*see* **Note 4**).

2. *For Fixed Cell Experiments Only*: Remove the medium and add 4% formaldehyde solution (prepared in Subheading 5.2, **step 2**) to each well and incubate for 15 min at room temperature (*see* **Note 12**).

3. *For Fixed Cell Experiments Only*: Remove fixative and gently wash the fixed cells 2–3 times with PBS.

4. *For Fixed Cell Experiments Only*: Add the 0.1% Triton® X-100 solution (prepared in Subheading 5.2, **step 3**) to each well and incubate for 15 min at room temperature.

5. *For Fixed Cell Experiments Only:* Remove permeabilization solution and gently wash wells 2–3 times with PBS.

6. Prepare a 1× Hoechst 33342 staining solution by adding 5 μL of the 10 mg/mL stock solution to 10 mL PBS for fixed cells or complete media for live cells (See **Note 13**).

7. Add 100 μL of Hoechst 33342 staining solution (prepared in Subheading 5.3, **step 6**) to each well and incubate for 15 min at room temperature.

8. Optimize the exposure conditions and scan samples on any high content imaging instrument. The protocol used here was developed on a Thermo Scientific ArrayScan™ V^{TI} platform with the General Intensity Measurement Tool BioApplication (Fig. 2c) (Table 2) (*see* **Note 7**).

6   Notes

1. Plates should be selected for an available form factor and thickness for the HCI platform being used and cells should be plated uniformly at lower densities for good separation of cells and accurate segmentation of objects (Subheadings 3.1, **item 5**, 3.3, **step 1**, 4.1, **item 5**, 4.3, **step 1**, 5.1, **item 5**, and 5.3, **step 1**).

2. Other fixation methods such as methanol have not been validated for use with HCS CellMask™ stains (Subheading 3.1, **item 6**). Literature searches may be helpful to determine if other fixation methods have been tested for these dyes.

3. While this study used a 40× objective, procedures outlined here are applicable to 10× and 20× magnifications (Subheading 3.1, **item 11**, Subheading 4.1, **item 9**, and 5.1, **item 10**).

4. Store any unused HCS CellMask™ stain at −20 °C, protected from light. For optimal results, use frozen aliquots within 6 months of preparation. Avoid freeze/thaw cycles (Subheadings 3.2, **step 2** and 3.3, **step 9**).

5. For assays where there may be apoptotic or dead cells, fixative can be added to the existing media in a 2× solution (Subheading 3.3, **step 2** and 5.3, **step 2**).

6. Other incubation times and Triton-X concentrations may be used. The researcher should use the optimal permeablization conditions for the cell type and fluorophore being used (Subheading 3.3, **step 4**).

7. Optimize the labeling conditions and software parameters for segmentation so at least an average of 95% of cells per field are segmented correctly (Table 2) (Subheadings 3.3, **step 9**, 4.3, **step 6**, and 5.3, **step 8**) (Fig. 3).

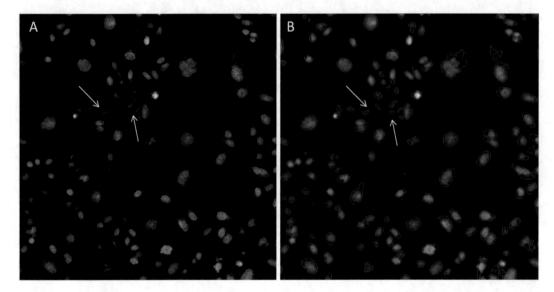

Fig. 3 Improper labeling conditions can affect segmentation of cells. Improper labeling conditions for Hoechst 33342 results in only half of some nuclei being labeled (*arrows* in **a**) which prevents proper segmentation (*arrows* in **b**) (see **Note 7**)

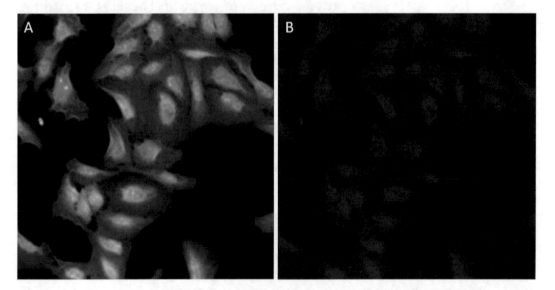

Fig. 4 Dyes "bleeding through" into other channels can affect assay and data integrity. Images from 60 ms exposures in the Texas Red (**a**) and TRITC (**b**) channels showed HCS CellMask Red "bleeding through" into the shorter wavelength TRITC channel obscuring dim targets (See Notes 8, 9, and 10)

8. When multiplexing HCS CellMask™ stains, CellTracker™ dyes, or Hoechst 33342 with other fluorophores, first perform small-scale optimization labeling conditions to ascertain fluorescence compatibility prior to large-scale screening. In a

96-well plate format, serial dilutions can be used effectively to optimize staining and labeling conditions (Fig. 4).

9. Given the brightness of these stains, "bleed through" into neighboring fluorescence channels can easily be mitigated by reducing the final concentration of the HCS CellMask™ reagent down to as low as 1 μg/mL in the staining solution. HCS CellMask™ reagents bleed through more prominently into channels with emission wavelengths shorter than the ideal emission wavelength of the dye, so it is advisable to assign less-abundant targets to longer wavelength emission channels when designing experiments (Fig. 4).

10. HCS CellMask™ stains are compatible only with fixed and permeabilized cells. When combining with antibody-based probes, labeling with HCS CellMask™ should be performed as the last step to avoid competitive blocking of epitopes intended for binding by primary antibodies (Fig. 4).

11. Any unused CellTracker™ Deep Red working solution should be discarded (Subheading 4.2, **step 2**).

12. Other fixation methods such as methanol may be used instead of formaldehyde (Subheadings 5.1, **item 6** and 5.3, **step 2**).

13. Store any unused Hoechst 33342 staining solution at 4 °C, protected from light (subheading 5.3, **step 6**).

References

1. Hill AA, LaPan P, Li Y et al (2007) Impact of image segmentation on high-content screening data quality for SK-BR-3 cells. BMC Bioinformatics 8:340

2. Giuliano KA, Haskins JR, Taylor DL (2004) Advances in high content screening for drug discovery. Assay Drug Dev Technol 1(4):565–577

3. Zhou X, Wong STC (2006) High content cellular imaging for drug development. IEEE Signal Process Mag 23(2):170–174

4. Buchser W., Collins M., Garyantes T. et al. (2012) Assay development guidelines for image-based high content screening, high content analysis and high content Imaging. Assay Guidance Manual. http://www.ncbi.nlm.nih.gov/books/NBK100913/. Accessed 1 Oct 2012

5. Miller EH, Harrison JS, Radoshitzky SR et al (2011) Inhibition of ebola virus entry by a C-peptide targeted to endosomes. J Biol Chem 286(18):15854–15861

Chapter 3

Tools to Measure Cell Health and Cytotoxicity Using High Content Imaging and Analysis

Bhaskar S. Mandavilli, Robert J. Aggeler, and Kevin M. Chambers

Abstract

High content screening (HCS)-based multiparametric measurements are very useful in early toxicity testing and safety assessment during drug development, and useful in evaluating the impact from new food supplements and environmental toxicants. Mitochondrial membrane potential, plasma membrane permeability, oxidative stress, phosphoplipidosis, and steatosis are a few of the important markers routinely studied for the assessment of drug-induced liver injury and toxicity. Mitochondrial dysfunction leads to oxidative stress and cell death. Liver injury from drug-induced phospholipidosis and steatosis is routinely studied in hepatotoxicity investigations to determine the risk factors and fate of drugs or chemical compounds as some drugs can lead to defects in lipid metabolism and accumulation of lipids in lysosomes. In this chapter, we describe fluorescent reagents and the protocols for the measurement of various parameters such as mitochondrial membrane potential, plasma membrane permeability, oxidative stress, phospholipidosis, and steatosis using high content imaging-based methodologies and instrumentation.

Key words Cytotoxicity, Mitochondrial membrane potential, Plasma membrane integrity, Oxidative stress, Reactive oxygen species

1 Introduction

Mitochondrial dysfunction and cell death are implicated in hepatotoxicity and are shown to be implicated in several other disease pathways including cancer, neurodegeneration, and aging [1–4]. Mitochondrial dysfunction can lead to increased oxidative stress resulting in apoptosis, necrosis, or autophagy [5–8]. These important biomarkers can be effectively used to measure chemical or drug-induced toxicity in cells [9–16]. Permeant fluorescent cationic dyes like tetra-methyl-rhodamine (TMRM or TMRE), JC-1, and MitoTrackers are organic dyes routinely used to measure the depolarization of mitochondrial membrane potential in living cells [17–20]. These fluorescent dyes accumulate in mitochondria based on membrane potential and are useful tools for measuring

Paul A. Johnston and Oscar J. Trask (eds.), *High Content Screening: A Powerful Approach to Systems Cell Biology and Phenotypic Drug Discovery*, Methods in Molecular Biology, vol. 1683, DOI 10.1007/978-1-4939-7357-6_3,
© Springer Science+Business Media LLC 2018

Table 1
Comparison of CellROX® reagents to existing probes for oxidative stress measurements

	H2-DCFDA	DHE	CellROX® Deep Red Reagent	CellROX® Orange Reagent	CellROX® Green Reagent
Add in complete media	No	Yes	Yes	Yes	Yes
Fixable	No	No	Yes	No	Yes
Detergent resistant	No	No	No	No	Yes
Multiplexibility	Yes	No	Yes	Yes	Yes
GFP compatible	No	No	Yes	Yes	No
RFP compatible	Yes	No	Yes	No	Yes
Photostability	*	N/A	***	****	**

Stars indicate the degree of photostability

mitochondrial dysfunction. Upon mitochondrial dysfunction, the cell undergoes oxidative stress and eventually cell death [5–8].

Another parameter that is very useful in the study of drug-induced toxicity is oxidative stress. Oxidative stress is the imbalance between the production of reactive oxygen species (ROS) and the ability of cells to destroy these oxidative species by antioxidant enzymes. Oxidative stress can be caused by various pathways including mitochondria and cell membrane-based NADPH oxidases [21]. ROS include reactive molecules like superoxide anion (O_2^-) and hydroxyl radical ($\cdot OH^-$) and hydrogen peroxide (H_2O_2). Both superoxide and hydroxyl radical are highly reactive and cause cytotoxicity by damaging DNA, RNA, proteins, and lipids [22]. H_2O_2 causes toxicity by oxidation of thiol groups in proteins. CellROX™ reagents are sensitive reagents that react with a broad range of ROS enabling a sensitive detection of oxidative stress in cells. CellROX™ Deep Red and Green are formaldehyde fixable, thus making them very useful reagents for HCS assays [23–30]. CellROX® reagents exhibit differential reactivity with various reactive oxygen species and different stimuli serve as positive controls for different CellROX® reagents (Table 1). Dihydroethidine is another routinely used fluorescence-based method for oxidative stress, but it reacts with only superoxide anion in cells and is not formaldehyde fixable.

Drug-induced phospholipidosis and steatosis are precursors to nonalcoholic steatohepatitis, liver fibrosis, and cirrhosis of the liver and are important early indicators in drug-induced hepatotoxicity [31–35]. Phospholipidosis can result from drug-induced inactivation of phospholipases in lysosomes and leads to accumulation of phospholipids. Steatosis is the accumulation of neutral lipids resulting in fatty deposits in cells. Drug-induced phospholipidosis and steatosis can be efficiently measured using fluorescent stains for phospholipids and neutral lipids [36]. Nile Red is routinely used to label neutral lipids but the emission profile of the dye is too broad for effective implementation of a multiplexed assay [37]. However, the combination of HCS LipidTOX™ steatosis reagents provides a labeling strategy to perform high content imaging multiplexed assays to characterize phospolipidosis and steatosis, and to measure the effects on lipid metabolism that can be triggered by drugs or other compounds. In this book chapter, we describe the protocols to measure multiple parameters of cell health with examples using specific positive control compounds. (1) Mitochondrial health and cytotoxicity assay using valinomycin as a positive control compound in HepG2 cells. Valinomycin is an ionophore which causes mitochondrial membrane depolarization and cell death (Fig. 1). (2) Oxidative stress assay using CellROX™ reagents using menadione or TBHP as positive controls. The cells were

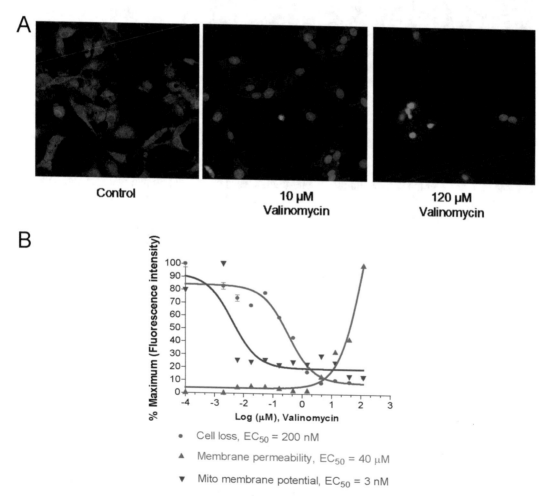

A

Control

10 µM
Valinomycin

120 µM
Valinomycin

B

% Maximum (Fluorescence intensity)

Log (µM), Valinomycin

● Cell loss, EC$_{50}$ = 200 nM

▲ Membrane permeability, EC$_{50}$ = 40 µM

▼ Mito membrane potential, EC$_{50}$ = 3 nM

Fig. 1 HepG2 cells were plated on collagen-coated 96-well plates and treated with various doses of valinomycin for 24 h before staining with Image-iT® DEAD Green™ and MitoHealth stains for 30 min. The cells were then fixed and counterstained with Hoechst nuclear stain. Imaging and analysis was done on a Thermo Scientific Cellomics® ArrayScan® VTI. The images at select concentrations of valinomycin are shown in panel (**a**). (**b**) The log concentrations of valinomycin were used to calculate EC$_{50}$ values for cell loss, mitochondrial membrane potential, and membrane permeability

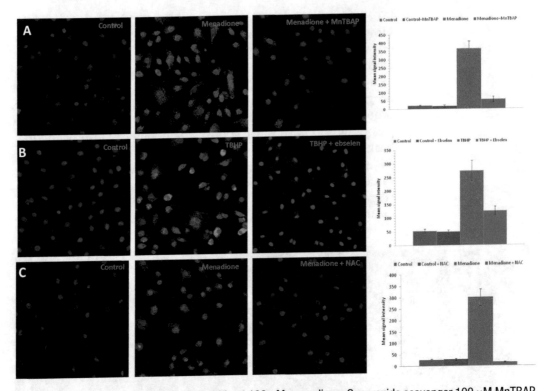

Fig. 2 (**a**) BPAE cells were treated with or without 100 μM menadione. Superoxide scavenger 100 μM MnTBAP was added to some of the control and menadione-treated wells for the last 30 min of incubation. The cells were stained with CellROX™Deep Red Reagent for 30 min. (**b**) BPAE cells were treated with or without 200 μM tert-butyl hydroperoxide (TBHP). The peroxide scavenger 10 μM ebselen was added to some of the control and TBHP-treated cells. The cells were then stained with CellROX™ Orange Reagent and incubated for 30 min. (**c**) BPAE cells were treated with or without 100 μM menadione. The antioxidant 50 μM *N*-Acetyl cysteine (NAC) was added to some of the control and menadione-treated wells for the last 30 min of incubation. The cells were then stained with CellROX™ Green Reagent The cells were then washed and analyzed on a Thermo Fisher Cellomics ArrayScan® VTI. Reduction of the signal in antioxidant treated cells confirmed indication of oxidative stress by CellROX™ reagents

also incubated with or without antioxidants to demonstrate the specificity of the assay (Fig. 2). (3) Phospholipidosis and steatosis assay using the pharmaceutical drugs, propranolol and cyclosporin A to induce phospholipidosis and steatosis respectively (Fig. 3).

1.1 Measurements of Mitochondrial Dysfunction and Cell Death

The HCS Mitochondrial Health and cytotoxicity assay described in this chapter is an optimized assay to measure mitochondrial membrane potential and cell death using high content imaging in both live and formaldehyde-fixed cells. The kit has three different components for multiparametric measurements of nuclear morphology (Hoechst 33342), mitochondrial membrane potential (MitoHealth stain) and plasma membrane permeability, a measure of cell death (Image-iT® DEAD™ Green) [38].

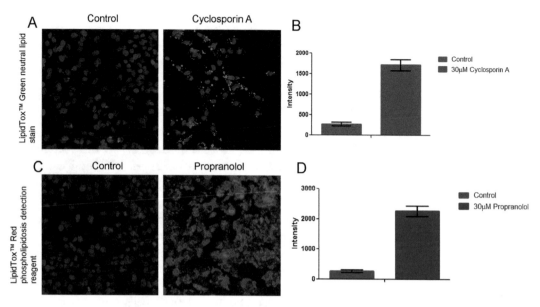

Fig. 3 HepG2 cells were plated at 2500 cells per well on a 96-well collagen-coated plate. The next day, media was replaced with fresh media containing 30 μM propranolol, 30 μM cyclosporin A, or vehicle and LipidTOX™ Red phospholipidosis detection reagent (1:1000) and incubated for 48 h. Cells were then fixed with 4% formaldehyde and labeled with 5 μg/mL HCS NuclearMask™ Blue stain in media for 15 min**. The cells were washed with DPBS before labeling with LipidTOX™ Green neutral lipid stain (1:1000) for 30 min. Imaging was performed on a Thermo Scientific Cellomics® ArrayScan® VTI. As expected, propranolol treatment resulted in increased LipidTOX™ Red phospholipidosis staining (**c** and **d**) while cyclosporine A treatment resulted in increased LipidTOX™ Green neutral lipid staining (**a** and **b**)

1.2 Materials

1. HCS Mitochondrial health Kit
   The components of the kit include:
   (a) 1 mM Image-iT® DEAD Green™ viability stain (in DMSO).
   (b) MitoHealth stain (solid).
   (c) 10 mg/mL Hoechst 33342.
   (d) Dimethylsulfoxide (DMSO).
2. Valinomycin, positive control compound.
3. HepG2 cell line.
4. Minimal essential medium (MEM).
5. Fetal bovine serum (FBS).
6. TrypLE™ express.
7. Sodium pyruvate.
8. Essential amino acids.
9. Formaldehyde.
10. Laminar flow hood.

11. Cell culture incubator.

12. 96-well cell culture microplates.

13. High content imaging instrument.

1.3 Reagent Preparation

1. To make 30 mM stock solution of valinomycin, add 750 μL of DMSO to a vial containing 25 mg of valinomycin.

2. Prepare MitoHealth stain (component B) stock solution by adding 30 μL of DMSO (component D) to the contents of the vial. Store any unused solution at −20 °C, desiccated and protected from light. Valinomycin stock solution may form precipitate when frozen, precipitate can be dissolved by thawing the sample in a 37 °C incubator.

3. Prepare the counterstain/fixation solution by adding 3 mL of 16% formaldehyde and 6 μL of Hoechst 33342 (Component C) to 9 mL PBS.

1.4 Method

This protocol is developed for HepG2 cells and has also been successfully tested in HeLa and A549 cells.

1. Seed HepG2 cells into a collagen-coated 96-well microplate at a density of 7500 cells/well in 100 μL of MEM supplemented with 10% FBS, sodium pyruvate and essential amino acids (complete growth medium), and place the plates in an incubator overnight at 37 °C with 5% CO_2 (see **Note 1**).

2. Prepare a suitable 10-point dilution series (see Subheading 1.5) of valinomycin from the 30 mM stock solution (prepared in Subheading 1.2) in a complete growth medium beginning at a maximum concentration of 120 μM. Valinomycin should be prepared as a 5× solution of drug in the complete growth medium (see **Note 2**).

3. Add 25 μL of valinomycin stock (prepared in Subheading 1.4, **step 2**) to each of the wells containing 100 μL of complete growth medium and incubate for 24 h at 37 °C with 5% CO_2. The total volume in each well is 125 μL at this point (see **Note 3**).

4. After 24 h incubation with valinomycin, prepare the cell staining solution by adding 10.5 μL of the MitoHealth stain solution (prepared in Subheading 1.2), 0.6 μl of Hoechst 33342 and 2.1 μL of Image-iT® DEAD Green™ viability stain to 6 mL of complete medium.

5. Add 50 μL of the cell staining solution (prepared in **step 5)** to each well and incubate the plate at 37 °C in 5% CO_2 for 30 min. The total volume in each well should be 175 μL at this point (see **Note 4**).

6. Remove the medium and wash the cells 3× with PBS (see **Note 5**).

7. The cells can be imaged at this point. The protocol used here was developed on a Thermo Scientific Cellomics® ArrayScan® VTI using the Compartmental Analysis Bioapplication. Cytoplasmic spot intensity measurements are made for mitochondrial membrane potential changes, while nuclear intensity changes are measured for cytotoxicity indicating plasma membrane permeability with the Image-iT® DEAD Green™ viability stain. Nuclear morphology and condensation measurements can also be made to look at the cytotoxic effects of drugs (*see* **Note 6**).

8. If performing antibody staining, the cells are fixed with 4% formaldehyde for 15 min and permeabilized with 0.25% of Triton X-100 for 10 min (*see* **Note 7**).

9. Cover plates with microplate seals and scan on a high content imaging instrument (Fig. 1).

1.5 ***Notes***

1. The method described used HepG2 cells, was also effective in HeLa and A549 cells exposed to valinomycin, and can be adapted to any other cell line and drug.

2. Valinomycin tends to precipitate when frozen, especially at high concentrations. Make sure to vortex the vial rigorously once the solution is thawed or sonicate. 10-point dose response was done in the experiment described here but the dose response can be expanded or reduced as needed.

3. DMSO concentrations higher than 1% in the incubation media with live cells may alter the biological response. When adding drugs or test compounds at this step, consider the additional 0.06% DMSO introduced with the cell staining solution. If other organic solvents are used for preparing drug stock solutions, determine their compatibility with cells.

4. When incubating at room temperature, make sure the plates are not exposed to light.

5. If washing manually take care not to touch the bottom of the wells with the pipette tip when removing the medium. The pipetting has to be slow when adding PBS to reduce the cell loss. For automatic washes care should be taken to add and aspirate the solutions at slower speed.

6. The approximate excitation and emission wavelength maxima for Hoechst 33342: 350/461 nm bound to DNA (DAPI filter); Image-iT® DEAD Green™ viability stain: 488/515 nm (FITC filter); MitoHealth stain: 550/580 nm (TRITC filter).

7. If samples are fixed, scan the plates within 24 h post fixation. If imaging cells live, scan the plates within 2–3 h after completion of staining.

2   Oxidative Stress

The method described here is based on CellROX™ reagents that measure general oxidative stress in cells (Table 1). The reagent comes in three different colors: green, orange, and deep red. CellROX® Green Reagent is a nucleic acid-binding dye, that upon oxidation, binds predominantly to DNA and thus, its signal is localized primarily in the nucleus. In contrast, the signals from CellROX® Deep Red and CellROX® Orange Reagents are localized primarily in the cytoplasm. The fluorescence resulting from CellROX® Reagents can be easily measured by high content imaging and analysis [39]. CellROX™ reagents work well in live cells and the cells can be fixed if using Deep Red and Green reagents for longer stability of signal. There is a basal level of ROS production in cells and it may vary from cell line to cell line. The staining workflow is simple and the reagent can be applied to cells in complete growth media or buffer. CellROX® Reagents are more photostable when compared to traditional ROS detection dyes like dicholoro-dihydro fluoresceins and dihydroethidium (Table 1). All three dyes have minimal background in control cells and give good signal-to-noise values upon treatment with positive control compounds (Fig. 2). Menadione can be used to generate positive controls for CellROX® Green and Deep Red Reagents while TBHP is used for the CellROX® Orange Reagent.

2.1   Materials

1. Bovine pulmonary artery endothelial cells (BPAE) (*see* **Note 1**).
2. Fetal bovine serum (FBS).
3. Dulbecco's Minimal Essential Medium (DMEM).
4. TrypLE™ express.
5. CellROX® Reagents.
6. Phosphate-buffered saline (PBS).
7. Hoechst 33342.
8. Menadione.
9. Tert-Butyl hydroperoxide (TBHP).
10. Laminar flow hood.
11. Cell Culture incubator.
12. 96-well cell culture plates.
13. High content imaging instrument.

2.2   Reagent Preparation

1. Prepare 50 mM stock solution of menadione in DMSO. Dilute the menadione stock solution into the complete growth medium to a final concentration of 100 μM.
2. Prepare 100 mM stock solution of tert-butyl hydroperoxide (TBHP) (*see* **Note 5**) by adding 1–77 μL of PBS. Dilute the stock solution of TBHP in the complete growth medium to a final concentration of 200 μM TBHP.

2.3   Method

1. Seed BPAE cells in a 96-well plate at a density of 10,000 cells/well in 100 μL volume of DMEM supplemented with 20% FBS (complete growth medium). Incubate plates overnight at 37 °C with 5% CO_2 (*see* **Note 1**).

2. Remove the medium from the cells. When using CellROX® Green and Deep Red Reagents, add menadione to half the 96-well plate. Treat the other half of the plate with an equal amount of DMSO without menadione for vehicle controls. For oxidative stress measurements using CellROX® Orange Reagent, add 200 μM TBHP to the one half of the plate. Incubate the cells for 1 h for menadione or 2 h at for TBHP at 37 °C with 5% CO_2 (*see* **Notes 2–5**).

3. Prepare 10 μM CellROX® reagents (2×) by adding 40 μL of the reagent to 10 mL of complete growth media. Add 2 μl of Hoechst 33342 to the CellROX® working solution. Add 100 μL of the CellROX® reagent to each well for the last 30 min of the drug incubation. The final concentration of CellROX® Reagents on cells is 5 μM (*see* **Notes 6** and **7**).

4. Rinse the cells 3× with PBS and then finally add 100 μL of PBS to each well.

5. Scan the plates live or fixed and analyze on a high content imaging instrument using a 20× objective (Fig. 2). The protocol used here was developed on a Thermo Scientific ArrayScan® VTI using Compartmental Analysis bioapplication (*see* **Note 8** and **9**).

2.4   Notes

1. BPAE cells were used as they respond well to oxidative stress and give good signal/noise.

2. CellROX® Green is fluorescent upon oxidation and binding to DNA or RNA and thus localized in nucleus with some staining of mitochondrial DNA. CellROX® Deep Red and Orange reagent signals localize more in the cytoplasm as spots.

3. CellROX® Deep Red is amenable to formaldehyde fixation. CellROX® Green is amenable to both formaldehyde fixation and detergent permeabilization (Table 1). Imaging live cells is better to obtain good S/N values. After formaldehyde fixation, there is some increase in background levels leading to a decreased S/N.

4. When using antioxidants (Fig. 2), pretreat cells with these compounds for 30 min before adding the drug inducing oxidative stress.

5. Menadione produces superoxide by redox cycling [40] and TBHP is routinely used as an oxidant [41]. CellROX® reagents do not react with hydrogen peroxide and this is not a good positive control compound.

6. Scan the plates within 3 h of staining for CellROX® Deep Red and Orange reagents. CellROX® Green is compatible with labeling of intracellular targets using antibodies or other probes which require permeabilization of cells. The plates can be stored at 4 °C and read within 4 days after staining.

7. CellROX® reagents are compatible with other dyes for multiplexing and care should be taken that there is no spectral overlap when performing a multiplex assay.

8. Avoid prolonged exposure of the CellROX® dyes to light and air to prevent oxidation.

9. The approximate fluorescence excitation and emission maxima: CellROX® Green reagent: 508/525 nm (oxidized product bound to nucleic acid), works well with regular FITC filter; CellROX® Orange reagent: 545/565 nm (oxidized product), works well with regular TRITC filter; CellROX® Deep Red Reagent: 644/665 nm (oxidized product), works well with Cy5 filter; Hoechst 33342: 352/461 nm, works well with any DAPI filter.

3   Phospholipidosis and Steatosis Measurements

The method described here is based on the HCS phospholipidosis and steatosis kit. The kit supplies LipidTOX™ Red phospholipid stain and LipidTOX™ Green neutral lipid stain, which can be used sequentially for the analysis of phospholipidosis and steatosis, respectively, or can be used in isolation for single-parameter analysis using high content imaging [33–35].

3.1 Materials

1. HCS LipidTOX™ Phospholipidosis and Steatosis detection kits; The kit has following components
   (a) Red phospholipidosis detection reagent (1000× stock).
   (b) Green neutral lipid stain (1000× stock).
   (c) Hoechst 33342 (1000× stock).
   (d) Propranolol.
   (e) Cyclosporin A.
   (f) Dimethylsulfoxide (DMSO).

2. HepG2 cells.

3. Minimal essential medium.

4. Fetal bovine serum (FBS).

5. TrypLE™ express.

6. Sodium pyruvate.

7. Essential amino acids.

8. Laminar flow hood.

9. Cell culture incubator.

10. 96-well thin-bottom plates.

11. 16% formaldehyde.

12. High content imaging instrument equipped with either wide field or confocal capabilities with 20× objective.

3.2 Reagent Preparation

All the components of the kit except positive control compounds, cyclosporine A and propranolol are in the solution.

1. Make a 10 mM propranolol (mol. Wt. 295.80) stock solution in DMSO provided in the kit. Aliquot in smaller volume (20 µL) and freeze them at −20 °C and the stock solutions can be stored for 1 month without any loss in function.

2. Make a 10 mM cyclosporine A (mol. wt. 1202.61) stock solution in DMSO provided in the kit. Aliquot in smaller volume (20 µL) and freeze them at −20 °C and the stock solutions can be stored for 1 month without any loss in function.

3.3 Method

1. Seed HepG2 cells in a 96-well plate at a density of 5000 cells/well in 100 µL volume of MEM supplemented with 10% FBS, sodium pyruvate and essential amino acids (complete growth medium). Incubate plates overnight at 37 °C with 5% CO_2.

2. Dilute the 1000× LipidTOX™ Red phospholipidosis reagent 1:500 in the complete growth medium to make a 2× solution.

3. Dilute the 10 mM stock solution (prepared in Subheading 3.2) of propranolol in growth media to a final concentration of 60 µM. Mix the solution vigorously to avoid precipitation of the compound upon dilution. A volume of 50 µL/well is required for this protocol (*see* **Note 1**).

4. Dilute the 10 mM stock solution of cyclosporine A (prepared in Subheading 3.2) in the growth media to a final concentration of 60 µM. Mix the solution vigorously to avoid precipitation of the compound upon dilution. A volume of 50 µL/well is required for this protocol (*see* **Note 1**).

5. Remove media from the wells of the plate.

6. Add 50 µL/well of LipidTOX™ Red phospholipidosis detection reagent (prepared in Subheading 3.2) to all the wells.

7. To the appropriate wells containing LipidTOX™ Red phospholipidosis detection reagent, add either propranolol (prepared in **step 3**) or cyclosporine A (prepared in **step 4**). Add the same amount of DMSO as drug-treated wells to the vehicle control wells. Return plates to the incubator for 24 h at 37 °C with 5% CO_2.

8. Prepare 4% formaldehyde in PBS containing Hoechst 33342 (prepared from the 1000× Hoechst solution from the kit).

9. Remove the incubation medium from the wells and add 100 μL of formaldehyde with Hoechst 33342 to each well and incubate at room temperature for 15 min. Formaldehyde or paraformaldehyde should work well. Commercially available formaldehyde solutions are pretty stable as stock solutions and dilutions should be made just before use. If making fresh stock solution in the lab, make sure that pH is not in acidic range. pH changes can affect the fixative properties of the formaldehyde. Under fixation of the cells may result in loss of cells and improper staining.

10. Remove the fixative solution from the wells and rinse 3× with PBS.

11. Prepare LipidTOX™ Green neutral lipid stain by diluting the 1000× reagent in PBS to make a 1× solution.

12. Remove PBS from cells and add 100 μL of LipidTOX™ Green neutral lipid stain to each well. Incubate the plates at room temperature for 30 min.

13. Rinse the cells 3× with PBS and then finally add 100 μL of PBS to each well.

14. Seal the plates and analyze them on a high content imaging instrument (Fig. 3). The protocol used here was developed on a Thermo Scientific Cellomics® ArrayScan® VTI (*see* **Notes 2–4**).

3.4 Notes

1. Make sure the DMSO concentration does not exceed 1% final concentration on cells. When adding drug, consider the DMSO concentration of the cell staining solution that is added at the end of the incubation with the drug.

2. Image the plates as soon as possible after processing or leave them at 4 °C and use within 1 week. The staining starts becoming weaker after 1 week.

3. The approximate excitation and emission maxima for LipidTOX™ Red phospholipidosis detection reagent: 595/615 nm, Texas Red filter set; LipidTOX™ Green neutral lipid stain: 495/505 nm, FITC filter set; and Hoechst 33342: 352/461 nm, DAPI filter set.

4. Signal for phospholipidosis or steatosis is measured as cytoplasmic spot intensities.

References

1. Lee J, Giordano S, Zhang J et al (2012) Autophagy, mitochondria and oxidative stress: crosstalk and redox signaling. Biochem J 441 (2):523–540

2. Tait SWG, Green DR (2010) Mitochondria and cell death: outer membrane permeabilization and beyond. Nat Rev Mol Cell Biol 11:621–632

3. Kamogashira T, Fujimoto C, Yamasoba T et al (2015) Reactive oxygen species, apoptosis, and mitochondrial dysfunction in hearing loss. Biomed Res Int 2015:1–7

4. Lin MT, Meal MF (2006) Mitochondrial dysfunction and oxidative stress in neurodegenerative diseases. Nature 443:787–795

5. Pieczenick SR, Neustadt J (2007) Mitochondrial dysfunction and molecular pathways of disease. Exp Mol Pathol 83:84–92

6. Eckert A, Keil U, Marques CA et al (2003) Mitochondrial dysfunction, apoptotic cell death, and Alzheimer's disease. Biochem Pharmacol 66(8):1627–1634

7. Jiang T, Harder B, de la Vega R et al (2015) p62 links autophagy and Nrf2 signaling. Free Radic Biol Med 88:199–204

8. Go KL, Lee S, Zendejas I et al (2015) Mitochondrial dysfunction and autophagy in hepatic ischemia/reperfusion injury. Biomed Res Int 2015:1–14

9. Persson M, Loye AF, Mow T et al (2013) A high content screening assay to predict human drug-induced liver injury during drug discovery. J Pharmacol Toxicol Methods 68:302–313. (MTO and TOTO3)

10. Dykens JA, Will Y (2007) The significance of mitochondrial toxicity testing in drug development. Drug Discov Today 12:777–785

11. Trask JO, Moore A, LeCluyse EL (2014) A micropatterned hepatocyte coculture model for assessment of liver toxicity using high-content imaging analysis. Assay Drug Dev Technol 12:16–27. (MTO, YOYO)

12. Abraham VC, Towne DL, Waring JF (2008) Application of a high-content multiparameter cytototoxicity assay to prioritize compounds based on toxicity potential in humans. J Biomol Screen 13:527–537. (TMRE, YOYO)

13. Tomida T, Okamura H, Satsukawa M (2015) Multiparametric assay using HepaRG cells for predicting drug-induced liver injury. Toxicol Lett 236:16–24

14. O'Brien PJ, Irwin W, Diaz D et al (2006) High concordance of drug-induced human hepatotoxicity with in vitro cytotoxicity measured in a novel cell-based model using high content screening. Arch Toxicol 80:580–604

15. Tolosa L, Gomez-Lechon J, Donato TM (2015) High-content screening technology for studying drug-induced hepatotoxicity in cell models. Arch Toxicol 89:1007–1022. (TMRM)

16. Adler M, Ramm S, Hafner M (2015) A quantitative approach to screen for nephrotoxic compounds *In Vitro*. J Am Soc Nephrol 10:1681–1693. (CRDR)

17. Liu Y, Batchuluun B, Ho L (2015) Characterization of zinc influx transporters (ZIPs) in pancreatic β cells: roles in regulating cytosolic zinc homeostasis and insulin secretion. J Biol Chem 290:18757–18759. (CRDR)

18. Delgado T, Carroll PA, Punjabi AS (2010) Induction of the Warburg effect by Kaposi's sarcoma herpesvirus is required for the maintenance of latently infected endothelial cells. Proc Natl Acad Sci U S A 107:10696–10701. (image-it dead green)

19. Perry SW, Norman JP, Barbieri J (2011) Mitochondrial membrane potential probes and the proton gradient: a practical user guide. BioTechniques 50:98–115

20. Huang S (2002) Development of a high throughput screening assay for mitochondrial membrane potential in living cells. J Biomol Screen 7:383–389

21. Iannetti EF, Willems PHGM, Pellegrini M et al (2015) Toward high-content screening of mitochondrial morphology and membrane potential in living cells. Int J Biochem Cell Biol 63:66–70

22. Scheiber M, Chandel NS (2014) ROS function in redox signaling and oxidative stress. Curr Biol 24:R453–R462

23. Sirenko O, Hesley J, Rusyn I (2014) High-content assays for hepatotoxicity using induced pluripotent stem cell-derived cells. Assay Drug Dev Technol 12:43–54

24. Saunders DN, Falkenberg KJ, Simpson KJ (2014) High-throughput approaches to measuring cell death. Cold Spring Harb Protoc 10:591–601

25. Giuliano KA, Gough AH, Taylor DL et al (2010) Early safety assessment using cellular systems biology yields insights into mechanisms of action. J Biomol Screen 15:783–797. (TMRE and DHE)

26. Lannetti EF, Willems PHGM, Pellegrini M et al (2015) Toward high-content screening of mitochondrial morphology and membrane

potential in living cells. Int J Biochem Cell Biol 63:66–70

27. Alileche A, Goswami J, Bourland W (2012) Nullomer derived anticancer peptides (NulloPs): differential lethal effects on normal and cancer cells *in vitro*. Peptides 38:302–311. (HCS mitoHealth Kit/PI/DHE)

28. Wilson MS, Graham JR, Ball A (2014) Multiparametric high content analysis for assessment of neurotoxicity in differentiated neuronal cell lines and human embryonic stem cell-derived neurons. Neurotoxicology 42:33–48. (mitotracker red)

29. Wilson J, Berntsen HF, Zimmer KS (2016) Effects of defined mixtures of persistent organic pollutants (POPs) on multiple cellular responses in the human hepatocarcinoma cell line, HepG2, using high content analysis screening. Toxicol Appl Pharmacol 294:21–31. (CRDR)

30. De Raad M, Teunissen EA, Lelieveld D et al (2012) High-content screening of peptide-based non-viral gene delivery systems. J Control Release 158:433–442. (Live Dead)

31. Selvaratnam J, Paul C, Robaire B (2015) Male rat germ cells display age-dependent and cell-specific susceptibility in response to oxidative stress challenges. Biol Reprod 93:1–17. (CellEvent and CellROX)

32. Becker B, Clapper J, Harkins KR et al (1994) In Situ screening assay for cell viability using a dimeric cyanine nucleic acid stain. Anal Biochem 221:78–84

33. Bauch C, Bevan S, Woodhouse H (2015) Predicting in vivo phospholipidosis-inducing potential of drugs by a combined high content screening and in silico modeling approach. Toxicol In Vitro 29:621–630

34. Van de Water FM, Havinga J, Ravesloot WT (2011) High content screening analysis of phospholipidosis: validation of a 96-well assay with CHO-K1 and HepG2 cells for the prediction of *in vivo* based phospholipidosis. Toxicol In Vitro 25:1870–1882

35. Billis P, Will Y, Nadanaciva S (2014) High-content imaging assays for identifying compounds that generate superoxide and impair mitochondrial membrane potential in adherent eukaryotic cells. Curr Protoc Toxicol 25:1.1–1.25

36. Labbe G, Pessayre D, Fromenty B (2008) Drug-induced liver injury through mitochondrial dysfunction: mechanisms and detection during preclinical safety studies. Fundam Clin Pharmacol 22:335–353

37. Donato MT, Gomez-Lechon MJ (2012) Drug-induced liver steatosis and phospholipidosis: cell-based assays for early screening of drug candidates. Curr Drug Metab 13:1160–1173

38. Willebrods J, Pereira IVA, Maes M et al (2015) Strategies, models and biomarkers in experimental non-alcoholic fatty liver disease research. Prog Lipid Res 59:106–125

39. Greenspan P, Mayer EP, Fowler SD (1985) Nile Red: a selective fluorescent stain for intracellular lipid droplets. J Cell Biol 100:965–973

40. Criddle DN, Gillies S, Baumgartner-Wilson HK et al (2006) Menadione-induced reactive oxygen species generation via redox cycling promotes apoptosis of murine pancreatic acinar cells. J Biol Chem 281:40485–40492

41. Vessey DA, Lee KH, Blacker KL (1992) Characterization of the oxidative stress initiated in cultured human keratinocytes by treatment with peroxides. J Invest Dermatol 97:442–446

Chapter 4

Cell-Based High Content Analysis of Cell Proliferation and Apoptosis

Bhaskar S. Mandavilli, Michelle Yan, and Scott Clarke

Abstract

High content imaging-based cell cycle analysis allows multiplexing of various parameters including DNA content, DNA synthesis, cell proliferation, and other cell cycle markers such as phosho-histone H3. 5'-Ethynyl-2'-deoxyuridine (EdU) incorporation is a thymidine analog that provides a sensitive method for the detection of DNA synthesis in proliferating cells that is a more convenient method than the traditional BrdU detection by antibody. Caspase 3 is activated in programmed cell death induced by both intrinsic (mitochondrial) and extrinsic factors (death ligand). Cell cycle and apoptosis are common parameters studied in the phenotypic analysis of compound toxicity and anti-cancer drugs. In this chapter, we describe methods for the detection of s-phase cell cycle progression by EdU incorporation, and caspase 3 activation using the CellEvent caspase 3/7 detection reagent.

Key words Cell cycle, Apoptosis, 5-Ethynyl-2-deoxyuridine, Click chemistry

1 Introduction

Cell proliferation and apoptosis are cellular parameters commonly measured in cancer drug screening [1]. The signaling pathways that control cell cycle progression and apoptosis are tightly regulated to prevent aberrant cells with mutations from proliferating [2–4]. Cell proliferation is commonly measured using the Ki 67 antibody, a marker associated with cell proliferation. Ki 67 is present in all the active phases of cell cycle and is not a good marker to differentiate between different phases of cell cycle except in resting cells (G_0) [5, 6]. Nascent DNA synthesis in replicating cells is a marker for S phase and is traditionally measured with synthetic thymidine analogs such as bromodeoxyuridine (BrdU). BrdU is incorporated into the DNA of actively dividing cells and then detected using an anti-BrdU antibody, which typically requires acid denaturation of the DNA to provide access for the antibody [7]. Alternately, the nascent DNA synthesis can be measured using the incorporation

Paul A. Johnston and Oscar J. Trask (eds.), *High Content Screening: A Powerful Approach to Systems Cell Biology and Phenotypic Drug Discovery*, Methods in Molecular Biology, vol. 1683, DOI 10.1007/978-1-4939-7357-6_4,
© Springer Science+Business Media LLC 2018

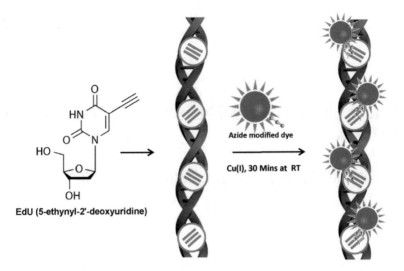

Fig. 1 Detection of incorporated EdU with AlexaFluor® Azide using copper-based click reaction

of 5′-ethynyl-2′-deoxyuridine (EdU) into the DNA followed by copper-based click reaction chemistry using an azide conjugated to a fluorophore [8–11]. The Click-iT EdU HCS assay is an alternative to traditional methods using BrdU [6, 12] for the detection and quantitation of newly synthesized DNA and to classify cells in the G1-S phase of the cell cycle [13]. Azides and alkynes bind together in a copper-catalyzed "click" reaction with high specificity to form stable, inert products. In this application, the alkyne is EdU and the azide is a modified Alexa Fluor® dye (Fig. 1).

Apoptosis is the process of programmed cell death that is characterized by distinct morphological features and activation of biochemical pathways [14, 15]. Apoptosis can be induced by either non-receptor-mediated intrinsic pathways [16, 17], or death receptor regulated extrinsic pathways [18–20]. Caspase 3 is a common effector proteolytic enzyme for both apoptotic pathways that is cleaved by upstream caspase 8 or 9 [14, 15]. Caspase 3 has no activity unless it is cleaved upon the initiation of apoptosis [15]. Activation of caspase-3 is considered an essential event during apoptosis, making it a definitive biomarker for the analysis of apoptotic cells that is regularly used for the preclinical toxicity screening of anti-cancer compounds [21].

2   Materials

2.1   Chemicals

1. Aphidicolin.

2. Click-iT®EdU 488 HCS assay kit. Components of the kit are:

    (a) 5-Ethynyl 2′-deoxyuridine (EdU).

    (b) Alexa Fluor® 488 azide.

      (c) Click-iT® EdU reaction buffer.

      (d) Copper sulfate ($CuSO_4$).

      (e) Click-iT® EdU buffer additive.

      (f) Click-iT® reaction rinse buffer.

      (g) HCS NuclearMask™ Blue stain.

3. Phosphate-buffered saline (PBS).

4. 16% formaldehyde.

5. Triton X-100.

6. CellEvent® Caspase 3/7 Green Detection Reagent.

7. Hoechst® 33342.

8. Staurosporine.

2.2 Cell Culture

1. 96-well cell culture plates.

2. U-2 OS cells.

3. McCoys Medium.

4. TrypLE™ express.

5. HeLa cells.

6. Fetal bovine serum (FBS).

7. Essential amino acids.

8. Minimal essential medium (MEM).

9. Deionized water.

10. Laminar flow hood.

11. Cell Culture incubator.

12. High content imaging instrument.

3 Methods

3.1 Cell Proliferation

The method described here is based on Click-iT EdU HCS 488 assay kit which contains all the components to quantitate the proliferative capacity of cells in a population based on nascent DNA synthesis. Using this method in conjunction with a suitable nuclear DNA stain, it is possible to measure both DNA content and cell proliferation based on EdU incorporation during DNA synthesis. The protocol described here shows the application of the HCS-based EdU cell proliferation assay in measuring the effects of DNA synthesis inhibitor, aphidicolin in Human bone osteosarcoma epithelial (U-2 OS) cells (Fig. 2).

3.2 Reagent Preparation

1. To make a 10× stock solution of the Click-iT® EdU buffer additive (component E) add 2 mL of deionized water to the vial and mix until fully dissolved. The solution can be stored at

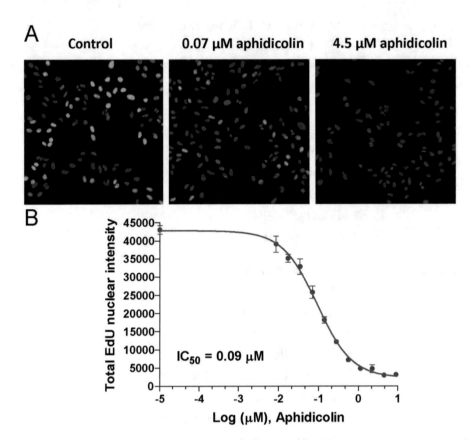

Fig. 2 U-2 OS cells were treated with the indicated amounts of aphidicolin for 3 h, and the effects on the DNA synthesis were assayed after 1 h incubation with 10 μM EdU. Cells were then fixed and permeabilized, and EdU incorporated into newly synthesized DNA was detected using the fluorescent Alexa Fluor® 488 azide from the Click-iT® EdU Alexa Fluor® 488 Assay (**a**). Quantitative analysis was performed using the Thermo Scientific Cellomics® ArrayScan® VTI and the compartmental analysis image analysis algorithm. The EC$_{50}$ value was derived by plotting the total EdU nuclear intensity against the indicated concentrations of aphidicolin (**b**)

−20 °C and is stable for 1 year at this temperature. If the solution develops brown color, it has degraded and should be discarded.

2. Prepare 1× Click-iT® EdU reaction buffer (component C) by diluting the 10× stock buffer in deionized water. E.g., for 40 mL of 1× Click-iT® EdU reaction buffer, mix 4 mL of the 10× Click-iT® EdU reaction (component C) buffer with 36 mL of the deionized water.

3.3 Labeling Cells with EdU

1. Seed U-2 OS cells in a 96-well plate at 60% confluence and incubate overnight at 37° C in a CO$_2$ incubator (*see* **Note 1**).

2. Treat the cells in the growth medium with different concentrations of aphidicolin, the highest concentration being 1 μM, and incubate at 37 °C for 3 h (Fig. 2).

Table 1
The click reaction cocktail

Reaction components	Number of plates			
	1	2	5	10
1× Click-iT® EdU reaction buffer (prepared in step)	10.3 mL	20.6 mL	51.5 mL	103 mL
CuSO₄ (Component)	480 µL	960 µL	2.4 mL	4.8 mL
Alexa Fluor® azide (Component)	30 µL	60 µL	150 µL	300 µL
1× Click-iT buffer additive (prepared in step)	1.2 mL	2.4 mL	6 mL	12 mL
Total volume	*12 mL*	*24 mL*	*60 mL*	*120 mL*

3. Remove media containing aphidicolin from the cells.

4. Treat cells with 10 µM EdU and incubate at 37 °C for 1 h (*see* **Note 1**).

5. Wash the cells 1× with PBS.

6. Fix cells by adding 100 µL of 4% formaldehyde and incubate for 15 min at room temperature.

7. Remove fixative from the cells and then wash the cells 3× with PBS. Samples may be stored at 4 °C at this time or proceed with permeabilization step.

8. Permeabilize cells by adding 100 µL of 0.1% Triton X-100 for 15 min at room temperature.

3.4 EdU Detection

1. Prepare the Click-iT reaction cocktail (Table 1) (*see* **Notes 2** and **3**).

2. Remove the permeabilization buffer from each well and wash 2× with PBS.

3. Remove the wash solution and add 100 µL of click reaction cocktail (prepared in **step 1**) to each well.

4. Incubate for 30 min at room temperature, protected from light.

5. Remove the reaction cocktail and wash the cells once with Click-iT® reaction rinse buffer.

6. Remove the rinse buffer and wash the cells 2× with PBS.

7. At this stage, samples may be labeled with other probes such as antibodies, or proceed with nuclear staining and imaging. Remember to protect cells from light during these procedures.

3.5 Nuclear Staining

The following protocol is based upon 50 µL of HCS Nuclear-Mask™ Blue (Hoechst 33342) staining solution per well.

1. Dilute HCS NuclearMask™ Blue stain solution 1:2000 in PBS to obtain a 1× NuclearMask™ Blue stain working solution.

2. Remove any wash solution from the cells.

3. Add 50 μL of 1× HCS NuclearMask™ Blue stain solution to each well. Incubate for 30 min at room temperature protected from the light (*see* **Note 4**).

4. Remove the HCS NuclearMask™ Blue stain solution and wash each well 2× with PBS.

5. Remove the wash solution and add 100 μL of PBS, seal the plates with the plastic film.

6. Scan the plates using an automated imaging (HCS) platform. The data shown in Fig. 1 were generated on a Thermo Scientific ArrayScan VTI using a 20× objective to acquire images and the compartmental analysis bioapplication for image analysis (*see* **Note 5**).

3.6 Apoptosis

The CellEvent® Caspase-3/7 Green Detection Reagent is intrinsically non-fluorescent as the DEVD peptide inhibits the ability of the dye to bind to DNA. However, after the activation of caspase-3/7 in apoptotic cells, the DEVD peptide is cleaved enabling the dye to bind to DNA and produce a bright, fluorogenic response (Fig. 3). The fluorescence emission of the dye when bound to DNA is ~530 nm and can be observed using a standard "FITC" filter set (Fig. 3) [22–35]. The method described here uses a protein kinase inhibitor and apoptosis inducer, staurosporine to measure activation of caspase 3/7 using CellEvent® Caspase-3/7 Green Detection Reagent (Fig. 4).

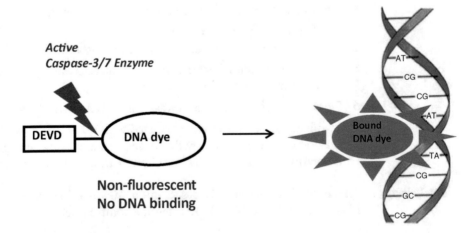

Fig. 3 Detection of cleaved caspase 3/7 by CellEvent™ Caspase-3/7 Green Detection Reagent

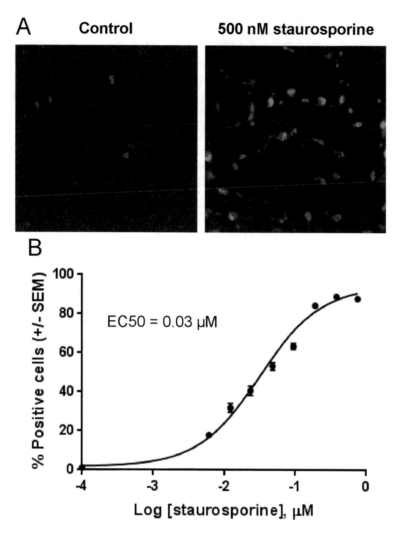

Fig. 4 (**a**) HeLa cells were treated with either vehicle (*left*) or 500 nM staurosporine for 4 h. The cells were then stained with 1 μg/mL Hoechst 33342 and 5 μM CellEvent™ Caspase-3/7 Green Detection Reagent at 37 °C for 30 min. (**b**) HeLa cells were treated with different concentrations of staurosporine (0–0.75 μM) for 4 h and then stained with Hoechst 33342 and 5 μM CellEvent™ Caspase-3/7 Green Detection Reagent for 30 min at 37 °C. The plate was scanned using the Thermo Scientific Cellomics® ArrayScan® VTI. The mean signal intensities were plotted against staurosporine concentrations and EC_{50} values calculated

1. Seed HeLa cells on a 96-well plate at a density of 5000 cells per well in 100 μL of MEM supplemented with 10% FBS (complete growth medium).

2. Incubate the plates overnight at 37 °C with 5% CO_2.

3. Dilute the stock solution of stauropsorine in a complete growth medium to a final concentration of 500 nM (*see* **Note 6**).

4. Prepare a 500 μM stock solution of staurosporine in DMSO by mixing 0.5 mg of the drug in 1.072 mL DMSO. Aliquot the

stock solution immediately and store at $-20\ °C$. The solution is stable for 6 months.

5. Remove the medium from the cells and add 100 μL of complete growth medium (prepared in **step 4**) to half the plate. Treat the other half of the plate with a medium containing DMSO without staurosporine for vehicle controls and incubate for 4 h at $37\ °C$ with 5% CO_2.

6. Add CellEvent™ Caspase-3/7 Green Detection Reagent at a final concentration of 5 μM to control and/or treated cells and incubate for 30 min at $37\ °C$ (*see* **Note 7**).

7. Label cells with Hoechst® 33342 (final concentration of 1 μg/ mL) for the final 10 min of CellEvent™ labeling at $37\ °C$ with 5% CO_2 (Fig. 4).

8. Labeled cells can be preserved with a formaldehyde-based fixative at this stage. Fixation with 4% formaldehyde for 15 min at RT is recommended, but this can be altered based on the cell type.

9. Remove the medium from the cells and wash the cells $3×$ with PBS and scan the plates using an automated imaging platform. For the experiments detailed in this section, the images were acquired on a Thermo Scientific Cellomics® ArrayScan® VTI with a $20×$ objective and analyzed using the Compartmental analysis Bioapplication (*see* **Notes 8–10**).

4   Notes

1. The protocol described above used U-2 OS cells as an example, but the assay can be used in any other cell line. The EdU concentration and pulsing time should be optimized for each cell line.

2. The copper catalyzed click reaction can destroy fluorescent proteins. If multiplexing with fluorescent proteins such as GFP, use AlexaFluor picolyl azide instead of the AlexaFluor azide described in the method above [36].

3. After the Click-iT® reaction, the cells can be labeled with antibodies for multiplex analysis.

4. Please use caution when using HCS NuclearMask™ Blue stain as it is a known mutagen.

5. The protocol described here used the Thermo Scientific Cellomics® ArrayScan® VTI platform using a $20×$ objective and the compartmental analysis bioapplication. Channel 1 is used to segment the nuclear region using the Hoechst 33342 signal. The EdU signals were measured in the nuclear region of

channel 2 using a 488 nm excitation filter capturing 25% of the maximum intensity of the signal.

6. Gentle handling of cells after exposure to toxic drugs that induce apoptosis is recommended to minimize cell loss. The aspiration of medium or PBS from cells has to be done slowly without touching the bottom of the well. Alternately, the cells can be fixed by adding 4× concentrated (16%) formaldehyde on the top of the medium at the end of the drug incubation to minimize cell loss.

7. Initial testing between 2 and 10 μM of CellEvent™ Caspase-3/7 Green Detection Reagent is recommended; however, optimal concentrations may be more or less depending on the cell model.

8. CellEvent™ Caspase-3/7 Green Detection Reagent signal is predominantly localized in the nucleus. Measure signal intensity only in nuclear area to avoid measuring any cytoplasmic background and to improve the signal-to-noise (S/N) of the assay.

9. The Hoechst 33342 staining can be very intense in apoptotic cells due to chromatin condensation. This feature can be used as another parameter for apoptosis along with reduced nuclear area.

10. Excitation filters at 386 nm (DAPI) and 485 nm (FITC) were used for the measurement of Hoechst 33342 in channel 1 and CellEvent Caspase 3/7 reagent in channel 2 respectively. The caspase signal was measured in the nuclear region of the cell. The signal intensities are measured at 25% of maximum intensity in both the channels. % responders for caspase 3 activation are plotted in the graph. To calculate % responders the threshold intensity values are based on the maximum intensity values of negative control wells and all the cells that are significantly above these intensity values are considered responders for caspase 3 activation.

References

1. Towne DL, Emily EN, Comess KM et al (2012) Development of a high-content screening assay panel to accelerate mechanism of action studies for oncology research. J Biomol Screen 17(8):1005–1017

2. Gou M, Hay BA (1999) Cell proliferation and apoptosis. Curr Opin Cell Biol 11:745–752

3. Pucci B, Maragaret K, Giordano A (2000) Cell cycle and apoptosis. Neoplasia 2(4):291–299

4. Pietenpol JA, Stewart ZA (2002) Cell cycle checkpoint signaling: cell cycle arrest versus apoptosis. Toxicology 181-182:475–481

5. Scholzen T, Gerdes J (2000) The Ki-67 protein: from the known and the unknown. J Cell Physiol 182(3):311–322

6. Gasparri F, Mariani M, Sola F (2004) Quantification of the proliferation index of human dermal fibroblast cultures with the ArrayScan™ high-content screening reader. J Biomol Screen 9(3):232–243

7. Rothaeusler K, Baumgarth N (2006) Evaluation of intranuclear BrdU detection procedures for use in multicolor flow cytometry. Cytomtery A 69:249–259

8. Salic A, Mitchinson TJ (2008) A chemical method for fast and sensitive detection of DNA synthesis *in vivo*. Proc Natl Acad Sci U S A 105(7):2415–2420

9. Buck SB, Bradford J, Gee K (2008) Detection of S-phase cell cycle progression using 5-ethynyl-2′-deoxyuridine incorporation with click chemistry, an alternative to using 5-bromo-2′-deoxyuridine antibodies. BioTechniques 44(7):927–929

10. Diermeier-Daucher A, Clarke ST, Hill D (2009) Cell type specific applicability of 5-ethynyl-2′-deoxyuridine (EdU) for dynamic proliferation assessment in flow cytometry. Cytometry 75A:535–546

11. Limsirichaikul S, Niimi A, Fawcett H (2009) A rapid non-radioactive technique for measurement of repair synthesis in primary human fibroblasts by incorporation of ethynyl deoxyuridine (EdU). Nucleic Acids Res 37:1–10

12. Crosby LM, Luellen C, Zhang Z (2011) Balance of life and death in alveolar epithelial type II cells: proliferation, apoptosis, and the effects of cyclic stretch on wound healing. Am J Physiol Lung Cell Mol Physiol 301: L536–L546

13. Robertson FM, Ogasawara MA, Ye Z (2010) Imaging and analysis of 3D tumor spheroids enriched for a cancer stem cell phenotype. J Biomol Screen 15(7):820–828

14. Elmore S (2007) Apoptosis: a review of programmed cell death. Toxicol Pathol 35 (4):495–516

15. Cullen SP, Martin SJ (2009) Caspase activation pathways: some recent progress. Cell Death Differ 16:935–938

16. Cai J, Yang J, Jones DP (1998) Mitochondrial control of apoptosis: the role of cytochrome C. Biochim Biophys Acta 1366:139–149

17. Saelens Z, Festjens N, Wande Valle L (2004) Toxic proteins released from mitochondria in cell death. Oncogene 23:2861–2874

18. Chinnaiyan AM, O'Rourke K, Tewari M et al (1995) FADD, a novel death domain-containing protein, interacts with the death domain of Fas and initiates apoptosis. Cell 81:505–512

19. Chicheportiche Y, Bourdon PR, Xu H et al (1997) TWEAK, a new secreted ligand in the tumor necrosis factor family that weakly induces apoptosis. J Biol Chem 272:32401–32410

20. Peter ME, Krammer PH (1998) Mechanisms of CD 95 (Apo-1/Fas)-mediated apoptosis. Curr Opin Immunol 10:545–551

21. Uttamapinant C, Tangpeerachaikul A, Grecian S et al (2012) Fast, cell-compatible click chemistry with copper-chelating azides for bimolecular labeling. Angew Chem Int Ed 51:5852–5856

22. Segawa K, Kurata S, Yanagihashi Y et al (2014) Caspase-mediated cleavage of phospholipid flippase for apoptotic phosphatidylserine exposure. Science 344:1164–1168

23. Divya KR, Liu H, Ambudkar SV et al (2014) A combination of curcumin with either gramicidin or ouabain selectively kills cells that express the multi-drug resistance-linked ABCG2 transporter. J Biol Chem 289:31397–31410

24. Bolanos JMG, Da Silva CMB, Munoz PM et al (2014) Phosphorylated AKT preserves stallion sperm viability and motility by inhibiting caspases 3 and 7. Reproduction 148:221–235

25. Hristov G, Martilla T, Durand C (2014) SHOX triggers the lysosomal pathway of apoptosis via oxidative stress. Hum Mol Genet 23:1619–1630

26. Anoop Chandran P, Keller A, Weinmann L (2014) Inflammation, extracellular mediators, and efgfector molecules: the TGF-β-inducible miR-23a cluster attenuates IFN levels and antigen-specific cytotoxicity in human CD8[+] T cells. J Leukoc Biol 96:633–645

27. Munoz MP, Ferrusola CO, Vizuete G et al (2015) Depletion of intracellular thiols and increased production of 4-hydroxynonenal that occur during cryopreservation of stallion spermatozoa lead to caspase activation, loss of motility, and cell death. Biol Reprod 93 (6):1–11

28. Funk J, Biber N, Scneider M (2015) Cytotoxic and apoptotic effects of recombinant subtilase cytotoxin variants of shiga toxin-producing *Escherichia coli*. Infect Immun 83:2338–2349

29. Fuchs R, Schwach G, Stracke A (2015) The anti-hypertensive drug prazosin induces apoptosis in the medullary thyroid carcinoma cell line TT. Anticancer Res 35:31–38

30. Sweetwyne MT, Gruenwald A, Niranjan T (2015) Notch 1 and Notch 2 in podocytes play differential roles during diabetic nephropathy development, Diabetes 64:4099–4111

31. Kular J, Tickner JC, Pavlos NJ (2015) Choline kinase β mutant exhibit reduced phosphocholine, elevated osteoclast activity, and low bone mass. J Biol Chem 290:1729–1742

32. Dong DJ, Jing YP, Liu W (2015) The steroid hormone 2—hydroxyecdysone up regulates Ste-20 family serine/threonine kinase Hippo to induce programmed cell death. J Biol Chem 290:24738–24746

33. Tsunoda T, Ishikura S, Doi K (2015) Establishment of a three-dimensional floating cell culture system for screening drugs targeting KRAS-mediated signaling molecules. Anticancer Res 35:4453–4459

34. Sergin I, Bhattacharya S, Emanuel R et al (2016) Inclusion bodies enriched for p62 and polyubiquitinated proteins in macrophages protect against atherosclerosis. Sci Signal 9:ra2

35. Carpio MA, Michaud M, Zhou W (2015) BCL-2 family member BOK promotes apoptosis in response to endoplasmic reticulum stress. Proc Natl Acad Sci U S A 112:7201–7206

36. Antczak C, Takagi T, Ramirez CN et al (2009) Live-cell imaging of caspase activation for high-content screening. J Biomol Screen 14 (8):956–969

Chapter 5

Tools to Measure Autophagy Using High Content Imaging and Analysis

Nick J. Dolman, Brent A. Samson, Kevin M. Chambers, Michael S. Janes, and Bhaskar S. Mandavilli

Abstract

Macroautophagy, hereafter referred to as autophagy, is a predominately pro-survival catabolic process responsible for the degradation of long-lived or aggregated proteins, invading microorganisms and damaged or redundant intracellular organelles. Removal of these entities is achieved through encompassment of the target by the autophagosome and subsequent delivery to the lysosome. The use of fluorescence microscopy is a common method to investigate autophagy through monitoring the spatial and temporal recruitment both of autophagosomal markers and cargo to the autophagosome. In this section, we will discuss the use of high content imaging (HCI) and analysis in the study of autophagy with reference to commonly used markers of autophagosomal formation.

Key words High content imaging, High content screening, Autophagy, Autophagosome, LC3B, ATG5, CRISPR

1 Introduction

The process of autophagosome formation, cargo recruitment, and fusion with the lysosome is a highly complex, dynamic, and transient process [1]. The stimulation of autophagy leads to the formation of an isolation membrane which is then followed by recognition and engulfment of cargo by the expanding isolation membrane. Closure of this membrane to form the autophagosome is followed by fusion with the lysosome (Fig. 1a). For a recent review of the molecules involved in autophagy, please *see* the manuscript by Mizushima and Komatsu [1]. The elucidation of these steps, and the signaling cascades that regulate them, have relied heavily upon fluorescence imaging of cells. The ability of fluorescence imaging to track these features in space and over time has led to the widespread use of fluorescence imaging in the field of autophagy. For a recent review of methods to measure autophagy,

Paul A. Johnston and Oscar J. Trask (eds.), *High Content Screening: A Powerful Approach to Systems Cell Biology and Phenotypic Drug Discovery*, Methods in Molecular Biology, vol. 1683, DOI 10.1007/978-1-4939-7357-6_5,

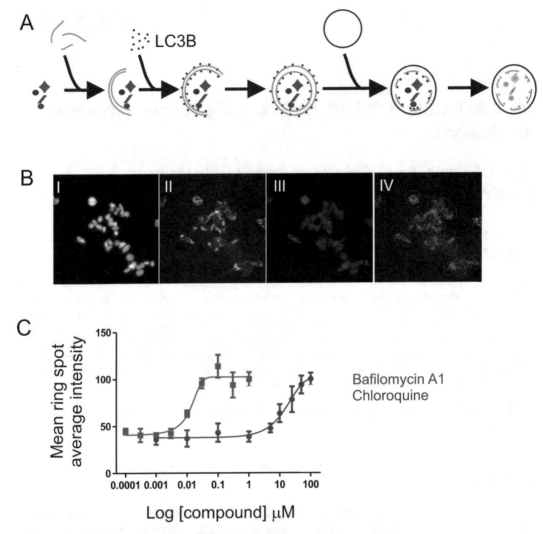

Fig. 1 Markers of autophagy. (**a**) Stages of autophagy described in the text, including the recruitment of LC3B to the autophagosomal membrane. (**b**) Segmentation of IBMK cells expressing GFP-LC3. Nuclei, labeled with Hoechst 33342 (I) autophagosomes labeled with GFP-LC3 (II), GFP-positive puncta are clustered around the nuclei (III), rings (*green*) encapsulate each cell and mark the boundaries within which LC3B positive spots (*red*) are segmented (IV). (**c**) Dose-dependent accumulation of autophagosomes in response to chloroquine diphosphate or Bafilomycin A1, reported by anti-LC3B immunolabeling, in A549 cells

see Klionsky et al. [2]. Automated high content imaging (HCI) offers the advantages of fluorescence imaging with unparalleled throughput [3, 4] and has played an important role in many studies of autophagy [5, 6]. HCI approaches have been used to examine genome-wide RNAi screens to identify novel regulatory proteins [7], as well as factors involved in selective autophagy [8], promotion of cancer cell survival [9], regulation of autophagy by small-molecules [10], interplay between antioxidants and autophagy during neurodegeneration [11], and regulators of mitophagy [12] to name but a few examples. The preferred means to assess autophagy is to monitor the recruitment of microtubule-associated protein 1 light chain 3 (MAP1LC3), commonly abbreviated to LC3, to the autophagosome [2, 13]. LC3 is the mammalian homolog of the yeast protein ATG8 and the four isoforms of LC3 (A, B, B2, C) form one family of these homologs [13]. The second family contains GABA$_A$-associated receptor proteins (GABARAP) of which there are three members [13].

Each of these molecules can be used in the same manner to monitor autophagy [13]. In this chapter we will present data using LC3 as a marker; however, these approaches are applicable to the other molecules described above. Under conditions of low autophagy, LC3 is distributed throughout the cytoplasm in a non-lipidated form (LC3-I). Upon induction of autophagy, LC3 is conjugated to phosphatidylethanolamine and it is this lipidated form of LC3 that associates with the autophagosomal membrane [13]. The recruitment of LC3 can be reported using a number of means, for example, antibodies against LC3 or fluorescent protein (FP) chimeras of LC3. Both these approaches have their advantages and disadvantages and the method employed should be carefully considered a function of the particular experimental goals and circumstances. Generally speaking, the use of FP constructs involves additional work upstream of the actual autophagy measurements, either establishing a stable cell line or transient expression of an FP biosensor. This can be offset by a reduced workflow circumventing the process of immunolabeling of targets and, of course, enables live-cell imaging of autophagosome and autolysosome formation. Antibodies against LC3 label endogenous LC3 and are therefore not prone to potential artifacts introduced through overexpressing an FP-based reporter [14]. However, fixation protocols that preserve autophagosomal staining by anti-LC3 need to be confirmed.

The use of antibodies involves a more complex multi-step protocol after the treatment of cells. We describe the materials and methods required to perform HCI assays of autophagy using antibodies against LC3B. Many other markers of autophagy, both upstream and downstream of LC3 recruitment, are amenable to imaging but are beyond the scope of this chapter. For these additional reporters two recent reviews should be consulted [2, 13]. This chapter describes the materials and methods required to monitor autophagy using HCI and provides detail on the relevant fluorescent probes and sensors that form an integral part of these experiments.

2 Materials

2.1 Autophagosomal Quantification with an Antibody Against LC3B

Labeling cells with antibodies against LC3 allows for the detection of autophagosome formation through detection of endogenous LC3. Primary antibody staining can also be detected using a broad range of fluorophores coupled to secondary antibodies facilitating the analysis of multiple markers within the same sample. The antibody used in the following protocol recognizes residues at the amino terminal end of LC3B from human, rat, and mouse cells. This antibody does not show a preference for either the cleaved and lipidated (LC3B-II) or the non-cleaved form of LC3B (LC3B-I). Possible artifacts introduced through the fixation and permeabilization should be considered [15] along with the extra time and multiple steps required for immunolabeling. This is especially relevant if automation/automated liquid handling is desired, for example when screening through large compound or siRNA libraries.

1. Greiner CELLSTAR® 96-well plates, black polystyrene wells flat bottom (with micro-clear bottom) (*see* **Note 1**).
2. A549 Cells, ATCC CCL-185 (*see* **Note 2**).
3. Minimal essential medium (MEM).
4. Fetal Bovine Serum (FBS).
5. Rabbit polyclonal anti LC3B (*see* **Note 3**).
6. Goat-anti Rabbit IgG conjugated to Alexa Fluor™ 647 dye (*see* **Note 4**).
7. Hoechst 33342 dye.
8. Chloroquine diphosphate.
9. Bafilomycin A1.
10. 16% Formaldehyde, EM Grade, Methanol free.
11. Triton X-100.
12. Dulbecco's phosphate-buffered saline (DPBS).
13. Heat inactivated normal goat serum.
14. Bovine serum albumin.

2.2 Specificity of Autophagosomal Labeling

The accumulation of LC3-positive vesicles (or other family members) is accepted as indicative that autophagy has been induced and forms the basis of most assays [13]; however, LC3 puncta can form independent of autophagy. For this reason, it is important to confirm the specificity of the chosen labeling approach. Pharmacological inhibitors of phosphatidylinositol 3-kinase (for example wortmannin or LY294002) can be used. However, these compounds inhibit multiple isoforms of phosphatidylinositol 3-kinase and may not show truly specific actions upon autophagy [16]. A more definitive approach is genetic deletion that results in the loss

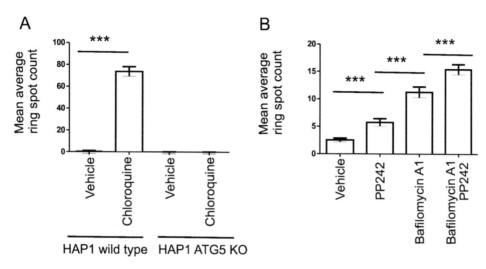

Fig. 2 Specificity and source of LC3 labeling. (**a**) Hap1 cells, parental or ATG5 KO, immunolabeling with anti-LC3B. Parental Hap1 cells showed chloroquine diphosphate-induced accumulation of LC3B positive puncta, as revealed by a significant increase in the mean average "ring spot" count. In Hap1 cells where CRISPR/Cas9 was used to knock out ATG5 the chloroquine diphosphate-induced increase in mean average ring spot is absent, confirming the specificity of this method to detect autophagosomes formed by canonical, ATG5 dependent, autophagy. (**b**) Source of autophagosomes. PP242 increases the number of LC3B-positive puncta via stimulation of autophagy as the average number of LC3-positive puncta (mean average "ring spot" count) in the presence of Bafilomycin A1 is significantly higher than either PP242 or Bafilomycin A1 alone. This indicates that PP242 is not acting to inhibit lysosomal activity and is therefore stimulating autophagy. ***p, 0.005, (Student's t-test)

of key autophagy genes. We have successfully knocked down ATG4B, ATG5, or ATG7 using siRNA (data not shown) and observed attenuated LC3B labeling. More recently, the description of CRISPR/Cas9 systems to "edit" genomes has provided an invaluable tool to effectively delete genes from target cells [17]. Here, we describe the use of CRISPR/Cas9 to delete a region within the coding sequence of ATG5 which completely ablates the LC3B labeling pattern that is otherwise observed in response to compounds that block autophagic flux (Fig. 2a and *see* **Note 5**).

1. Greiner CELLSTAR® 96-well plates black polystyrene wells flat bottom (with micro-clear bottom) (*see* **Note 1**).

2. Parental Human haploid cells (Hap1).

3. CRISPR-edited Human haploid cells (Hap1) lacking ATG5 (ATG5 KO) (*see* **Note 5**).

4. Iscove's Modified Dulbecco's Medium (IMDM).

5. Fetal calf serum (FCS).

6. Penicillin/streptomycin.

7. Materials listed in Subheading 2.1 to label cells with anti-LC3B.

2.3 Source of LC3B Labeling

The use of knock-out models confirms the specificity of LC3 labeling to the process of autophagy (*see* above). If it is necessary to identify which process gave rise to the increase in LC3 labeling, experiments should be performed to measure stimulation of autophagy or inhibition of lysosomal breakdown as the full flux through the pathway is completed. This can be achieved by performing the experimental manipulation in the presence and absence of a known lysosomal modulator, such as Bafilomycin A1 or chloroquine diphosphate [18]. Lysosomal inhibitors impair the ability of the lysosome to degrade cargo and therefore cause an increase in LC3 puncta. If an experimental manipulation stimulates the production of LC3 positive puncta, then in the presence of the lysosomal inhibitor, the level of staining seen will be higher than with the lysosomal modulator on its own. On the other hand, if the manipulation also acts to inhibit lysosomal function, then no additive level of staining is seen [18] (*see* **Note 6** and Fig. 2b).

1. Materials listed in Subheading 2.1 (in the case of labeling cells with anti-LC3B).

2. Bafilomycin A1 (*see* **Note 6**).

3. PP242 (mTOR inhibitor) (*see* **Note 6**).

3 Method

3.1 LC3B Antibody: Cell Culture

1. Prepare complete growth media by adding FBS to MEM to a final concentration of 10%.

2. A549 cells are seeded in a 96-well plate at a density of 5000 cells per well in 100 μL of complete growth media (prepared in **step 1**).

3. Incubate plates overnight at 37 °C in a humidified atmosphere with 5% CO_2.

3.2 Experimental Manipulation: Compound Dose-Response

1. Prepare a serial dilution of chloroquine diphosphate and Bafilomycin A1 covering the range of 90 μM to 1 nM and 10 μM to 100 pM, respectively, in complete growth media.

2. Prepare vehicle controls, for Bafilomycin A1 this is DMSO and for chloroquine diphosphate this is water.

3. Aspirate complete growth media from each well.

4. Pipette 100 μL of the designated Bafilomycin A1 or chloroquine diphosphate concentration into each column [2–11]. Column 1 will be treated with vehicle whereas column 12 is a media-only control.

5. Incubate plates overnight at 37 °C in a humidified atmosphere with 5% CO_2.

3.3 Immunolabeling A549 Cells with LC3B Antibody

1. Prepare 10 mL of 4% formaldehyde solution (in PBS warmed to 37 °C).

2. Prepare 10 mL of permeabilization buffer (0.2% Triton X-100 solution in DPBS).

3. Prepare 10 mL of blocking buffer by adding bovine serum albumin and heat inactivated normal goat serum to a final concentration of 3% (w:v) and 5% (v:v) respectively.

4. Aspirate complete growth media from each well.

5. Immediately replace with 100 μL of 4% formaldehyde solution (prepared in **step 1**).

6. Incubate the plate at 37 °C for 15 min.

7. Aspirate the 4% formaldehyde from each well.

8. Immediately replace with 100 μL of permeabilization buffer.

9. Incubate at room temperature for 15 min with gentle rocking.

10. Aspirate permeabilization buffer.

11. Wash three times in DPBS (use 100 μL per well).

12. Incubate cells in 100 μL of blocking buffer for 1 h at room temperature with gentle rocking.

13. Just prior to the end of the blocking step, prepare the primary antibody solution by adding 6 μL of rabbit polyclonal LC3B antibody to 12 mL of blocking buffer.

14. Aspirate blocking buffer.

15. Incubate cells in 100 μL primary antibody solution for 1 h at room temperature with gentle rocking.

16. Aspirate primary antibody solution.

17. Wash three times in DPBS (use 100 μL per well). For each wash, the DPBS may be left on for 5 min with gentle rocking to ensure complete removal of residual (unbound) primary antibody.

18. During the final wash step, prepare the secondary antibody solution by diluting 6 μL of goat anti-rabbit Alexa Fluor™ 647 conjugate in 12 mL (to a final concentration of 1 μg/mL) of blocking buffer. Nuclear counterstaining can be performed at the same time as secondary antibody labeling by adding 1.2 μL of Hoechst 33342 to the 12 mL of blocking buffer (to a final Hoechst 33342 concentration of 1 μg/mL) containing the secondary antibody.

19. Aspirate wash solution.

20. Add 100 μL of secondary antibody solution (prepared in **step 18**) to each well and incubate at room temperature for 30 min with gentle rocking.

21. Wash three times in DPBS (use 100 μL per well). For each wash, the DPBS may be left on for 5 min with gentle rocking to ensure complete removal of residual (unbound) primary antibody.

22. Image the cells using a HCI platform. The data in Fig. 1 was acquired using the CellInsight™ Cx5 high content screening platform with a 20× air objective lens (*see* **Note 7**) using the autophagy bio application in HCSStudio™ 3.0. Use filters appropriate for Hoechst 33342 and Alexa Fluor™ 647 (*see* **Note 8**). Automated image acquisition of samples takes place after focusing criteria, exposure times, background removal, object identification, segmentation, validation, and selection thresholds have been set. Depending on the goal of the researcher, the thresholds and ranges are typically set with regard to images from wells which contain extreme biological treatments or control compounds as a point of reference for the rest of the samples.

23. For the images and datasets included here, we constructed a two-channel image analysis protocol (in the HCS Studio 3.0 software) using the SpotDetector.v4 BioApplication. Channel one images were of cells labeled with Hoechst 33342 and channel two were images of cells labeled with anti-LC3. Background fluorescent intensities were removed from both channels one and two. Nuclei were identified in channel one by applying smoothing followed by a fixed thresholding level and an intensity based segmentation method. Once nuclei were robustly identified in channel one with a circular overlay at the nuclear membrane, spots representing LC3 positive autophagosomes were identified in channel two with a ring overlay in the cytoplasm. To achieve this, a circular ("circ") primary mask was used as the center of the ring region of interest (ROI) for which LC3 spots were identified in channel two. The distance of the ring mask was expanded to encapsulate the bulk of the cell cytoplasm. We have found that the ring captures the most spots when there is no distance between the edge of the nuclear membrane "circ" region and where the cytoplasmic "ring" begins. Spot identification used a box method with a triangle thresholding. The exact values of the nuclear and spot identification thresholds will vary depending on cell type and the specific fluorescent labeling protocol.

24. A number of parameters extracted for the digital images by the image analysis algorithm can be presented (Fig. 3 and *see* **Note 9**). Dose-dependent accumulation of anti-LC3B (Alexa Fluor™ 647) positive puncta following treatment with either chloroquine diphosphate or Bafilomycin A1 is shown in Fig. 1c as the mean average intensity of spots within the ring.

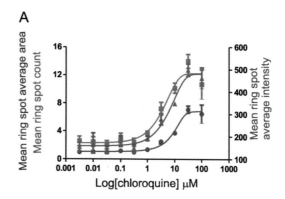

Fig. 3 HCS parameters. (**a**) Autophagy can be quantified with several different readouts. LC3B puncta can be measured for average area (*purple*), number (*green*), or average intensity (*red*). All three parameters show the same dose-dependent relationship with increasing concentrations of chloroquine diphosphate

3.4 Hap1 Cell Culture

1. Prepare complete growth media by adding FBS and penicillin/streptomycin to IMDM to a final concentration of 10% and 1% respectively.

2. Parental or ATG5 KO Hap1 cells are seeded in a 96-well plate at a density of 5000 cells per well in 100 μL of complete growth media (prepared in **step 1**).

3. Incubate plates overnight at 37 °C in a humidified atmosphere with 5% CO_2.

3.5 Confirmation of Reporter Specificity Using CRISPR Edited Hap1 Cells

1. Prepare a 30 μM chloroquine diphosphate solution by diluting the 30 mM stock 1 in 1000 in complete growth media.

2. Prepare vehicle control by adding 1 μL of sterile distilled water to 1 mL of complete growth media

3. Aspirate complete growth media from each well.

4. Pipette 100 μL of the 30 μM chloroquine diphosphate or vehicle control into each column [2–11]. Column 1 will be treated with vehicle whereas column 12 is a media-only control.

5. Incubate plates overnight at 37 °C in a humidified atmosphere with 5% CO_2.

6. Label with anti-LC3B as described in Subheading 3.3, **steps 1–21**.

7. Image the cells using a HCI platform. The data in Fig. 1 were acquired using the CellInsight™ Cx7 high content screening platform with a 40× air objective lens (*see* **Note 7**) using the autophagy bio application in HCS Studio™ 3.0. Use filters appropriate for Hoechst 33342 and Alexa Fluor™ 647 (*see* **Note 8**). Image acquisition and analysis procedures followed

similar steps to those described in Subheading 3.3, **steps 22–24**. Channel one images were of cells labeled with Hoechst 33342 and channel 2 images were of cells labeled with anti-LC3.

8. The specificity of LC3B immunolabeling is depicted in Fig. 2a as the mean average number of spots detected within the ring.

3.6 Confirmation of the Source of Autophagosomes

1. Prepare complete growth media by adding FBS to MEM to a final concentration of 10%.

2. A549 cells are seeded in a 96-well plate at a density of 5000 cells per well in 100 μL of complete growth media (prepared in **step 1**).

3. Incubate plates overnight at 37 °C in a humidified atmosphere with 5% CO_2.

3.7 Experimental Manipulation: Addition of Inducers and Inhibitors

1. Prepare complete growth media containing vehicle (DMSO) or 1 μM Bafilomycin.

2. Aspirate complete growth media from each well.

3. Add 100 μL of either vehicle or Bafilomycin A1 to the appropriate wells.

4. Incubate plates overnight at 37 °C in a humidified atmosphere with 5% CO_2.

4. Prepare complete growth media containing 20 μM PP242 or both PP242 and 1 μM Bafilomycin A1.

5. Aspirate complete growth media from the wells to be treated with PP242 or both PP242 and Bafilomycin A1.

6. Add 100 μL of complete growth media containing 20 μM PP242 or both PP242 and 1 μM Bafilomycin A1 to the appropriate wells.

7. Incubate plates for 1 h at 37 °C in a humidified atmosphere with 5% CO_2.

8. Follow the steps for anti LC3B staining in Subheading 3.3, **steps 1–21**.

9. The protocol described here was developed on a Thermo Scientific CellInsight™ Cx5 high content screening platform using a 20× air objective (*see* **Note 7**) and the autophagy bioapplication in HCS Studio™ 3.0. Use filters appropriate for Hoechst 33342 and Alexa Fluor™ 647 (*see* **Note 8**). Image acquisition and analysis procedures followed similar steps to those described in Subheading 3.3 **steps 22–24**. Channel one images were of cells labeled with Hoechst 33342 and channel two images were of cells labeled with anti-LC3.

10. Discrimination of stimulation versus inhibition is shown in Fig. 2b as mean average number of spots counted within the ring.

4 Notes

1. In the protocol described above (Subheadings 3.1–3.4), we have used Greiner CELLSTAR® 96-well plates black polystyrene wells flat bottom (with micro-clear bottom). We have also used 96-well plates from Perkin Elmer and Nunc. Plates selected should be suitable for fluorescence imaging.

2. We describe the use of A549 cells (Subheadings 3.1 and 3.6), an immortalized cell line derived from a human lung carcinoma. These cells show exceptional response to induction of autophagy through rapamycin analogs (for example PP242) and nutrient starvation. We have also performed the assays described above in HeLa, U-2 OS, HEK293, CHO-M1, and HepG2 cells.

3. Many homologs of LC3B exist [13]. If the LC3B antibody fails to give a signal in the cell system under study, then antibodies against these other homologs may be evaluated (Subheadings 3.1–3.3 and 3.5–3.7).

4. The method described (Subheadings 3.1–3.3 and 3.5–3.7) used goat anti-rabbit Alexa Fluor™ 647 conjugate as a secondary antibody. We have also performed this method successfully with Alexa Fluor™ 488, Alexa Fluor™ 555, Alexa Fluor™ 568, and Alexa Fluor™ 594 conjugated secondary antibodies.

5. The knockout of ATG5 in these cells (Subheadings 3.4 and 3.5) was carried out using CRISPR/Cas9 genome editing [17]. A 20 nucleotide guide RNA (AACTTGTTTCACGCTA-TATC) was used to make an 8 base-pair deletion in exon 2 of the ATG5 gene (Chromosome 6). ATG5 deletion was confirmed using sequencing (performed as part of the QC protocol by the supplier) and western blot (data not shown). We have also used siRNA-mediated knockdown of the genes ATG4B, ATG5, and ATG7 as a tool to inhibit the formation of autophagosomes (data not shown).

6. Application of a known lysosomal modulator (for example Bafilomycin A1, Subheadings 3.6 and 3.7) aids the discrimination of compounds that cause an increase in autophagosome number through the stimulation of autophagy (for example rapamycin, serum starvation or blocking proteasomal degradation) versus those that inhibit autophagic flux (hydrolase

inhibitors or modulators of lysosomal pH). Incubation of cells with either PP242 (to inhibit mTOR) or Bafilomycin A1 causes a significant increase in LC3B staining (Fig. 2b). When co-applied, the increase in LC3B labeling is significantly higher than either compound alone (Fig. 2b). Therefore, PP242 acts to increase LC3B labeling via a mechanism distinct from inhibiting flux. By using drugs such as Bafilomycin A1 (or chloroquine diphosphate) that act at defined points in the pathway, it is possible to determine if an unknown compound or gene is acting on the lysosome to cause an increase in LC3 labeling through inhibition of autophagic flux or by stimulating the formation of autophagosomes [18] (Subheadings 3.6 and 3.7).

7. We primarily use a 20× objective for autophagy assays (Subheadings 3.1–3.7). The magnification afforded by the 20× is sufficient to accurately identify spots while giving a large enough number of cells per field for valid statistical analysis. The use of 40× (Subheadings 3.5), 60×, 63×, and 100× objective lenses enables identification of spots but will compromise scan times. Use of lower magnification lenses (for example 10×) is not ideal for capturing "spot" features.

8. Image cells using filters appropriate for Hoechst 33342 (excitation and emission maxima at 350 nm and 461 nm respectively) and Alexa Fluor™ 647 (excitation and emission maxima at 650 nm and 670 nm respectively). On the systems we have used (CellInsight™ CX5 high content screening platform, CellInsight™ CX7 high content screening platform, and ArrayScan™ VTI High Content Platform) the filter set "BGRFR" is able to efficiently excite and collect fluorescence from these fluorophores (Subheadings 3.1–3.7).

9. Cytoplasmic spot intensity, area, or number can be measured for autophagosome formation. These measurements showed similar dose-dependent trends (Fig. 3) following application of drugs that inhibit autophagic flux (chloroquine diphosphate, Bafilomycin A1); however, care should be taken to measure all three when performing novel manipulations of autophagy and to validate the fidelity of a given parameter under those conditions (Subheadings 3.1–3.7).

References

1. Mizushima N, Komatsu M (2011) Autophagy: renovation of cells and tissues. Cell 147:728–741

2. Klionsky DJ, Abdelmohsen K, Abe A et al (2016) Guidelines for the use and interpretation of assays for monitoring autophagy (3rd edition). Autophagy 12:1–122

3. Feng Y, Mitchison TJ, Bender A et al (2009) Multi-parametric phenotypic profiling: using cellular effects to characterize small-scale compounds. Nat Rev Drug Discov 8:567–578

4. Zanella F, Lorens JB, Link W (2010) High content screening: seeing is believing. Trends Biotechnol 28(5):237–245

5. Nyfeler B, Bergman P, Wilson CJ et al (2012) Quantitative visualization of autophagy induction by mTOR inhibitors. Methods Mol Biol 821:239–250

6. Joachim J, Jiang M, McKnight NC et al (2015) High-throughput screening methods to identify regulators of mammalian autophagy. Methods 75:96–104

7. Mcknight NC, Jefferies HBJ, Alemu EA et al (2012) Genome wide siRNA screen reveals amino acid starvation-induced autophagy requires SCOC and WAC. EMBO J 31:1931–1946

8. Ovredahl A, Sumpter R Jr, Xiao G et al (2011) Image-based genome-wide siRNA identifies selective autophagy factors. Nature 480:113–117

9. Buchser WJ, Laskow TC, Pavlik PJ et al (2012) Cell-mediated autophagy promotes cancer cell survival. Cancer Res 12:2970–2979

10. Zhang L, Yu J, Pan H et al (2007) Small molecule regulators of autophagy identified by an image-based high-throughput screen. Proc Natl Acad Sci U S A 104:19023–19028

11. Underwood BR, Imarisio S, Fleming S et al (2010) Antioxidants can inhibit basal autophagy and enhance neurodegeneration in models of polyglutamine disease. Hum Mol Genet 19:3413–3429

12. Hasson SA, Kane LA, Yamano K et al (2013) High-content genome-wide RNAi screens identify regulators of parkin upstream of mitophagy. Nature 504:3675–3688

13. Klionsky DJ, Abdalla FC, Abeliovich H et al (2012) Guidelines for the use and interpretation of assays for monitoring autophagy. Autophagy 8:445–544

14. Miyawaki A (2011) Proteins on the move: insights gained from fluorescent protein technologies. Nat Rev Mol Cell Biol 23:656–668

15. Schnell U, Dijk F, Sjollema KA et al (2012) Immunolabeling artifacts and the need for live-cell imaging. Nat Methods 9:152–158

16. Mizushima N, Yoshimori T, Levine B (2011) Methods in mammalian autophagy research. Cell 140:313–326

17. Wright AV, Nuñez JK, Doudna JA (2016) Biology and applications of CRISPR systems: harnessing nature's toolbox for genome engineering. Cell 164:29–44

18. Hancock MK, Hermanson SB, Dolman NJ (2012) A quantitative TR-FRET plate reader immunoassay for measuring autophagy. Autophagy 8:1227–1244

Part II

Selecting Microtiter Plates, Identify Imaging Artifacts, RNAi Specificity, Data Handling, Kinetic Assays, and Ultra-High Throughput HCS

Chapter 6

Guidelines for Microplate Selection in High Content Imaging

Oscar J. Trask

Abstract

Since the inception of commercialized automated high content screening (HCS) imaging devices in the mid to late 1990s, the adoption of media vessels typically used to house and contain biological specimens for interrogation has transitioned from microscope slides and petri dishes into multi-well microtiter plates called microplates. The early 96- and 384-well microplates commonly used in other high-throughput screening (HTS) technology applications were often not designed for optical imaging. Since then, modifications and the use of next-generation materials with improved optical clarity have enhanced the quality of captured images, reduced autofocusing failures, and empowered the use of higher power magnification objectives to resolve fine detailed measurements at the subcellular pixel level. The plethora of microplates and their applications requires practitioners of high content imaging (HCI) to be especially diligent in the selection and adoption of the best plates for running longitudinal studies or larger screening campaigns. While the highest priority in experimental design is the selection of the biological model, the choice of microplate can alter the biological response and ultimately may change the experimental outcome. This chapter will provide readers with background, troubleshooting guidelines, and considerations for choosing an appropriate microplate.

Key words Microplates, Microtiter, Multiwell, Plates, Slides, Surface, Flat, Bottom, Autofocus

1 Introduction

A key consideration for choosing the appropriate microplate begins with the geometric and physical dimensions. Most manufacturers create microplates according to the recommended guidelines initially proposed by the Society for Biomolecular Screening (SBS) for a standardized microplate with a rectangular footprint dimensional size of 127.76 mm × 85.48 mm with standardized height dimensions, bottom outside flange dimensions, well positions, and well bottom elevations. These standards are now better known as the American National Standards Institute (ANSI)/Society of Laboratory Automation and Screening (SLAS) microplate standards [1–7]. The advantage of standardized microplates is interoperability with a diverse array of automated liquid handling devices,

Paul A. Johnston and Oscar J. Trask (eds.), *High Content Screening: A Powerful Approach to Systems Cell Biology and Phenotypic Drug Discovery*, Methods in Molecular Biology, vol. 1683, DOI 10.1007/978-1-4939-7357-6_6,
© Springer Science+Business Media LLC 2018

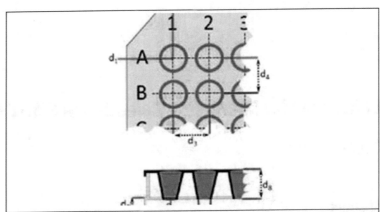

Dimensions: d_1 = well A1 offset distance from left side; d_2 = well A1 offset distance from top side; d_3 = well center to well center vertical distance; d_4 = well center to well center hortizontal distance; d_5 = distance from bottom of well to surface (skirt); d_6 = thickness of the bottom of plate; d_7 = surface area of bottom of well; d_8 = height of microplate distance from bottom surface to top surface (with or without lid)

Fig. 1 Microplate dimensions

robots, plate reader detection platforms, and HCS imagers. The specifications of microplates are available from the manufacturers and an illustration of the microplate is commonly included to show the length, width, and height of the microplate, the well positions, well bottom elevation, and the plate shirt height, outside flange dimensions, bottom thickness, volume, area, and distance between wells (Fig. 1). This information is typically presented in a figure form.

Microplates are available in a variety of well density formats from 6 well to 3456 wells and imbedded with colors of black, white, or transparent (clear). The color of the microplate commonly dedicates the type of application these are often suited toward, for luminesce-based assays, the microplate well wall and bottom are typically solid white. For fluorescence applications however, well walls are usually black and for imaging applications the well bottom surface is transparent to allow light transmission. Entirely transparent clear plates are used for a variety of applications from cell growth to chemical compound storage. Microplates and microscope slides are each designed with a fixed surface area and maximum volume capacity to accommodate cells and media respectively (Table 1). Most microplate wells have a circular geometry; however, higher density 384-well and 1536-well microplates are available in both round and square shapes. The bottom of higher density microplates such as a 96-well or greater is molded in shapes including flat (f-bottom), round (u-bottom), conical (v-bottom), or curve (c-bottom). Flat-bottom plates are the most commonly used microplates for microscopy and HCI applications; however, curved or round-bottomed plates are often used with suspension

Table 1
Reference table for microplates and chamber slide surface area and maximum volume per well

Plate type	Well shape	Est area (cm^2)	Maximum volume (mL)
6 well	Round	9.5	16.8–17.7
12 well	Round	3.8	6.76–6.90
24 well	Round	1.9	1.8–3.5
48 well	Round	0.95	1.5–1.6
96 well	Round	0.32	0.3–0.4
96 well half-area	Round	0.16	0.18–0.20
384 well	Square	0.056	0.105–0.145
384 well, low volume	Square	0.031	0.028–0.050
1536 well	Square	0.023	0.012–0.018
1 chamber side	Rectangle	9.4	4.5–5.0
2 chamber side	Square	4.2	2–2.5
4 chamber side	Rectangle	1.8	0.9–1.3
8 chamber side	Square	0.8	0.4–0.6

cells and more recently for tumor spheroid and organoid culture systems [8]. A variety of other microplate options that have been developed for specific biological applications include, but are not limited to, well-less microplates, half-area configurations (approximately 50% area of a normal well of the same multi-well configuration), filtration, micro-channels between wells, "ladders or stairsteps," evaporation wells or areas between wells, and barcode labeling for tracking bioassays and compound samples during high-throughput screening. Tissue culture treatment processes such as substrate coating with poly-D-lysine, poly-L-lysine, or extracellular matrix proteins, enhancements to promote 3-dimensional (3D) growth such as Hydrogel or Matrigel, and surfaces with ultra-low attachment properties are other options that will be discussed below. Since these microplate attributes are also covered by the ANIS/SLAS microplate standards, the majority of treated plates are compatible with commercial HCS imagers. One of the most important considerations of the microplate physical dimension elements for imaging applications is the plate thickness and skirt height. The challenge for HCI practitioners is that *not* all plate thicknesses match the working distance specifications of the microscope objective lens used in HCS imaging platforms. It is critical therefore that these variables are examined and carefully considered during the experimental design phase, so that HCI practitioners understand the capabilities and/or limitations of the HCS imager and the microplate combination they have selected. If higher numerical aperture (NA) objectives are required to distinguish and measure detailed biological information within the subcellular context of the cell, the corresponding plate thickness will need to match the minimum working distance (WD) of the objective lens selected. For example, if the WD of the objective lens is 0.17 mm then the microplate bottom interface should be approximately 170 μm thick to achieve optimal performance from the HCS imager optics. When using objective lenses with a WD of 0.17 mm, microplates having a bottom thickness > 200 μm will result in a loss of image quality with a corresponding reduction in resolution and distortion of the image. In a comparison of two different 20× objective lenses with a low NA of 0.4 and high NA of 0.8, microsphere beads seeded into a microplate exhibit "in-focus" and "out-of-focus" objects respectively. These differences have a statistically significant impact on the ability of the image analysis segmentation to identify individual objects that in this case resulted in a 43% decrease in the number of objects identified in images acquired with the larger 0.8 NA objective that would ultimately be translated to fewer individual cell measurements (Fig. 2). One consideration that provides some flexibility and a solution with respect to the variations in microplate bottom thicknesses that are available is the use of microscope objective lenses with a correction collar, which allows practitioners to "dial-in" the appropriate WD on the objective lens to match the thickness of the microplate bottom. However, objective lenses with correction collars typically have a lower NA specification, which reduces their light-gathering ability and resolution. In most cases, as the objective lenses WD increases, the NA decreases. Lower NA objective lenses reduce light throughput resulting in longer exposure times, and perhaps also reduced image quality. For example, in cells that were labeled with Alexa Fluor 488® Phalloidin, a bicyclic peptide probe to identify filamentous actin (F-actin) structures, and were excited with an LED light source, exposure times of 1.05 s and 0.072 s were required to achieve a 50% saturation of the camera detector with a 20×/0.4NA and 20×/0.8NA objective lens respectively. At a constant 0.1 s exposure, the F-actin fluorescence saturation levels were 23% with the 20×/0.4NA, and 87% with the 20×/0.8NA objective lens (Fig. 3).

In instances when HCS studies have not been conducted in the standard ANSI/SLAS microplates, such as on microscope slides, chamber slides, petri dishes, and other vessel configurations, careful consideration should be given to whether the sample holder is compatible with the HCS imager stage and objective lenses selected. The legacy of microscope slides (1″ × 3″) that do not fit in the typical microplate stage requires the use of slide adaptors that

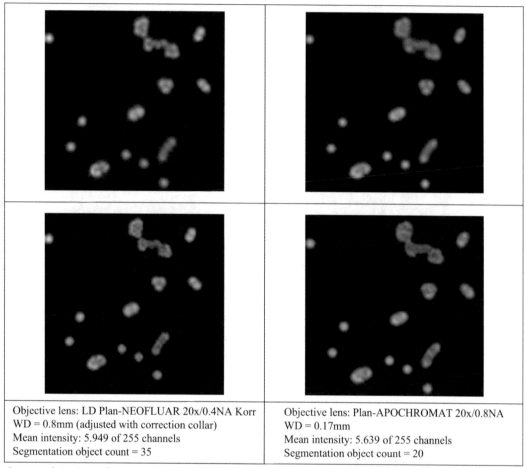

| Objective lens: LD Plan-NEOFLUAR 20x/0.4NA Korr WD = 0.8mm (adjusted with correction collar) Mean intensity: 5.949 of 255 channels Segmentation object count = 35 | Objective lens: Plan-APOCHROMAT 20x/0.8NA WD = 0.17mm Mean intensity: 5.639 of 255 channels Segmentation object count = 20 |

Images of 4 micron fluorescent beads seeded on polystyrene microplate with 0.8mm bottom thickness. Images captured on High Content Imager with two different objective lenses using LED light excitation and pentabandpass emission filter.

Fig. 2 Microscope objective lens working distance effects on focus, resolution, and segmentation

meet the ANSI/SLAS dimensional microplate format with each holder capable of housing up to four independent slides standard microscope slides or chamber slides per tray (Fig. 4).

As a general guideline, when measuring large organisms, e.g., Zebrafish, explant tissue, or large clusters of cells, then lower magnification objective lenses with lower NA are typically used with either petri dishes or microplates of lower well densities (≤ 96 well). However, when the investigation involves small objects within the cell such as nuclear foci, then consideration for thin bottom optical clear plastic or glass is recommended in combination with high NA objective lenses with magnifications $>20\times$.

 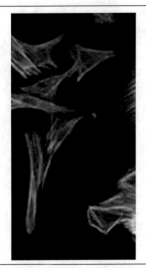

Objective lens: LD Plan-NEOFLUAR 20x/0.4NA Korr	Objective lens: Plan-APOCHROMAT 20x/0.8NA
Mean intensity: 22.1 of 255 channels	Mean intensity: 51.3 of 255 channels
Saturation: 23%	Saturation: 87%
Exposure time to achieve 50% saturation: 1.05 sec	Exposure time to achieve 50% saturation: 0.072 sec

Images of cells mounted on microscope slide labeled with 488 phalloidin. Images captured on High Content Imager with two different objective lenses using LED light excitation and pentabandpass emission filter.

Fig. 3 Microscope objective lens numerical aperture decreases exposure time and increases resolution

Representation of ANSI/SLAS standard microplate mold for holding up to 4 individual microscope slides

Fig. 4 Microscope slide holder

2 Plate Composition

The materials used to fabricate the bottoms of microplates typically are composed of glass, quartz, polymers, mainly polystyrene and in rare cases polypropylene or polyethylene, or other novel composite materials including cyclic olefin copolymers [9, 10]. The glass and quartz surfaces are one of the hardest materials available which translates into very thin (#2 coverslip thickness or better) rigid material with a flat uniform bottom. In comparison to traditional plastic polymer microplate materials, glass and quartz transmit light extremely well with minimum optical interference; therefore, the use of high NA objective lenses for image capture is possible. These materials are particularly useful when oil immersion objectives are required to achieve NA levels >1. The disadvantage of glass bottom plates is primarily their cost, and in certain cases poor cell attachment and/or changes in the in vitro biological responses when compared to plastic polymer microplates that have been tissue culture treated to promote cell attachment [11]. Additionally, glass bottom plates tend to be heavier and can be scratched easily, and therefore require delicate handling to prevent damage. The polymer-based materials, mainly polystyrene and newer composite cyclic olefin polymer materials, are most commonly used to fabricate the microplates used in HCS today. Unlike glass bottom plates, polymer-based materials are not as rigid and are available in a wide range of thicknesses from 100 μm to nearly 1 mm. Polymer-based composite materials may contain constituents that autofluorescence or contribute to optical interference due to deflection or reflection of transmitted light, and therefore their performance in imaging assays "may" be diminished compared to glass materials [12–14]. However, polymer-based microplates are typically less expensive than glass bottom plates. Another important consideration is that the materials used to fabricate microplates may exert a biological effect. For example, it is known that some plastic material resins contain bisphenol A (BPA) which is a selective estrogen receptor modulator that can promote an estrogenic response that may alter biological outcomes [15–18]. Whenever possible these materials should be avoided. In rare instances, the adhesives or chemical residues produced during the microplate manufacturing process may result in poor cell attachment, adverse health effects of the cells, or even cytotoxicity. Contact your plate manufacturer if evidence supports these findings.

3 Substrates to Enhance Cell Attachment

Most microplates are sterilized and plastic materials are tissue cultured treated to create and/or provide a hydrophilic surface to facilitate cell attachment. However, not all cells attach to plastic or glass materials and additional surface materials are required to enhance attachment and promote cellular function. There are a number of substrate materials available to coat the bottom surface of the microplates to mediate attachment and spreading of cells, and in some instances to enhance biological functions and more closely mimic in vivo responses [19, 20]. The simplest substrates used to improve cell attachment to microplate surfaces incorporate poly-lysine in the form of either poly-D-lysine (PDL) or poly-L-lysine (PLL). Poly-lysine is a positively charged hydrophilic synthetic amino acid that neutralizes negatively charged ions found on plastic and glass surfaces as well as in cell membranes [21–24]. For some cells that otherwise will only attach loosely to the bottom of a microplate, the use of PDL or PLL treatments will increase cell adherence and reduce the loss of cells due to cytotoxic insults, cell proliferation, or during washing steps prior to cell fixation with aldehyde or alcohol reagents. Likewise, extracellular matrix (ECM) protein substrates are used to not only enhance cell attachment, but they have a much broader application in recapitulating an in vivo like environment for many cell models [25]. The function of ECM proteins is to provide nutrients for structural support, maintenance of cell health and proliferation, and signaling [26]. The most common ECM protein substrates used in HCS imaging assays are Collagen-I, Collagen-IV, Fibronectin, and Laminin. Each ECM substrate has a unique biological function that practitioners should consider during the experimental design phase of assay development, because the selection of an individual ECM protein or a combination of multiple ECM proteins can dramatically alter the biological outcomes of experiments. For instances, in cell spreading and motility or migration, the adhesion molecules expressed by cells and the matrix molecule interaction from each ECM protein substrate or combination thereof can alter the overall cell movement, duration, direction, and distance traveled [27, 28]. Matrigel is a basement membrane containing a unique heterogeneous cocktail of proteins including ECM and growth factors isolated from a primary mouse sarcoma tumor that liquefies at 4 °C and solidifies into a gel at 37 °C [26]. Matrigel's utility to support growth in primary cell culture models, ex vivo tissues, and promoting 3D cell formation is well established [29, 30]. Hydrogels are hydrophilic polymers that exhibit plasticity due to their high water retention resembling the natural environment for cells and tissues and are frequently used as substrates for 3D cell culture models [31, 32].

4 Plate Flatness and Focusing

Since not all microplates are created equal to the same surface flatness across rows and columns or even within a well, achieving accurate autofocusing during microplate scanning is challenging.

The process of autofocusing is a critical aspect of capturing high quality images from an automated high content imaging platform. There are two types of autofocus routines used, laser based autofocus and image base autofocus. Laser-based autofocus is achieved through reflection of a light source, typically a near-IR laser by the microplate bottom interface of the plate and typically an offset is used to identify objects in the well. In most image-based autofocusing, an algorithm is used to determine the contrast intensity differences between background and object; without a cell or object, image-based autofocus undoubtedly fails. In both the autofocusing processes, the microplate bottom surface materials can affect the outcome. Even though the specifications and quality assurance of manufactured microplates are accessible to customers, it is critical that any new microplate is properly assessed to determine the performance of the high content imager autofocus system, because the "in-focus" variance in well to well uniformity in microplates can be quite different.

The z-axis coordinate is an HCI feature that is often overshadowed and not commonly used by practitioners, but can provide meaningful insight into the performance of autofocusing and microplate well flatness. A surface graph plot of the z-axis coordinates from 384-well microplate shows the variations of well to well positioning at the autofocus plane providing a representation to quickly determine overall performance of plate flatness or suspicious areas on the plate that should be investigated for out-of-focus images (Fig. 5). When microplate well flatness is nonuniform the impact can be amplified when using higher NA and higher magnification objective lenses. The intra-well and intra-field image autofocus z-axis coordinates within a well should be determined as even a slight variance may generate out-of-focus images. To illustrate this point, a single well from a newly developed commercially available 384-well microplate was processed using a "one-time" laser-based autofocusing method in the center of the well. Then remaining images in the well were captured in a montage form according to the determined autofocus z-coordinates. As a control, other wells within the microplate were imaged to determine if the out-of-focus images were due to instrument drift from the objective lens or microscope stage during image capture, or if it was due to the microplate flatness. A montage image (12-fields $\times$ 13-fields) comprising of 156 individual fields of mouse embryonic fibroblast cells labeled with mitochondrial probe was captured with $20\times/0.75$NA objective. The findings showed in-focus images in the center of the well at the point of the autofocus z-coordinates, and to the right of center, but left of center showed out-of-focus areas (Fig. 6). These results were subsequently confirmed with another HCS imager and the microplate findings were reported to the vendor; the microplate design was later revised with a new microplate having improved flatness and no out-of-focus issues. To help circumvent this issue

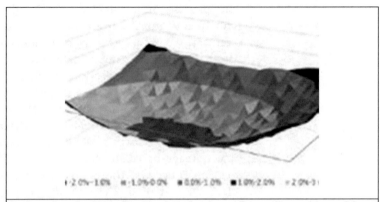

Surface plot of a 384-well microplate showing well to well z-positions of focus using a 10x/0.45NA objective lens. X-axis and y-axis represents microplate well positions and z-axis represents z-position of focus. Gray scale percentages represents the difference from the average mean z-position of all wells recorded during focusing.

Fig. 5 Microplate flatness variance

Montage image (12-fields x 13-fields) comprised of 156 individual fields captured with 20x/0.75NA objective lens from a single well of a 384-well microplate. Mouse embryonic fibroblast cells labeled with mitochondrial probe. The right and lower right areas are in-focus and left and upper left areas that are out-of-focus. Results replicated on two independent high content imagers.

Fig. 6 Intra-well image capture showing "in-focus" and "out-of-focus" field areas

when capturing multiple fields, autofocusing every few fields or every field may be necessary when using high NA or magnification objective lenses or when microplate flatness is in question. However, the disadvantages to autofocusing on every field include an increase in the time required to scan microplates, potential photo-damage occurring if using image-based autofocusing, and an increased frequency in autofocus failures. Alternatively, HCI practitioners can adjust the scan area within the well to prevent out-of-focus fields from being captured or evaluate other microplates that do not exhibit similar issues. Before out-of-focus images become a major problem, it is better to evaluate the overall performance of microplates in advance of launching any screening project. Furthermore, it is best to use the same lot number of commercial available microplates throughout a screening campaign or research project to exempt this variable from becoming an issue.

If unresolved issues from poor image capture, focusing, or microplate concerns remain there are potential solutions to overcome these obstacles as outlined in the troubleshooting guide chart.

Troubleshooting guide

Problem	Solution
Autofocus issues	• Confirm plate is locked into stage position correctly • Clean bottom of plate with ddH_2O, dry, then reimage • Be sure there are no scratches on bottom of plate; more common in glass bottom plates • Determine if cells are identifiable in scan area • Clean optics on imaging device • Confirm wells are filled with buffer or solution
Cell seeding is not uniform across well	• Use automated cell dispenser; if not available, then carefully move plate in x and y directions several times; allow cells to settle for 10–15 min at room temp before placing in incubator. • Confirm incubator is level
Evaporation	• Increase the humidity levels in the incubator • Increase liquid volume in wells • Use microplates with evaporation chambers • If feasible, decrease incubation times or replenish medium or buffer
Focus issues within a well; not all fields are in the same focus level	• Confirm plate is locked into stage position correctly • Confirm the microscope objective lens is planar and matches the plate bottom thickness

	• If substrate was used, then lifting of cells on substrate may have occurred in either uniform or non-uniformed areas within the well. • If using image or laser-based autofocusing, determine if field to field autofocusing is activated. Turn off and repeat. If results did not change and no ECM coating was used, contact microplate manufacturer to determine if microplate lot was recalled. • Clean imaging device optics
Plate leaks liquid	• Pre-warm microplates from 4 °C storage • Keep microplate in upright flat position • This could be due to wicking (*see* below) • Contact microplate manufacturer
Unhealthy cells	• Confirm preseeding viability • If substrate was used, plate cells without it to determine if cytotoxic • Contact manufacturer to determine if other customers were affected
Wicking of liquid from wells	• Zap plate with static discharge gun before cell seeding and before applying plate seals. • Use static free plate seals. • Contact microplate manufacturer

5   Conclusions

There are a number of different materials including glass, polymers, and co-polymers used in the manufacture of microplates, and each of these materials has desirable and undesirable traits. The initial selection of microplates by HCS practitioners is heavily influenced by budget constraints, but the impact of microplate selection based on biological function and responses, and the contribution to autofocusing and high-resolution image acquisition are equally important. The biological response in an experimental assay can be uniquely altered using microplate coating substrates, and undoubtedly, one of the most important considerations in choosing the appropriate microplate and substrate coatings are their overall effects on biological health and function. When an assay is developed, transferred, or miniaturized into a microplate, the biological performance must be maintained; otherwise, the quality and robustness of the assay including the signal-to-noise measurements of the experiment can inadvertently be lost and become burdensome to overcome, and very expensive. The selection of microplates for an HCS assay is often overlooked because other factors may confound the decision including costs, marketing materials from manufacturers, "word of mouth," or even historical experience or evidence. To achieve a successful outcome and prevent repeating unnecessary experiments, due diligence evaluation of microplates from the beginning is required.

6 Methods

All microplate fabrication materials have advantages and disadvantages; therefore, selection of the appropriate plate should undergo careful investigation to provide the best possible results. Please use this guideline method to help determine which microplate is optimal for your experiment.

6.1 Guideline Method for Microplate/Vessel Selection

1. Determine throughout requirement for screen or project. Choose the appropriate media vessel from slides to 1536-well microplates.

2. Determine microplate type from f-bottom, c-bottom, u-bottom, or v-bottom.

3. Determine microplate material (glass, polystyrene, or copolymers) and if subsequent surface coating (PDL, ECM, etc.) is required to maintain in vitro health or biological response.

4. Determine the image resolution required to address or answer a biological question.

   (a) Choose the appropriate magnification objective lens based on target protein identification.

   (b) Select the corresponding microplate bottom thickness to match the objective lens.

5. Evaluate and compare two or more microplates.

   (a) Criteria based on autofocus performance, image quality, and image analysis algorithmic statistical signal to noise.

6. Determine microplate availability and cost.

   (a) Be sure to stock enough microplates from a single manufacture lot for the entire project or screening campaign.

References

1. http://www.wellplate.com/ansi-slas-microplate-standards/ (as of April 2016)

2. http://www.slas.org/resources/information/industry-standards/ (as of April 2016)

3. ANSI/SLAS 1–2004: Microplates — Footprint Dimensions → http://www.slas.org/default/assets/File/ANSI_SLAS_1-2004_FootprintDimensions.pdf

4. ANSI/SLAS 2–2004: Microplates — height dimensions

5. ANSI/SLAS 3–2004: Microplates — bottom outside flange dimensions

6. ANSI/SLAS 4–2004: Microplates — well positions

7. ANSI/SLAS 6–2012: Microplates — well bottom elevation

8. Vinci M, Gowan S, Boxall F et al (2012) Advances in establishment and analysis of three dimensional tumor spheroid-based functional assays for target validation and drug evaluation. BMC Biol 10:29

9. Kohara T (1996) Development of new cyclic olefin polymers for optical uses. Macromol Symp 101(1):571–579

10. Niles WD, Coassin PJ (2008) Cyclic olefin polymers: innovative materials for high-density multiwell plates. Assay Drug Dev Technol 6 (4):577–590

11. Amstein CF, Hartman PA (1975) Adaption of plastic surfaces for tissue culture by glow discharge. J Clin Microbiol 2:46–54

12. Sultanova N, Kasarova S, Nikolov I (2009) Dispersion properties of optical polymers. Acta Phys Pol A 116:585–587

13. Bach H, Neuroth N (1998) The properties of optical glass, 2nd edn. Springer, New York. ISBN: 978-3-642-63349-2

14. Gliemeroth G (1982) Optical properties of optical glass. Proceedings of the conference on optical properties of glass and optical materials. J Non-Cryst Solids 47(1):57–68

15. Yang CZ, Yaniger SI, Jordan VC, Klein DJ, Bittner GD (2011) Most plastic products release estrogenic chemicals: a potential health problem that can be solved. Environ Health Perspect 119(7):989–996

16. Richter CA, Birnbaum LS, Farabollini F, Newbold RR, Rubin BS, Talsness CE, Vandenbergh JG, Walser-Kuntz DR, vom Saal FS. (2007) In vivo effects of bisphenol A in laboratory rodent studies. Reprod Toxicol 24(2):199–224

17. Welshons WV, Nagel SC, vom Saal FS. (2006) Large effects from small exposures. III. Endocrine mechanisms mediating effects of bisphenol A at levels of human exposure. Endocrinology 147(6 Suppl):S56–S69

18. Erickson BE (2008) Bisphenol A under scrutiny. Chemical and engineering news. American Chemical Society, Washington, DC, pp 36–39

19. Reid LM, Rojkind M (1979) New techniques for culturing differentiated cells: reconstituted basement membrane rafts. In: Jakoby WB, Pastan IH (eds) Cell culture, Methods in enzymology, Chapter 21, vol 58. Academic, New York, pp 263–278

20. Discher DE, Janmey P, Wang YL (2005) Tissue cells feel and respond to the stiffness of their substrate. Science 310(5751):1139–1143

21. Curtis ASG, Forrester JV, McInnes C, Lawrie F (1983) Adhesion of cells to polystyrene surfaces. J Cell Biol 97:1500–1506

22. Ramsey WS, Hertl W, Nowlan ED, Binkowski NJ (1984) Surface treatments and cell attachment. In Vitro 20:802–808

23. Shen M, Horbett TA (2001) The effects of surface chemistry and adsorbed proteins on monocyte/macrophage adhesion to chemically modified polystyrene surfaces. J Biomed Mater Res 57(3):336–345

24. Mazia D, Schatten G, Sale W (1975) Adhesion of cells to surfaces coated with polylysine. Applications to electron microscopy. J Cell Biol 66(1):198–200

25. Xu C, Inokuma MS, Denham J, Golds K, Kundu P, Gold JD, Carpenter MK (2001) Feeder-free growth of undifferentiated human embryonic stem cells. Nat Biotechnol 19 (10):971–974

26. Benton G, George J, Kleinman HK, Arnaoutova I (2009) Advancing science and technology via 3D culture on basement membrane matrix. J Cell Physiol 221(1):18–25

27. Pierres A, Eymeric P, Baloche E, Touchard D, Benoliel A-M, Bongrand P (2003) Cell membrane alignment along adhesive surfaces: contribution of active and passive cell processes. Biophys J 84(3):2058–2070

28. Kuntz RM, Saltzman WM (1997) Neutrophil motility in extracellular matrix gels: mesh size and adhesion affect speed of migration. Biophys J 72(3):1472–1480

29. Hughes CS, Postovit LM, Lajoie GA (2010) Matrigel: a complex protein mixture required for optimal growth of cell culture. Proteomics 10(9):1886–1890

30. Arnaoutova I, George J, Kleinman HK, Benton G (2009) The endothelial cell tube formation assay on basement membrane turns 20. Angiogenesis 12(3):267–274

31. Camci-Unal G, Nichol JW, Bae H, Tekin H, Bischoff J, Khademhosseini A (2013) Hydrogel surfaces to promote attachment and spreading of endothelial progenitor cells. J Tissue Eng Regen Med 7(5):337–347

32. Mellati A, Dai S, Bi J, Jin B, Zhang H (2014) A biodegradable thermosensitive hydrogel with tuneable properties for mimicking three-dimensional microenvironments of stem cells. RSC Adv 4(109):63951–63961

Chapter 7

Quality Control for High-Throughput Imaging Experiments Using Machine Learning in Cellprofiler

Mark-Anthony Bray and Anne E. Carpenter

Abstract

Robust high-content screening of visual cellular phenotypes has been enabled by automated microscopy and quantitative image analysis. The identification and removal of common image-based aberrations is critical to the screening workflow. Out-of-focus images, debris, and auto-fluorescing samples can cause artifacts such as focus blur and image saturation, contaminating downstream analysis and impairing identification of subtle phenotypes. Here, we describe an automated quality control protocol implemented in validated open-source software, leveraging the suite of image-based measurements generated by Cell-Profiler and the machine-learning functionality of CellProfiler Analyst.

Key words Cell-based assays, High-content screening, Image analysis, Microscopy, Quality control, Machine learning, Open-source software

1 Introduction

The use of automated microscopy combined with image analysis methods has enabled the extraction of quantitative image-based information from cells, tissues, and organisms while speeding analysis and reducing subjectivity (*see* refs. 1, 2). Any number of high-content assays can be quantified by combining high-resolution microscopy with sophisticated image analysis techniques in order to create an automated workflow with a high degree of reproducibility, fidelity, and robustness (*see* ref. 3). Analyzing experiments that are comprised of tens to millions of images allows for quantitative modeling of biological processes and discerning complex and subtle phenotypes.

However, reliable downstream processing of such datasets often depends on robust exclusion of images that would otherwise be erroneously scored as screening hits or inadvertently ignored as false negatives. Abnormalities in image quality can degrade otherwise high-quality microscopy data and, in severe cases, even render

Paul A. Johnston and Oscar J. Trask (eds.), *High Content Screening: A Powerful Approach to Systems Cell Biology and Phenotypic Drug Discovery*, Methods in Molecular Biology, vol. 1683, DOI 10.1007/978-1-4939-7357-6_7,
© Springer Science+Business Media LLC 2018

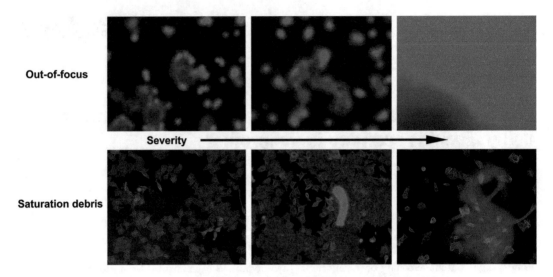

Fig. 1 Examples of HCS images containing artifacts. Out-of-focus (*top row*) and saturation debris (*bottom row*) examples are shown. Images are taken from the Broad Bioimage Benchmark Collection (BBBC) at http:/www. broadinstitute.org/bbbc/BBBC021/. These images come from a compound mechanism of action assay consisting of MCF-7 cells labeled with fluorescent markers for DNA (*red*), β-tubulin (*green*), and actin filaments (*yellow*)

some experimental approaches infeasible. In our experience, as many as 5% of the fields of view in a routine screen can be affected with such artifacts to varying degrees. For high-throughput assays, manual inspection of all images for quality control (QC) purposes is not tractable; therefore, the development of QC methodologies must be similarly automated to keep up with the increasing demands of modern imaging experiments.

This chapter outlines a protocol for the characterization of images for common artifacts that confound high-content imaging experiments, including focus blur and image saturation (Fig. 1). The protocol uses the open-source, freely downloadable software packages, CellProfiler and CellProfiler Analyst. CellProfiler has been validated for a diverse array of biological applications, typically for generating features on a per-cell basis (*see* refs. 4, 5). Likewise, CellProfiler Analyst has been previously used for per-cell classification of phenotypes (*see* refs. 4, 6). The workflow described below expands our prior work using CellProfiler and CellProfiler Analyst validating image-based metrics for QC (*see* ref. 7) and provides a step-by-step protocol that leverages the functionality of both of these packages for QC purposes.

2   Materials

2.1   High-Content Fluorescent Images for Assessment

1. Either single channel or multichannel fluorescent images acquired on a microscopy platform, conventionalor automated, may be analyzed. CellProfiler is capable of handling both

fluorescence and transmitted-light (e.g., bright-field) images; however, this protocol assumes that a fluorescent assay is being evaluated (*see* **Note 1**).

2. More than 120 file formats are readable by CellProfiler, including TIF, BMP, and PNG; standardized HCS image data formats such as OME-TIFF are also supported. Some file formats are more amenable to image analysis than others (*see* **Note 2**).

3. For most screening applications, images are captured by an automated microscope from multi-well plates, such that each image is annotated with unique plate, well, and site metadata identifiers. Using this metadata will enable some features in CellProfiler Analyst, as described below.

4. The images may be contained in a single folder or in a set of folders or subfolders. While hundreds of images may be analyzed on a single computer, such a computing solution is insufficient for the thousands or millions of images characteristic of large-scale screens. In the latter case, a CellProfiler analysis can be run on a computing cluster, taking advantage of the hardware infrastructure to process any number of images in parallel (*see* Subheading 8.1).

5. An example screening image set (BBBC021v1) is available from the Broad Bioimage Benchmark Collection (BBBC) at http://www.broadinstitute.org/bbbc/BBBC021/ (*see* refs. 8, 9). These images come from a small molecule mechanism of action assay consisting of MCF-7 cells labeled with fluorescent markers for DNA, β-tubulin, plasma membrane, and actin filaments.

2.2 A Desktop or Laptop Computer

1. A Mac, PC, or Linux computer with at least 4 GB of RAM, a 2 GHz processor and a 64-bit processor is recommended. If the images are stored remotely, a fast Internet connection is recommended for rapid image loading.

2. A single image set such as those in the example set demonstrated here will be processed in <1 min/image on a single computer with a 2.67 GHz processor and 8 GB RAM.

3. Large image sets (greater than ~1000 images) will likely require a computing cluster (*see* **Note 3**).

2.3 CellProfiler and CellProfiler Analyst Software

1. Both applications are free and open-source (BSD license).

2. The CellProfiler image analysis software package is available as a distributable installation package for Windows and Mac and can be downloaded at http://cellprofiler.org/. This protocol uses CellProfiler version 2.2.0. Researchers who wish to implement their own image analysis algorithms or run CellProfiler on UNIX/Linux or a computing cluster will want to download

the source code (*see* **Note 4**). All versions are free and open-source (BSD license).

3. The CellProfiler Analyst software package is available for Windows and Mac as a distributable installer package at http://cellprofiler.org/. This protocol uses CellProfiler version 2.2.0.

4. For both packages, follow the installation instructions from their respective download pages. If difficulties on this step are encountered, visit the online forum (http://forum.cellprofiler.org/) to search if the problem has been previously encountered and resolved, or post the issue to the forum.

3   Methods

The protocol begins with configuring the input and output file locations for the CellProfiler program and constructing a modular QC "pipeline". Image processing modules are selected and placed in the pipeline and the modules' settings are adjusted appropriately according to the specifics of the HCS project (for example, spatial scales for blur measurements, and the channels used for thresholding; see the section "Configuring the `MeasureImageQuality` module" below). The pipeline is then run on the images collected in the experiment to assemble a suite of QC measurements, including the image's power log-log slope, textural correlation, percentage of the image occupied by saturated pixels, and the standard deviation of the pixel intensities, among others. These measurements are used within the machine-learning tool packaged with CellProfiler Analyst to automatically classify images as passing or failing QC criteria determined by a classification algorithm. The results can either be written to a database for further review, or the classifier can be used to filter images within a later CellProfiler pipeline so that only those images which pass QC are used for cellular feature extraction. An overview of the workflow is shown in Fig. 2.

3.1   Starting CellProfiler and Loading a Pipeline

1. Start CellProfiler by selecting CellProfiler from the Start Menu (Windows) or Applications folder (Mac), or from the command line (any OS). The CellProfiler welcome screen and graphical user interface will appear (Fig. 3a).

2. Download an example quality control (QC) pipeline from http://pubs.broadinstitute.org/bray_methodsmolbiol_2016/. From the main menu bar, select *File > Import > Pipeline from File...* and browse to the location of the downloaded pipeline (or alternately, drag/drop the pipeline file into the CellProfiler pipeline panel). This will load the QC pipeline which can be adjusted as needed for other assays; the details on the specific settings are described in the following sections and associated notes.

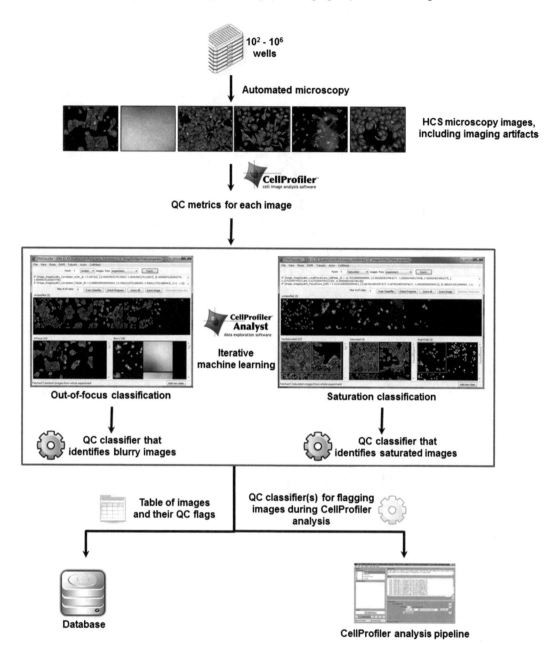

Fig. 2 Overall quality control workflow. A suite of image quality measures are obtained by CellProfiler from images collected by an automated microscope. These measurements are used as input into a supervised machine learning tool in CellProfiler Analyst; the researcher then trains the computer to classify images as out-of-focus or containing saturation artifacts. The classifier then scores all images from the experiment, with the QC results stored as metadata in a database, or the classifier incorporated into an analysis pipeline

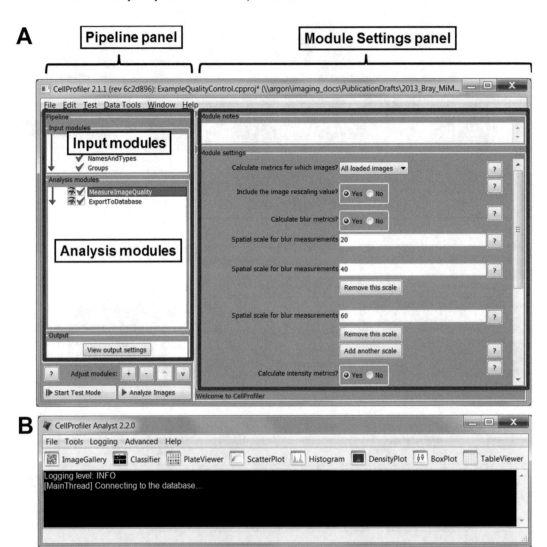

Fig. 3 Screenshots of the software packages described in the protocol. (**a**) The CellProfiler interface. The Pipeline panel is divided into three sections: the Input modules which specify information about the images to be processed, the Analysis modules which are executed sequentially to collect the measurements, and the Output, specifying the location of the output files. The Module settings panel provides the customizable settings for each selected module in the Pipeline panel. (**b**) The CellProfiler Analyst user interface. The Classifier tool (*icon on the upper left*) is used to train a classifier to distinguish between images of various types; other icons launch tools used for data visualization and exploration

3. Because CellProfiler is usable on a wide variety of assays, only the modules and associated settings relevant to the quality-control protocol are listed and described in this protocol. The remaining module settings are not mentioned and should be adjusted to suit each specific assay set as needed; each module

has extensive documentation to assist with fine-tuning the settings (*see* **Note 5**). Other modules can also be added and positioned if further image measurements are desired (*see* **Note 6**).

3.2 Configuring Image Input for CellProfiler

1. Select the `Images` module, the first module in the Input modules section of the pipeline panel. A box will appear in the module settings panel prompting for files to be placed into the box. Drag-and-drop the desired image files (or folders containing the image files), as described in the Materials section, into this box; the box will update and show a listing of the collected files. These files will be used as input into the QC pipeline. Adding entire folders of files is acceptable even if some of the contents are not to be processed; these can be filtered in the next step.

2. Adjust the "Filter images?" drop-down below the file list box to specify what files in the file list are passed downstream for further processing. The default setting is "Images only," which is sufficient for most data sets. If only a subset of files are to be used as input (e.g., only process those files with the extension "TIF"), select "Custom" from the drop-down box and then define rules for filtering the files for processing (*see* **Note 7**).

3.3 Specifying Image Metadata (Optional)

1. If there is information (metadata) that is associated with the images, such as experiment, plate, and well identities, select the `Metadata` module (the second module of the Input modules) and select for "Yes" for "Extract metadata?" This module should definitely be used if information about the well layout is contained in the image filename or folder name. Configure the module according to the settings listed below.

2. *Metadata extraction method*: Select "Extract from file/folder names" if the metadata information is contained within the image filename or path, then select "File name" or "Folder name" from the *Metadata source* setting that appears. Select "Import metadata" if it is contained in a comma-delimited file (CSV) of values, then browse to the file location from the setting that appears.

3. *Regular expression*: This setting may require adjustment to match the nomenclature applied by the acquisition software (*see* **Note 8**). However, the default of ""^(?P<Plate>.*)_(?P<Well>[A-P][0–9]{2})_s(?P<Site>[0–9])_w(?P<ChannelNumber>[0–9])" is sufficient for a number of commercial systems (*see* **Note 9**).

4. If additional metadata needs to be included, click the "Add another extraction method" button to reveal additional

settings which can then be adjusted to include further metadata sources such as sample treatment information.

5. Click the "Update" button below the horizontal divider to display a table where each row shows an input image's filename and whatever associated metadata is available, including plate layout identifiers such as plate, well, and site, as well as sample and treatment information.

3.4 Specifying the CellProfiler Name and Type of Image Channels

1. Select the NamesAndTypes module, the third module in the Input modules section of the pipeline panel. This module is used to assign a user-defined name to particular images or channel(s), and define their relationship to one another. Configure the module according to the settings listed below.

2. *Assign a name to*: Select "Images matching rules" to select a subset of images from the Images module as belonging to the same channel.

3. *Select the rule criteria*: From the drop-down and edit boxes, select an identifier and the value for this identifier in order to distinguish a subset of images as a unique channel. In the example pipeline, the settings are specified as: "Metadata", "Does", and "Have ChannelNumber matching" in the three drop-down menus, and "1" is entered in the edit box. This combination of settings will identify those images which have the ChannelNumber metadata identifier specified as "1" and ignore all others. If no metadata was gathered from the Metadata module, then other image characteristics such as filename, extension, and image type may be used to identify a unique channel.

4. *Name to assign these images*: Enter a suitably descriptive name to identify the image for later use in the pipeline; downstream modules will then refer to the image by this name for processing. For example, "DNA" can be used to indicate that the first wavelength corresponds to DNA-stained images.

5. *Select the image type*: Select the image format that corresponds to this channel (*see* **Note 10**).

6. Press the "Add another image" button if the assay involves multiple channels; additional settings will be revealed so that further matching rules and names can be given to additional channels. Any number of channels may be specified using this method.

7. *Image set matching method*: This step associates multiple image channels with each other, for each field of view. If the Metadata module was used to specify the identifiers for the channel, select "Metadata" to display a panel containing a column for each channel given above, and a row of drop-down menus with available metadata identifiers. For each row, match the

metadata identifiers so that the channels are properly matched together (*see* **Note 11**).

8. Press the "Update" button below the horizontal divider to display a table where each row displays a unique metadata combination, and the image names are listed as columns. When the pipeline executes during the analysis run, each set of images specified in a row will be loaded and processed as an individual image set. Check the listing for any errors, e.g., image channel mismatches.

3.5 Configuring the *MeasureImage Quality* Module

1. Select the `MeasureImageQuality` module, located in the Analysis modules panel below the Input modules. This module measures features that indicate image quality, including measurements of blur (poor focus), intensity, and saturation. Configure the module according to the settings listed below.

2. *Calculate metrics for which images?* Select "All loaded images" in order to calculate QC metrics for all channels that were specified in the `NamesAndTypes` module. Choose "Select..." to select a subset of these channels.

3. *Calculate blur metrics?* Select "Yes" for this setting to calculate a set of focus blur metrics, one for each channel specified above (*see* **Note 12**).

4. *Spatial scale for blur measurements.* Enter a number specifying the size(s) of the relevant features, in pixels. For a given amount of focus blur, the degradation of image quality will depend in part on the size of the cellular features imaged. For example, nuclei that are typically 20 pixels in diameter may not be as affected by a small amount of blurring as thin actin filaments that are only 5 pixels wide. Since the size of the features can vary over a wide range in HCS, it is often helpful to specify several spatial scales in order to capture differing amounts of blur; $0.5\times$, $1\times$, and $2\times$ the size of a given structure of interest are good starting points. CellProfiler will measure the blurriness metrics for all specified spatial scales, for all selected input images; click the "Add another scale" button to include additional spatial scales. Later in the analysis, blur measurements resulting from each scale can be examined and assessed for utility.

5. *Calculate intensity metrics?* Select "Yes" for this setting to calculate a set of intensity measurements, one for each channel specified above (*see* **Note 13**).

6. *Calculate saturation metrics?* Select "Yes" for this setting to calculate a set of saturation metrics, one for each channel specified above (*see* **Note 13**).

1. Select the ExportToDatabase module in the Analysis modules panel. This module exports measurements produced by a pipeline directly to a database, which will be accessed by Cell-Profiler Analyst. Configure the module according to the settings listed below.

2. *Database type*: Select "MySQL" if remote access to a MySQL database is available; enter the host, username, and password information in the settings below. Select "SQLite" to instead store the data on hard drive space, such as a personal computer or networked storage space; this option does not require setting up or configuring a database (*see* **Note 14**).

3. *Experiment name*: Enter a name for the experiment; this can be any descriptor that uniquely identifies the analysis run. This name will be registered in the database and linked to the tables that ExportToDatabase creates.

4. *Database name*: Enter the name of the database that will store the collected measurements. CellProfiler will create the table if it does not exist already, or produce a warning prior to over-writing an existing table. Often this will be the same as *Table prefix* (*see* below).

5. If using "MySQL" for the database type, press the "Test connection" button to confirm the connection settings entered above.

6. *Overwrite without warning?* This setting will determine whether the database tables used to store the measurements will be created based on the researcher's response to a prompt at the beginning of the analysis run ("Never"), existing tables will be reused and measurements added or overwritten as needed ("Data only"), or the tables will be created without prompting ("Data and schema"). Be very careful with this setting, as it will enable overwriting existing data with the same database name.

7. *Add a prefix to table names?* Select "Yes" to this setting to uniquely specify the names of the tables created in the analysis run. If so, a setting labeled *Table prefix* will appear for entering the chosen identifier for the analysis run. This text will be prepended onto the default table name of "Per_Image" created in the database specified above; using a unique identifier allows multiple data tables to be written to the same database rather than over-writing the default table name with each run. For example, a QC pipeline run on the BBBC images described above (*see* Subheading 2) could use the prefix "BBBC021_QC" to distinguish the database table from QC runs performed on other BBBC images.

8. *Create a CellProfiler Analyst file?* Select "Yes" to this setting to create a configuration file (the "properties" file, described in

more detail in Subheading 3.9) that will be used by CellProfiler Analyst to access the images and measurements. Additional settings will appear upon selecting this option and are specified below.

9. *Access CPA images via URL?* Select "Yes" to this setting if the images are stored remotely and can be accessed via HTTP. If so, a setting labeled *Enter an image url prepend if you plan to access your files via http* will appear for entering a URL prefix. This prefix will be prepended onto all image locations during the analysis run. For example, if this setting is given as "http://some_server.org/images" and the path and file name in the database for a given image are "some_path" and "file.png," respectively, then CellProfiler Analyst will open "http://some_server.org/images/some_path/file.png".

10. *Select the plate type:* If using a multi-well plate assay, select the plate format from the drop-down box. Permissible types are 6, 24, 96, 384, 1536, and 5600 (for certain cell microarrays).

11. *Select the plate metadata:* If using multi-well plates, select the metadata identifier corresponding to the physical plate ID, otherwise leave as "None".

12. *Select the well metadata:* If using multi-well plates, select the metadata identifier corresponding to the well ID of the physical plates, otherwise leave as "None".

13. *Select the classification type:* Choose "Image" for this setting to enable image-based classification.

14. *Calculate the per-image mean values of object measurements?* Select "No" for this setting, because no objects are identified or measured in this pipeline.

15. *Export measurements for all objects to the database?* Select "None" from the drop-down box, because no objects are identified or measured in this pipeline. Note that a red triangle indicating module error on the setting "*Which objects should be used for locations?*" will disappear once this selection is made.

16. *Export object relationships?* Select "No" to this setting, because no objects are identified or measured in this pipeline.

17. *Write image thumbnails directly to the database?* Select "Yes" to this setting to write a miniature version of each image to the database. This is not necessary for the protocol described here, but may be helpful if using the PlateViewer tool in CellProfiler Analyst to explore images in a multi-well format.

18. *Select the images for which you want to save thumbnails:* Select the channels to be saved as thumbnails; use Ctrl-Click (Windows) or Command-Click (Mac) to select multiple channels.

3.7 Configuring the CellProfiler Output Settings

1. Click the button "View output settings" located at the bottom of the pipeline panel.

2. In the module settings panel, set the *Default Output Folder* to a folder that will contain the output. It is best to avoid spaces or special characters in naming the output folder. If this folder does not exist, it should be created beforehand using a file manager tool (e.g., Windows Explorer, Mac Finder), or by clicking the "New folder" button to the right of the Default Output Folder edit box.

3. Disable the creation of alternative-format output files by selecting the "Do not write MATLAB or HDF5 files" from the *Output file format* drop-down box; this additional output is not needed. Note that the regular-format output will still be produced by the ExportToDatabase module in the pipeline; this will be used for QC purposes.

3.8 Running the QC Pipeline

1. If running on a computing cluster, the CreateBatchFiles module must be added to the pipeline (*see* **Note 15**).

2. From the main menu bar, select *Window > Hide all windows on run*. By default, a window displaying the module results is typically opened for each module during the analysis run. Disabling these windows is recommended for the analysis run, as they will unnecessarily add to the overall run-time, and for the QC pipeline are rather uninformative.

3. Click the "Analyze images" button to begin the analysis processing run.

4. Upon starting the analysis, each image (or collection of images if multiple wavelengths are available) is processed by each module in the pipeline, in order.

3.9 Starting CellProfiler Analyst

1. Start CellProfiler Analyst (CPA) as instructed in the installation help.

2. A dialog box will appear requesting a "properties file." This file was produced by the CellProfiler QC pipeline and has the extension ".properties"; it is placed into the Default Output Folder specified by CellProfiler. This file contains the location of the database containing the QC measurements, image locations, and other associated information. Browse to the location of the properties file created above.

3. Once the properties file is loaded, the CPA interface will then appear (Fig. 3b). CellProfiler Analyst provides an interface with icons to launch a variety of tools.

3.10 Using the Classifier Tool to Detect Blurred Images

1. The exploration tools in CPA (e.g., PlateViewer, ScatterPlot and Histogram) are recommended for use if evaluation of a single QC measurement is sufficient to pass or fail an image (*see*

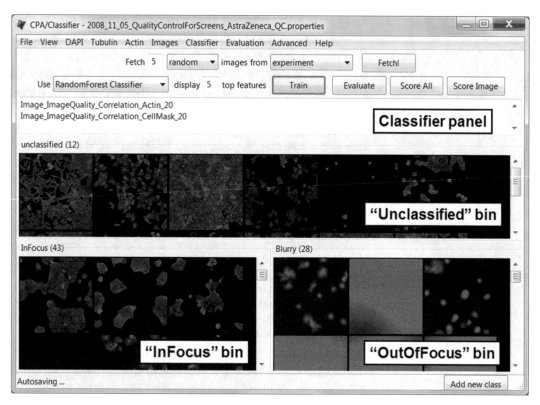

Fig. 4 Example screenshot of the Classifier tool. The panel for adjusting the type and number of images to retrieve ("fetch") is at the *top*. The classifier panel contains the top-scoring QC image features that Classifier has determined are best to distinguish the images in different bins. The "Unclassified" bin contains images that have yet to be sorted into a classification bin. The "InFocus" and "OutOfFocus" bins contain the training set images, that is, images designated by the user as belonging to one of the two classes; these samples will be used to generate the classifier sufficient to distinguish the classes

Note 16). However, if this is not the case, the following steps describe how machine-learning methods may be applied in order to automatically discern which measurements and cutoffs best apply to detect a given type of QC problem (a given QC "class").

2. Click the Classifier icon in the CPA interface to start the machine-learning tool Classifier (Fig. 4). Classifier trains the computer to discriminate user-defined image classes by iteratively applying a supervised machine-learning approach to the CellProfiler-generated QC measurements.

3. Right-click inside one of the bins located at the bottom to display a popup menu of options. Select the "Rename class" option and rename the bin to "InFocus"; this bin will contain examples of images that, by visual inspection, are properly focused. Rename the other bin to "OutOfFocus"; it will hold examples that fail QC due to blurriness.

4. In the top portion of the Classifier window, enter the number of images Classifier should retrieve (or "fetch"). The default of 20 images is a good starting point.

5. Click the "Fetch" button. Images will be randomly selected from the entire image set, and image tiles will begin to appear in the "Unclassified" bin (*see* **Note 17**).

6. Use the mouse to drag and drop the unclassified images into one of the two classification bins, "InFocus" or "OutOfFocus" (Fig. 4). Continue fetching and sorting images until at least ten examples populate each bin (*see* **Note 18**). If no examples of aberrant images are fetched after checking a few dozen, use the exploration tools to assist in finding a few examples based on some of the QC metrics measured by CellProfiler (*see* **Note 16**). The collection of images that have been annotated by classifying them (and are thus located in the lower bins of Classifier) is referred to as the *training set* (*see* **Note 19**).

7. In the top portion of the Classifier window, select the desired classifier from the drop-down box. We recommend using the default settings as a starting point (*see* **Note 20**).

8. Click the "Train" button. Depending on the classifier selected, the large text field near the top of the CPA interface will then populate with a list of the most important features selected by the initial classifier based on the training set; the machine learning algorithm is attempting to differentiate between the samples in each bin based on a combination of QC metrics measured by CellProfiler.

9. In the fetch controls (top part of the window), select "OutOfFocus" from the left-most drop-down menu. Click the "Fetch" button: Classifier will select examples that it deems as out-of-focus based on the current classifier and display them in the Unclassified bin. If the blurriness to be deemed as aberrant is fairly subtle, it may be helpful to fetch from the "InFocus" class images to make sure that only normal images are returned.

10. Correct any misclassifications you see (i.e., in-focus images classified as "OutOfFocus") by sorting them into the appropriate bins.

11. Click the "Train" button to revise the classifier based on the updated training set.

12. Repeat the above process of fetching images, sorting them into their appropriate classes, and re-training to improve the classifier until the results are sufficiently accurate (*see* **Note 21**). Two approaches for checking the classifier accuracy are provided under the "Evaluation" menu item: a confusion matrix displaying the fraction of images falling into the actual versus predicted class or a classification report displaying the precision,

recall, and F1-score for each class (*see* **Note 22**). Once an evaluation selection has been made, press the "Evaluate" button to generate the statistics, and *see* if more training samples are needed.

3.11 Using the Classifier Tool to Detect Saturated Images

1. Open another Classifier tool from the CPA main interface.

2. Right-click inside one of the bins located at the bottom to displays a popup menu of options. Rename the bin to "Non-Saturated". Rename the other bin to "Saturated".

3. Add additional bins if more classes are needed to discern subtleties between artifacts (*see* **Note 23**). Press the "Add new class" button at the bottom-right corner of the Classifier tool. At the prompt, give the new bin a descriptive name and a third bin will appear next to the others.

4. However many bins are needed to distinguish the desired classes, proceed with same procedure as in Subheadings 10.4–10.12 above, identifying images with various types of saturation (bright debris, whole-well fluorescence, etc.).

3.12 Saving the QC Results for Later Use

1. Save the training sets (for both the saturated and out-of-focus classifications) for future refinement, to regenerate a classifier across CPA sessions (but *see* **step 2**), and as an experimental record by selecting *File > Save Training Set* from the menu bar. It is advisable to do so periodically during the creation of a training set, but certainly before proceeding to scoring the experiment because scoring may take a long time for large screens.

2. Likewise, save the classifier generated by CPA (for both the saturated and out-of-focus classifications) by selecting *File > Save Classifier Model* (*see* **Note 24**). This classifier may also be used as part of a downstream CellProfiler analysis workflow; *see* Subheading 13 for details.

3. Click the "Score All" button to have Classifier score all images in the entire experiment. A dialog box will appear with scoring options; use the defaults of "Image" under "Grouping," and "None" under "Filter". Make sure the "Report enrichments?" box is unchecked before pressing the OK button, because this is only relevant for classifying individual objects.

4. Once scoring is completed, the results are presented in a Table-Viewer. Saving this table to a comma-delimited file (CSV) or to the original database can be done via *File > Save table to CSV* or *File > Save table to database*, respectively. In the latter case, the table can either be stored permanently or for the current session which means the table will be removed from the database when CPA is closed.

3.13 Using the QC Classification Results for CellProfiler Analysis (Optional)

1. The QC workflow described above is intended to form the initial steps of a larger data analysis workflow (*see* refs. 7, 10). Our laboratory typically runs the QC workflow prior to completing a full analysis of images using CellProfiler. A systematic microscopy error, for example, could be detected at this point and further downstream data processing could be aborted without further investment of valuable time. In the absence of such egregious problems, it is helpful to store the QC results as metadata alongside subsequent analysis results to allow for retrospective quality checks and to assist troubleshooting. Alternately, the QC results can be used to exclude aberrant images from full analysis: if using CellProfiler for post-QC data analysis, for example, the classifier produced by the above steps may be incorporated into the CellProfiler analysis pipeline with the `FlagImage` module. This module can mark images that pass or fail QC and can also skip further analysis of images that fail and proceed immediately to the next image. Use the following steps to modify an existing CellProfiler analysis pipeline to take advantage of the QC results.

2. Load or create an analysis pipeline designed to score the assay of interest; examples of analysis pipelines may be found at http://www.cellprofiler.org/examples.shtml. If other modules are needed in the pipeline, they may be added and arranged using the controls at the bottom of the pipeline panel (*see* **Note 8**).

3. Select and add the `MeasureImageQuality` module from the "File processing" module category. Generally, this module should be placed as the first of the analysis modules.

4. Give this module the same settings as in the QC pipeline. Failure to do so may result in an error, as the same sets of features are expected between the two pipelines.

5. Select and add the `FlagImage` module from the "Data tool" module category. Place it in the pipeline after the `MeasureImageQuality` module. Adjust the settings listed in the following steps.

6. *Name the flag's category*: Leave this as "Metadata".

7. *Name the flag*: Give the flag a meaningful name. For example, if using this module to detect out-of-focus images, the flag might be called "OutOfFocus".

8. *Skip image set if flagged*: Select "Yes" for this setting to skip downstream modules in the pipeline for any images that are flagged. This approach gives the option of omitting unnecessary analysis on aberrant images. By selecting "No", the analysis measurements are retained regardless of the QC flag, which may be helpful for later review.

9. *Flag is based on*: Select the option from the drop-down box corresponding to the classifier file saved from CPA. Additional settings will appear prompting you to specify the location and file name of the classifier to be applied to this data set.

10. Further settings will allow you to select which classes to flag when applying the QC criteria, with the flag is set if the image falls into the selected class. The module can also set the flag if the image falls into any of multiple classes, e.g., if you created classes in CPA for both out-of-focus images and images with low cell counts as well as a class of in-focus images with high cellular confluency.

11. If you have multiple classifiers that indicate other QC problems (such as saturation artifacts in addition to focal blur), you can press the "Add another flag" button to produce another collection of settings for a new metadata flag; repeat the above steps to provide additional QC criteria.

12. Alternately, you can combine QC criteria to produce a single metadata flag by pressing the "Add another measurement" button to produce another collection of settings for the same metadata flag. Repeat the above steps to provide additional QC criteria and under the *"How should measurements be linked?"* setting, indicate whether the flag should be set if all the conditions are met ("Flag if all fail"), or any of the conditions are met ("Flag if any fail").

13. After running the pipeline on the full image set (follow the instructions for the QC pipeline in Subheading 3.8 above), the results of the classifier will be stored as a per-image measurement, named according to the settings for the `FlagImage` module. For example, with the example given in (Fig. 4), the corresponding measurement will be named "Metadata_OutOfFocus", with an out-of-focus image receiving a value of "1" while an in-focus image will assigned a value of "0".

4   Conclusions

This protocol describes how a researcher can collect a suite of image-based quality metrics and use a machine-learning approach to distinguish between high-quality and aberrant images, all with the use of free, open-source software. Naturally, the best approach to remove artifacts is to prevent them from occurring in the first place during sample preparation and imaging. Simple steps include filtering the staining reagents before use to remove large particulates, and confirming the proper exposure settings for each channel prior to running an experiment (*see* ref. 11). While we have taken a supervised (i.e., human-guided) approach, unsupervised (i.e., purely computer guided) techniques to whole-image classification have also been described (*see* ref. 12). We have restricted our guidance to out-of-focus images and images containing saturation artifacts because these classes cover nearly all the artifacts we typically see in our own experience, but this basic approach may be used to identify any desired artifact, provided that their "phenotype" can be captured by one or several whole-image measurements. If the measurements provided in the `MeasureImageQuality` module turn out to be insufficient to capture the artifact in question, it may be helpful to include additional measurements in the pipeline directed towards the artifactual features, analogous to what is done in the screening domain by including image features specific to the phenotype of interest (*see* ref. 13).

5   Notes

1. It is essential that the fluorescence images be collected with a uniform protocol, in which the image acquisition settings (e.g., exposure time, magnification, gain) are kept constant throughout the entire experiment. Additional guidance on image acquisition can be found elsewhere (*see* ref. 11). This workflow can be adapted to handle brightfield images as well, with the following caveat: debris will not appear as a saturation artifact but rather as a dark region or smudge. In this case, by inverting the pixel intensities so that dark pixels become bright, and vice versa, the quality control metrics described above can be used without modification. This can be done using the `ImageMath` module (Category: Image Processing) in CellProfiler with "Invert" as the *Operation* setting.

2. We recommend the use of "lossless" image formats such as .TIF, .BMP, or .PNG. While "lossy" .JPG images are commonly used for photography, the smaller file size comes at the cost of artifacts that can hinder image analysis. For further reading, please *see* the online Assay Guidance Manual chapter on image-based high content screening (*see* ref. 11).

3. If processing a few hundred images, a stand-alone desktop is sufficient to complete the task in a matter of hours. For assays with thousands of images or more, the best practice is to use a computing cluster to parallelize and thus speed up processing. Suggested hardware specifications for a computing cluster are 64-bit architecture, with eight or more cores per compute node.

4. Although most users of the protocol described in this article will need source code, the source code is publicly available in Git repositories administered by GitHub, and can be downloaded from https://github.com/CellProfiler/CellProfiler/.

Information and resources for developers are available at https://github.com/CellProfiler/CellProfiler/wiki, including tips for running Cellprofiler on a cluster environment for large screens.

5. In addition to the Help menu in the main CellProfiler window, there are many "?" buttons in CellProfiler's interface containing more specific documentation. Clicking the "?" button near the pipeline window will show information about the selected module within the pipeline, whereas clicking the "?" button to the right of each of the module settings displays help for that particular setting. Additionally, the CellProfiler user manual is available online at http://www.cellprofiler.org/CPmanual/ (containing content copied verbatim from CellProfiler's help buttons), and a user forum (http://forum.cellprofiler.org/) is available for posting questions and receiving responses about how to use the software.

6. To add modules, click the "+" button below the pipeline panel (Fig. 3a). In the dialog box that appears, select the module category from the left-hand list. Select the module itself from the right-hand list. Double-click the module to add it to the pipeline, or click the "+ Add to Pipeline" button. Many modules can be added; click the "Done" button when finished. Modules can then be arranged in the pipeline by clicking the "^" or "v" buttons below the pipeline panel. Help is also available for each module by clicking the module to highlight it and then pressing the "?" button near the pipeline window.

7. By default, the Images module will pass all the files specified to later Input modules, in order to define the relationships between images and associated metadata (the Metadata module) and to have a meaningful name assigned to image types so other modules can access them (the NamesAndTypes module). Filtering the files beforehand is useful if, for example, a folder which was dragged-and-dropped onto the file list panel contains a mixture of images for analysis along with other files to ignore.

8. Often, the acquisition software of many screening microscopes will insert text into each image's file and/or folder name corresponding to the user-specified experiment name, plate, well, site, and wavelength number. For example, the BBBC images described above (*see* Subheading 2) use a common nomenclature, e.g., *Week1_150607_F10_s3_w1636CC6D1-0741-42BB-AF32-3785EB8BA086.tif*, where "Week1_150607" is the plate name, "F10" is the well, "s3" denotes site 3 in the well, and "w1" indicates that the first wavelength was acquired.

9. Regular expressions (regexp) are a versatile (albeit complex) text pattern-matching syntax. Patterns are matched using combinations of symbols and characters. By clicking the magnifying glass icon next to the regexp setting, a dialog is provided which shows a sample text string, a regexp, and the results of applying the regexp to the sample text; both the sample text and regexp can be edited by the researcher. CellProfiler's help text for the `Metadata` module provides an introduction to regular expressions. While regexp syntax is largely standardized, the Python programming language variant thereof is used here; a more in-depth tutorial can be found at http://docs.python.org/dev/howto/regex.html.

10. Raw grayscale images are recommended for fluorescence microscopy. If color images are acquired, select "Color image" for this setting, and, later, insert a ColorToGray module in the analysis portion of the pipeline. The `ColorToGray` module splits the original color image into its red, green and blue channels, each represented as grayscale image.

11. To use the metadata matching tool, select the metadata identifier that is required to uniquely match all the channels for each row. If multiple identifiers are needed, click the "+" button to add another row of metadata below the previous one, or the "−" button to remove a row. Click the up and down arrows to reorder the precedence that these identifiers are applied.

12. The QC metrics that are targeted to identify focal blur artifacts include: (a) Power spectrum slope: the image spatial frequency distribution, with lower values corresponding to increased blur; (b) Correlation: the image spatial intensity correlation computed at a given spatial scale offset, with lower values corresponding to decreased blur; (c) Focus score: the normalized image variance of the image, with lower values corresponding to increased blur; (d) Local focus score: the focus score computed in nonoverlapping blocks and averaged, with lower values corresponding to decreased blur. Details on robustness and prior validation of these metrics can be found elsewhere (*see* ref. 7).

13. The QC metrics that are targeted to identify saturation artifacts include: (a) Percent maximal: the percentage of the image occupied by saturated pixels; (b) Intensity standard deviation, which is useful for detecting images with very bright but subsaturated artifacts. Details on robustness and prior validation of these metrics can be found elsewhere (*see* ref. 7).

14. Measurements may reside in a MySQL or SQLite database. A MySQL database is recommended for storing large data sets (i.e., from more than 1000 images) or data that may need to be accessed from different computers. Consultation with the local

information technology staff on the details of setting up or accessing a database server is recommended. SQLite is another mode of data storage, in which tables are stored in a large, database-like file on the local computer rather than a database server. This is easier to set up than a full-featured MySQL database and is at least as fast, but it is not a good choice of storage if the data is to be accessed by multiple concurrent connections.

15. To prepare a pipeline for batch processing on a computing cluster, add the `CreateBatchFiles` module (Category: File Processing) as the last module of the pipeline and configure it according to the module instructions. Once done, click on the "Analyze images" button. CellProfiler will initialize the database tables and produce the necessary file for batch processing submission. Submit the batches to the computing cluster for processing; use the *Search Help...* function under the Help menu in CellProfiler to search for "batch processing" for details on cluster computing.

16. Click the PlateViewer icon in the CPA interface to launch a tool to view a single QC measurement as a per-well aggregate in a multi-well plate format. Use the ScatterPlot or Histogram tools for a more quantitative approach to reviewing single QC measurements. Details on the use of these tools for QC purposes can be found elsewhere (*see* ref. 7).

17. Each tile is a thumbnail of the full image; a small white square is displayed in the center of each tile as the mouse hovers over it. It may be that the image tile is too small to allow viewing a small or subtle artifact. One approach to handle this issue is opening the full image in a separate ImageViewer window by double-clicking the tile. From this window, the image can be placed into a bin by dragging and dropping the small white square in the image center. Another approach is to select "View" from the menu bar and in the dialog that appears, adjust the image zoom (indicated by the magnifying glass icon) by pulling the slider to the left, which will change the zoom of all the image tiles. Adjust until the image tiles are the desired size.

18. Images with no cells can usually be classified as in-focus for this purpose; enough residual cellular material often remains in such images for the microscope to maintain focus. Also, images with varying degrees of blurriness can all be included in the same bin for classification, as illustrated by the first two images in the "OutOfFocus" bin in Fig. 4.

19. Not all images in the "Unclassified" bin need to become part of the training set: if the classification of a particular image is uncertain, it can be ignored by leaving it in the "Unclassified" bin (or remove it by selecting it and pressing the Delete key).

Keep in mind, however, that Classifier will eventually be required to score all images in the experiment as one classification or the other, so the more information you provide in guiding this decision, the better.

20. The classifiers (except for Fast Gentle Boosting) are implemented using Python's scikit-learn package (*see* ref. 14). The values of the parameters used as input may be modified by selecting *Advanced > Edit Parameters*, but this is not recommended unless you are already comfortable with machine learning approaches.

21. The most accurate method to gauge Classifier's performance is to fetch a large number of images of a given class from the whole experiment, and evaluate the fraction of the images which correctly match the requested class. For example, if fetching 100 putative out-of-focus images reveals upon inspection that seven of the retrieved images are actually in-focus, then the classifier has a positive predictive value of roughly 93% (and thus a false positive value of 7%). Another approach is to click the "Evaluate" button to produce performance statistics (*see* **Note 22**); values closer to 1 indicate better performance. However, because the training set often includes a number of difficult-to-classify images (due to the recommended iterative training process), the accuracy reported by the "Evaluate" button should generally be considered the worst case scenario, that is, a lower bound on the true accuracy. The final approach is to open an image by double-clicking on an image tile and then select *Classify > Classify Image* to score the single image. While the results of this method cannot be extrapolated to other images, it can help improve a training set by identifying misclassified images to add to the classification bins; this can be done by left-clicking the full image and dragging-and-dropping it into the desired classification bin.

22. For both evaluation displays, the predictive ability is assessed using *cross-validation*, a technique in which the annotated set of images is split into a "training" subset to train the classifier to distinguish between classes, and a "test" subset to evaluate the accuracy. This procedure is repeated five times, with the training and test subsets randomly selected while preserving the percentage of samples for each class. The results are then aggregated to produce the evaluation displays. The confusion matrix shows a table in which the true classification of the images (rows) is shown versus the predicted classification (columns). Ideally, the table should have only non-zero values on the diagonal elements (i.e., where the row index = column index) and zeros elsewhere; this means that all images are correctly classified into their respective types. A large number of images in the off-diagonal elements indicates that the

classifier is "confusing" the classes with each other. The classification report displays a heatmap of three common metrics used in machine learning for each of the classes: the precision (how well the classifier avoided false positives, defined as the fraction of retrieved images for a given class that are correctly classified), the recall (how well the classifier obtained true positives upon request, defined as the fraction of correctly-classified images from the set of images retrieved) and the F1-score (a weighted average of the precision and recall; ranges from 0 to 1, with 1 as the best score). If using the classification report with the Fast Gentle Boosting classifier, the "Evaluate" button will produce a graph of the cross-validation accuracy for the training set, by estimating the classifier performance by training on a random subsample of the training set, then testing the accuracy on the samples not used for training. This value is plotted as an increasing number of image features are used. If the graph slopes upward at larger numbers of features, adding more features is likely to help improve the classifier. If the graph plateaus after a certain number of image features, then further features do not help improve accuracy. A downward slope may indicate more training examples are needed.

23. We have found that using only two classification bins to distinguish saturated from non-saturated images tends to fail for images that contain brightly fluorescing cells. This problem can be overcome by creating an additional class to distinguish bright, non-artifactual images from images containing actual saturated artifacts. If such images are unlikely to occur in a given assay, the creation and use of this extra bin can be omitted.

24. A saved classifier set can assist in initializing a QC classifier for a new experiment, as long as the stains and imaged channels are the same, as follows. Start a new CPA/Classifier session. Create and name the bins to match those from the previous Classifier session. Select *File > Load Classifier* from the Classifier menu to load your previously saved classifier. At this point, images can be fetched from the desired class without creating a training set first. The fetched images may have a large number of misclassifications, due to inter-experiment variability. If this is the case, the iterative workflow will still need to be followed as before.

Acknowledgements

This work was funded by the National Science Foundation (RIG DB-1119830 to M.A.B..) and the National Institutes of Health (R01 GM089652 to A.E.C.). We also thank Jane Hung and David Dao for offering helpful comments and suggestions during manuscript preparation.

References

1. Conrad C, Gerlich DW (2010) Automated microscopy for high-content RNAi screening. J Cell Biol 188:453–461

2. Thomas N (2010) High-content screening: a decade of evolution. J Biomol Screen 15:1–9

3. Niederlein A, Meyenhofer F, White D et al (2009) Image analysis in high-content screening. Comb Chem High Throughput Screen 12:899–907

4. Carpenter AE, Jones TR, Lamprecht MR et al (2006) CellProfiler: image analysis software for identifying and quantifying cell phenotypes. Genome Biol 7:R100

5. Kamentsky L, Jones TR, Fraser A et al (2011) Improved structure, function and compatibility for CellProfiler: modular high-throughput image analysis software. Bioinformatics 27:1179–1180

6. Jones TR, Carpenter AE, Lamprecht MR et al (2009) Scoring diverse cellular morphologies in image-based screens with iterative feedback and machine learning. Proc Natl Acad Sci U S A 106:1826–1831

7. Bray M-A, Fraser AN, Hasaka TP et al (2012) Workflow and metrics for image quality control in large-scale high-content screens. J Biomol Screen 17:266–274

8. Caie PD, Walls RE, Ingleston-Orme A et al (2010) High-content phenotypic profiling of drug response signatures across distinct cancer cells. Mol Cancer Ther 9:1913–1926

9. Ljosa V, Sokolnicki KL, Carpenter AE (2012) Annotated high-throughput microscopy image sets for validation. Nat Methods 9:637

10. Bray M-A, Carpenter A (2012) Advanced assay development guidelines for image-based high content screening and analysis. In: Sittampalam GS, Coussens NP, Nelson H et al (eds) Assay guidance manual. Eli Lilly & Company and the National Center for Advancing Translational Sciences, Bethesda, MD

11. Buchser W, Collins M, Garyantes T et al (2012) Assay development guidelines for image-based high content screening, high content analysis and high content imaging. In: Sittampalam GS, Coussens NP, Nelson H et al (eds) Assay guidance manual. Eli Lilly & Company and the National Center for Advancing Translational Sciences, Bethesda, MD

12. Rajaram S, Pavie B, Wu LF et al (2012) PhenoRipper: software for rapidly profiling microscopy images. Nat Methods 9:635–637

13. Logan DJ, Carpenter AE (2010) Screening cellular feature measurements for image-based assay development. J Biomol Screen 15:840–846

14. Pedregosa F, Varoquaux G, Gramfort A et al (2011) Scikit-learn: machine learning in python. J Mach Learn Res 12:2825–2830

Chapter 8

High-Content Screening Approaches That Minimize Confounding Factors in RNAi, CRISPR, and Small Molecule Screening

Steven A. Haney

Abstract

Screening arrayed libraries of reagents, particularly small molecules began as a vehicle for drug discovery, but the in last few years it has become a cornerstone of biological investigation, joining RNAi and CRISPR as methods for elucidating functional relationships that could not be anticipated, and illustrating the mechanisms behind basic and disease biology, and therapeutic resistance. However, these approaches share some common challenges, especially with respect to specificity or selectivity of the reagents as they are scaled to large protein families or the genome. High-content screening (HCS) has emerged as an important complement to screening, mostly the result of a wide array of specific molecular events, such as protein kinase and transcription factor activation, morphological changes associated with stem cell differentiation or the epithelial-mesenchymal transition of cancer cells. Beyond the range of cellular events that can be screened by HCS, image-based screening introduces new processes for differentiating between specific and nonspecific effects on cells. This chapter introduces these complexities and discusses strategies available in image-based screening that can mitigate the challenges they can bring to screening.

Key words Synthetic lethality, High-throughput screening, Phenotypic screening, Assay validation, Chemical biology

1 Introduction

Although genetic and chemical screening developed independently, they shared common limitations regarding what could be identified—generally factors that affect growth or survival because they were the easiest phenotypes to identify, although some extremely clever phenotypic screens in model organisms were critical to our understanding of development. Largely as a consequence of the advances in automation and data analysis, phenotypic screens have become manageable for many thousands of reagents, be they chemical or genetic. Chief among the technological improvements was the development of image-based screening, as it became an

Paul A. Johnston and Oscar J. Trask (eds.), *High Content Screening: A Powerful Approach to Systems Cell Biology and Phenotypic Drug Discovery*, Methods in Molecular Biology, vol. 1683, DOI 10.1007/978-1-4939-7357-6_8,
© Springer Science+Business Media LLC 2018

open toolbox for essentially any visual change in a cell, either endogenous or engineered through reporters such as fluorescent proteins. Either as true phenotypic screens (measuring changes in an intrinsic property of the cell, such as morphology) or an engineered molecular property (for example, measuring cyclin dynamics through expression of a fusion protein to GFP).

RNAi has had the strongest historical connection to image-based screens [1–5], RNAi screening is common in many multicellular organisms and some parasites and fungi. Image-based endpoints are also being used in other platforms. The screening of small molecules, either in drug discovery or for biological discovery (usually referred to as chemical biology, chemical genetics or chemical genomics has also used image-based endpoints [5, 6]. Screening for new drugs or tool compounds usually means that the target is not known at the time of screening, whereas chemical biology screens using libraries of kinase or GPCR active compounds means that at least some of the targets are known (but new ones are revealed fairly frequently). Most recently, the ability to precisely edit the mammalian genome using RNA-mediated endonucleases through the CRISPR/Cas9 system has developed into a new approach for genome-wide screening, facilitating its combination with imaging high-content analysis [7]. Some of the factors that are unique to CRISPR-based screens that can confound the interpretation of results are starting to be identified [8], but more will emerge over time.

This review highlights the major screening modalities in use currently and introduce the major technical challenges to running successful screens in each of these formats, and then cover aspects of HCS that can minimize or even directly address some of these challenges using strategies for increasing specificity in hits through increased attention to endpoints during the assay design and methods for detecting and eliminating false-positives and other anomalous data during data analysis. In general, this review will not cover approaches to mitigate artifacts through non-image-based methods, but will direct the reader to sources that do cover them well.

2   Caveats and Limitations of Popular Screening Platforms

As noted above, each screening approach has challenges, some are common to all and some are unique. Understanding how image-based screening can mitigate these limitations requires an understanding of the causes for each as function of the individual platforms. Following some discussion on how imaging screens can be used for innovative designs that intrinsically address some of these limitations, there will be a final discussion that returns to some of these specific issues.

2.1 RNAi Screening

RNAi screening uses short RNA sequences introduced into cells through transfection (as duplex RNAs via a carrier (such as a lipid), or by transduction using a viral or plasmid vector. The latter generally uses a lentiviral system to introduce the shRNA, which is then processed to produce a siRNA-like duplex. HCS-based RNAi screening is common for both of the major formats of RNAi [9–11]. shRNA-based systems also allow for some novel screen designs, such as pooling viruses and quantifying over- and under-represented RNAi sequences following selection pressure [12, 13], but these are usually selections based on survival of selected transductants rather than through a cytological phenotype. Some widely discussed problems with RNAi were recognized as it was emerging as a general screening platform. These include the potential to engage the dsRNA antiviral response, the inability to successfully target the intended gene and propensity to target unintended genes [14]. Several strategies for reducing these effects through improved RNAi sequence design and identify off-target effects have been very helpful [15, 16]. These strategies have improved RNAi library design and use, but some matters are still challenges to RNAi screens. These have been discussed in detail elsewhere [17], but are highlighted here:

1. *Penetrance*: The transfection process and RNAi effects can be highly inefficient when compared to alternative perturbations, so one readjustment is that the recognition that effects are typically more heterogeneous. Some cell lines are inherently more difficult to transfect that others (differences in viral-mediated transduction efficiency are minimized when a selection step is included in the screen design). Whether this heterogeneity is addressed directly in HCS, such as quantifying effects through comparing the distribution of the endpoint values, or not (taking the average intensity or localization ratio as one would for a small molecule treatment, but accepting a smaller assay window is more common), it is generally true that the effect of RNAi on a cellular process is less than what is observed through other modalities.

2. *Partial knockdown.* In addition to uneven knockdown across a population of cells, it is also true that many siRNAs are only partially effective at reducing target message levels under the best conditions. Sequence design improvements, as well as sequence scanning to identify optimal reagents, have reduced the list of genes that are difficult to knockdown, but false-negatives in screens do still exist due to poor knockdown of the target gene. In cases where developing a basic set of effective RNAi reagents is a problem, some groups (particularly within the shRNA platforms) have included extra reagents to help increase confidence for genes that can only be partially knocked down.

3. *Stress/toxicity.* As mentioned above, unintended effects on stress and inflammatory signaling were observed in some early RNAi experiments as a result of sequence-specific engagement of the cellular Toll receptors [18]. Other effects were the result of harsh transfection conditions. This was a particular problem for many screens in the cancer area, as cell death, apoptosis induction, or proliferation arrest endpoints are affected by harsh transfection/transduction conditions, synergizing with antiproliferative control RNAi reagents, artificially increasing the strength of the endpoints for positive controls. Methods that include careful monitoring of the cell health lead to better screens. Measurement of GAPDH levels is a popular method for general RNAi screening [19]. Morphology measures would be additional obvious candidates for image-based screens; this is discussed in depth below.

4. *Temporal complexity.* Small molecule and protein ligands typically engage their targets rapidly, and in some cases effects can be observed within seconds or minutes [11, 17]. RNAi functions indirectly on proteins through reduction of mRNA levels, so effects are not observed until the message has been depleted and the protein level have been reduced through normal turnover. Since protein turnover rates range from minutes to days, the knockdown of any gene could produce a phenotype anywhere along this continuum. As a general rule, endpoint data for RNAi screens are typically collected after 24–48 h. Proteins with faster turnover rates tend to have the stronger phenotypes. Those regulated through the cell cycle are among the most robust, such as the cyclins and many mitotic kinases; PLK1 is a particularly strong example, and in fact it is commonly used as a process control during many screens.

5. *Off-target effects.* If failing to significantly modulate the intended gene product was not enough, there is also the problem that RNAi tends to affect other genes too frequently. Why this happens is an active area of research, and some important principles have emerged. Probably the one with the most impact has been that siRNAs with high off-target effects tend to hit noncoding micro RNA (miRNA) target sequences, and in effect, can substitute for endogenous miRNAs [20, 21]. miRNAs are siRNA-like endogenous transcripts that regulate groups of genes posttranslationally [20]. RNAi reagents that mimic miRNAs can affect signaling or metabolic pathways through partial knockdown of several genes involved in a specific biological process. Most off-target effects, and effects on miRNA-regulated gene families in particular, occur through the seed sequence, a critical region for target gene recognition. siRNAs are 17–19 bp long, but nucleotides 2–7 are of greatest importance in target gene selection and this stretch is referred

to as the seed sequence [16]. Revised sequence selection algorithms for the design of RNAi reagents have been updated to avoid six-mer sequences that are found in miRNAs or are overrepresented in mRNA sequences generally [16, 22, 23].

6. *Plate (edge) effects.* In higher-throughout assay designs, given the sensitivity of cells to RNAi, the increased handling during the transfection or transduction process and the time needed to observe expression change-mediated effects, RNAi experiments tend to require more manipulations than other modalities. The traffic in and out of the incubator introduces fluctuations of temperature, humidity, and CO_2-levels during the course of the experiment. These end up affecting the phenotype on the wells where these changes are more severe - the outer wells of the plate. Such "edge-effects" can affect the endpoint measurements for these wells. Two common methods for limiting these effects are either to avoid them all together (through not using the outer wells), or to apply a location-based normalization; the most frequently used is to calculate a B-factor for each well using row- and column-wide adjustments of the raw data [24].

2.2 Small Molecule Screens

The many forms of screening small molecules, such as the identification of tool compounds to be used to characterize a biological process or screening known drugs for new uses (repurposing screens), owe much to the traditional screening of small molecules by biotechnology and pharmaceutical companies as they try to identify new drugs for a wide spectrum of diseases. In many regards, the drug development process, at least through the stage of demonstrating activity in animal models, is the process of addressing off-target effects (selectivity) and toxicity that are found in all screening methods. However, in drug discovery, these are addressed by a team of biologists and medicinal chemists over period of one or more years and coalesce around a chemical scaffold. This is clearly impractical for chemical biology/genetics, so instead they either work with compounds that have already been well-characterized or employ methods to identify and eliminate (rather than fix) compounds with significant liabilities. Some of the major complications to screening small molecule inhibitors are:

1. *Selectivity.* The chemists' term for off-target effects, it is pervasive in pharmacology. Some protein families share enough sequence and structural identity that compounds may react with several or even dozens of members. These are most common for proteases, protein kinases, and GPCRs, but this exists for all protein families. Additionally, there are "idiosyncratic" selectivity issues, frequently identified through chemoproteomics. This is when a small molecule is used to purify the proteins it interacts with. In many cases several unrelated

proteins (by function or family) can be purified and shown interact directly with the small molecule [25].

2. *Reactivity.* Chemical reactivity is much more complex than the cell stress that can occur during transfection. There are at least a dozen chemical motifs that occur relatively frequently in compound collections. Some of the most egregious motifs have been collectively termed PAINS, pan-assay interference compounds because they occur in many publications despite having non-drug like actions [26], but are instead the result of chemical reactivity that is related to the specific motifs incorporated into the compound. These are observed in both *in vitro* screens (using purified proteins) as well in cell-based screens. In general these compounds chemically react with proteins covalently or are redox-active, generating reactive species that will affect some proteins significantly, giving the impression of activity. As such, these do not show a single stress phenotype, like siRNAs.

3. *Solubility issues.* Compared to genetic manipulations, chemical compounds present logistical complications because of their highly unique properties. Chief among them is solvent-dependent solubility. Most screens make compound diversity a priority, and even focused libraries, such as "repurposing" libraries that facilitate new activities for clinical or FDA-approved compounds, collect molecules with widely different physical properties. When such libraries are treated systematically (resuspended in a common vehicle, diluted uniformly and measured at a common time point), idiosyncratic behavior occurs. One common effect is precipitation of some compounds during the dilution process, which rapidly changes their environment. Some compounds become insoluble during these rapid transitions and their effects on cells can be a product of their colloidal properties than pharmacological effects. Depending on the assay endpoints, these can result in false-positive or false-negative data.

2.3 CRISPR

CRISPR (Clustered Regularly Interspaced Short Palindromic Repeats) and the CRISPR-associated nuclease Cas9 together have been shown to edit the genome with high specificity and could therefore be used to engineer deletions in essentially any desired gene. The system uses a guide sequence (sgRNA) that is introduced into the cell to engage the Cas9 nuclease system at a desired target sequence. The system is very versatile (it can be used to introduce point mutations in genes, and the specificity is sufficient to enable the system to be used in genome-wide screens directly analogous to RNAi [7]. As the most recent addition to the options available for screening, its utility has been validated through several successful screens, but a complete understanding of limitations or confounding matters is emerging [27]. It is currently undergoing a period of vigorous methods development, including methods for introducing the Cas9 nuclease protein at the time of transfection of the sgRNA and the generation of murine lines that carry germ-level expression of the Cas9 gene. Some of the things that concern practitioners are:

1. *Penetrance/efficiency.* Similar to problems experienced in RNAi, CRISPR can have limited efficiency when guide RNAs are introduced into cells. Mali et al. [28] fond that efficiencies can be high, but in some commonly used cell lines it can be around 10% and iPSC cells were as low as 2%. Frequently a selection step can be added to enrich for the successful events, but some screen designs may be compromised by such a step.

2. *Off-target effects.* Early data suggests that off-target effects are minimal when compared to RNAi, but practitioners of RNAi recall that the extent of mistargeting by RNAi was grossly underappreciated in the beginning and strive to avoid a false sense of security. Nevertheless, studies are emerging that do continue to support this general appraisal [29], but they do occur [8].

3. *Unanticipated genetic adaptations and consequences.* Since subtle genetic changes, like point mutations in a target gene to eliminate translation initiation are considered the minimal perturbations and do not affect transcription, the potential for internal translation start sites can result in a successful, on-target, genetic change that does not completely eliminate protein function [8]. Another potential problem is genomic complexity. While the genome has been characterized for decades as a relatively small percentage of active genes encoded in a bed of generally nonfunctional DNA, the genome is in fact highly active including many transcripts that are transcribed but not translated into proteins [30]. Many of these sequences have been shown to play physiological roles and they can be embedded with other genes, making them unintended recipients of targeted genomic alterations. Unlike most of the challenges discussed here, this is generally preventable, but it can happen with some types of guide sequences.

3   Imaging Endpoints and Data Analysis Approaches That Can Improve on Screen Performance

The range of image-based assays can be appreciated through the contributions to this and other volumes on HCS [31–33]. However, HCS is more than an endpoint (or set of endpoints) in phenotypic screens [34].

3.1 Image-Based Endpoints That Increase the Specificity of Hit Criteria

For most researchers, the advantages of HCS-based assays are introduced through the concept of switching from imprecise and potentially artifactual whole-well assays to specific cell-based events. This is because general assays, such as cell death, can have many separate paths that lead to the same effect, but investigators are typically interested in a specific pathway, and a large hit list can slow progress on the project. The classic cases are false-positive hits in cell death or proliferation assays that result from extensive cell loss from nonspecific causes, such as cytoskeletal changes that cause the cells to round up and detach. An image-based endpoint can not only interrogate cells for a specific event, such as caspase or PARP cleavage as specific measures of apoptosis, but can do so on a per-cell basis, which can control for some variability in cell numbers and can also detect significant cell loss in a well. This specificity has had the effect of reducing the number of false-positives. A related example is the comparison between transcriptional reporter assays to measure the effect of a signaling pathway. For a promoter to be regulated by a transcription factor such as Foxo-3A, NF-κB or STAT3, the pathway regulating the transcription factor is obviously critical, but so can transcriptional co-regulators, epigenetic mechanisms and general chromatin structure. HCS endpoints specific to the transcription factor maintain focus on the relevant signaling events and can avoid complications from other causes of promoter activity.

3.1.1 Morphological Endpoints

Morphological endpoints are unique to imaging assays, particularly those that measure changes in the cytoskeleton or the shape of the cell or nucleus [35–37]. These can be highly relevant and informative endpoints, particularly for screens that investigate processes in developmental biology such as differentiation, motility and chemotaxis, stem cell biology, tumor metastasis or epithelial-mesenchymal transformation [38, 39]. These phenotypic endpoints link have been used in RNA screening to examine many of the most challenging biological and therapeutic problems, including neurotoxin sensitivity in Huntington's Disease [6], Tau hyperphosphorylation in Alzheimer's Disease [40], cholesterol accumulation in Neimann-Pick Type C cells [41] and the effect of genes identified in GWA studies for genetic contributions to Type II Diabetes through effects on glycogen accumulation in primary human hepatocytes [42].

3.1.2 Multiple Endpoints

HCS assays can typically quantify four or more separate channels, allowing for a configuration where one channel measures a nuclear/chromatin dye and leaves other channels to track separate proteins or functional endpoints. These can be configured for redundant positive endpoints, such as two separate points on the same pathway (phosphorylation of FAK and effects on the actin cytoskeleton is one example) or to configure a negative control in the primary screen, such as translocation of NF-κB but not STAT3 to differentiate between stress signaling pathways. These further enhance the specificity of the assay by adding additional information that can support initial hits from the single endpoint and imposing additional conditions that fit the goal of the screen [43, 44].

3.1.3 Integrated Multiparametric Endpoints

Given the extensive measurements made for any cellular fluorophore, there are many features available for extended data analysis. Such analyses can integrate dozens or even hundreds of features into feature maps that bring some of the techniques of systems biology into HCS [45–47]. These require a good biological rationale for such a complicated workflow, as well as the input of a biostatistician with good programming skills. Such approaches have been helpful in the toxicology area, where a well-differentiated approach to tracking idiosyncratic events are critical to monitor the potential for serious adverse clinical events [48, 49]. Another area where the use of integrated endpoints and cell-level analysis has been essential is stem cell biology [50]. While essentially a morphological or phenotypic screening approach, the heterogeneity of both colony morphology and the expression of stem cell-specific transcription factors requires the assessment of multiple features across a colony rather than the expression of one gene or cellular feature. Software applications are advancing the technology to allow phenotypic screens to be explored through machine-learning classifications [51–53].

3.1.4 Chemosensitization Screening Designs

Chemosensitization studies in oncology, comparing the effect of one drug on another to identify a synergy or antagonism between two therapeutics has been in use for years [54, 55]. RNAi screening has been an excellent platform for identifying potential combination strategies that associate new signaling pathways to mechanisms of drug resistance [56, 57]. The simplest design is to determine the IC10 or similar low dose of a compound (for example, a cytotoxic chemotherapeutic that is standard of care for some cancer treatment regimens, but suffers from acquired resistance or a substantial population of nonresponders) and screen an RNAi library for genes that increase the sensitivity of the cell line to the compound dramatically. Genes identified in such a screen indicate potential combination therapeutics strategies, through the development of a chemical inhibitor of a gene identified as sensitizing the initial treatment condition through RNAi knockdown. An example of such an analysis is shown in Fig. 1. After adjusting for subtle effects on the endpoint in the test condition, RNAi treatments can be plotted for the test and control conditions and will generally fall along the diagonal. siRNAs that fall above the line as plotted in the figure, are RNAi reagents that increase the effect of the test condition, or are synergistic, whereas those that fall below the line suppress the effects of the test treatment and are antagonistic. This kind of screen design can benefit from an image-based assay format, as treatment conditions that put stress on cells are likely to

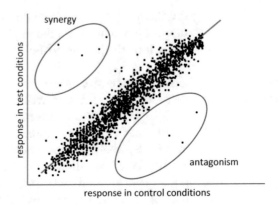

Fig. 1 Data analysis to identify RNAi interactions with cell culture treatments. Each RNAi reagent is used in a control and a test condition (typically the addition of a compound at a submaximal concentration) and the results for each reagent analyzed in a scatter plot. The majority of reagents have no specific effect, and therefore perform equivalently under the two conditions (after normalization). Those that show strong interactions will be visualized as far away from the diagonal line where most of the data falls

have both direct and indirect causes of synergy. A visual check of the hits could be an informal (nonstatistical) classification of hits into groups. Morphological changes or a cell cycle arrest could be valuable information in beginning the association of gene to phenotype during the hit assessment phase.

3.1.5 Mixed Culture Screening Designs

Mixed culture assays are an additional powerful method to assess an RNAi library, but they can be difficult to run for a true RNAi screen. Mixed culture assays are different from co-cultures in that the cell lines do not interact with each other, but are used to compare effects for different cells under the same treatment conditions. One classic design is to seed a cell line to be tested with a negative control line that is insensitive to changes in the primary endpoint, such as harboring a deletion in a gene that is critical for pathway function. Such lines would be labeled so that each can be identified as belonging to the test or control line, either through labeling the lines with fluorescent dyes/GFP variants or separated into compartments through micro-encapsulation [58].

3.1.6 Population or Cell-Level Analysis of High-Content Data

Although most high-content data is reduced to a well-level average or median value for cells in each well, it is valuable to remember that these summaries are derived from individual cellular measurements, and that these cell-level measurements can be evaluated directly. In this regard, HCS data shares some similarity with flow cytometry. HCS has fewer channels to track but can apply these cell-level measurements to morphological and subcellular events. The key point here is that HCS is unique from other plate-based methods in that samples do not need to treated at the well-level, including the

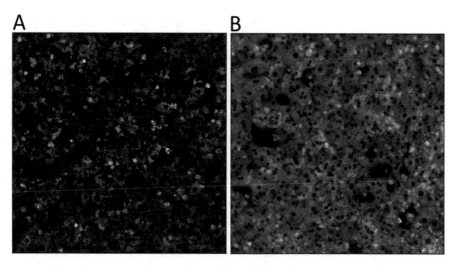

Fig. 2 Heterogeneity of glycogen accumulation in primary human hepatocytes. In (**a**) hepatocytes are treated with a siRNA that reduces glycogen accumulation, when compared to the treatment of hepatocytes in (**b**), where a different siRNA was used. A high level of heterogeneity at the cellular level can complicate a well-level assay statistic, but can be used at the cell level to measure changes in a population of cells

analysis of a screening window or the calculation of a Z'-factor coefficient based upon well-level summaries [59]. In one study, effects of siRNAs for genes associated with increased risk for Type II Diabetes were assessed in primary human hepatocytes [42]. In this case, the effect on glycogen accumulation, the primary endpoint, was measurable, but cellular heterogeneity was pronounced (*see* Fig. 2). In this case, the heterogeneity was normally distributed, so the effects of the siRNAs were assessed through a student's *t*-test measurement of the overall distribution of glycogen levels for the cells in each group, rather than as a well level *z*-score. This is a data-intensive approach to RNAi screening, but more efficient data analysis methods will enable sensitive primary screening strategies [51, 60].

4 Strategies for Identification of False-Positives or False-Negatives Through HCS Data Analysis

While the options available for running image-based screens outlined above implicitly help to address the challenges of screening in various formats, it is worth outlining a couple of explicit mitigation strategies.

4.1 Cellular Features That Can Be Used as a Counter Screen to Identify Off-Target or Nonselective Perturbants

In the absence of a specific counter screen endpoint such as described above, there are still options. Even the most basic HCS experiments capture more than a dozen unique measurements per cell. Monitoring some of these additional features may be a valuable reality check on whether the response could be based on anomalous behavior. In this context, morphological changes are a control instead of an endpoint. If changes in cell morphology should *not* occur following a change in the primary endpoint, such as nuclear fragmentation when a signaling pathway becomes activated, its co-occurrence could be a strong signal of an off-target/nonselective activity. Features that could be considered for this type of analysis include nuclear features (particularly common since chromatin staining is almost universally required for any HCS experiment). Such features that are typically available include nuclear area, variability in the nuclear intensity (a measure of chromatin compaction that can occur during apoptosis) or nuclear shape (such as perimeter/area ratio or length/width ratio). As an example, consider Fig. 3. In this case, the reduction of the primary endpoint, staining in the cytoplasm that can be observed in Fig. 3a, is reduced in cells treated with a different perturbation, shown in Fig. 3b. Other changes can also be used, including a loss of the actin cytoskeleton and reduced area or altered shape of the cell. Using morphological features that are analyzed along with the primary endpoint is a highly sensitive method for detecting confounding factors that could otherwise be interpreted as an active hit.

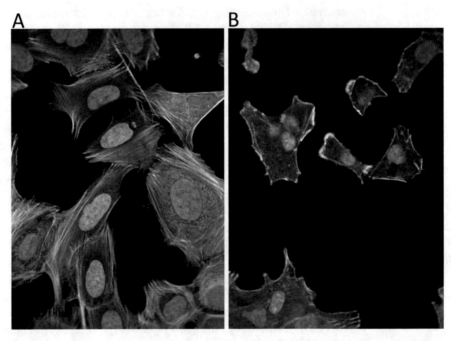

Fig. 3 Cellular morphology as a counter screen. In cases where the effect on a specific signaling pathway is desired, the incorporation of cellular features can function as an alert to nonspecific effects during the primary screen instead of during follow up screening. Original image was in three channels that included nuclei and actin staining, but was reduced to a monochromatic image for reproduction purposes. Changes in morphology, both nuclei and actin features, indicate that a reduction in the primary endpoint occurred through an indirect mechanism

A very subtle case of indirect influence on a screen endpoint is the case of "cell context effects." These types of problems difficult to detect, but can still have effects on the hits identified in a screen. This was the case for a set of studies by Pelkmans and colleagues, as they sought to understand the potential trends in genes identified by RNAi screening that effect viral infection for a number of human viral pathogens [61]. After running 45 RNAi screens against 17 pathogens, they noticed less concordance between the hit lists than was anticipated. An extensive statistical and visual analysis of the screens revealed that many of hits altered the colony morphology and growth patterns of the cells, features that most viruses are exquisitely sensitive towards. In fact, viral infection is a difficult process, and normal cellular heterogeneity affects infection patterns and rates. Starting with an intrinsically variable and heterogeneous process, RNAi screening introduced changes to colony growth that were far more nuanced that simple cell stress or death, but did affect infection rates through changes in colony growth and colony morphology. When the data was corrected for these indirect effects, hits between the different screens become much more concordant. The analytical process could only be achieved through a retrospective analysis of the images and through a cell-level analysis of the data, as it involves detecting and separating mixed-effects algorithmically, followed by validation studies.

4.2 Selective Cell Measurements to Compensate for Penetrance

Penetrance can be a significant problem for some screens, particularly those that use unusual cell types or culture conditions in RNAi and CRISPR screens. In these cases, it is possible to identify the treated cells by introducing a fluorescent marker during delivery of the reagent and then include only cells that display that marker during image analysis. The potential drawback to this approach is if there is a biological reason that facilitates transfection or transduction, for example damage to the cell or (de-) differentiation. In this case, the data will be biased for a subset of cells and may not be representative of the native population.

4.3 Measurement of Target Protein Levels to Address Partial Effects or Other Heterogeneous Behavior

In some cases, it may be desirable to track the extent or onset of a phenotypic effect as a function of target protein level. While not possible for primary screening, this can be helpful in evaluating hits from a screen. If the target protein can be labeled, then a relationship between protein level per cell and phenotype can be plotted. It can be useful to correlate target protein level changes to both support the relationship and gauge the extent of reduction that is necessary. This could be considered a case of correlation inferring causation, but the data could indicate important contextual information, such as whether high levels of inhibition are required to observe an effect (a potential red flag if considering developing a therapeutic). Since protein levels can vary widely across cells in native populations, it can also be possible to measure the effect of a protein level on a phenotype or response in untreated cells [36].

4.4 Analysis of Cell-Level or Field-Level Metadata to Identify Spurious Technical Problems During a Screen

Inspection of HCS data, particularly the cell-level data where the measurements of each cell are listed in a (very long) table, shows that the data is a mixture of actual measurements for the cell features and additional numerical data on the image. These latter data vary by instrument, but records technical aspects of the image that may include the distance between the plate surface and the objective, the exposure time, the number of cells identified per field, the field number itself, and so on. To some extent, these data could be considered metadata-data about the experiment, although the term is typically limited to logistical data such as well number and objective magnification. These data are highly redundant, since these do not change much per field or well…but that is exactly the point. When problems in image acquisition occur, they result in large changes to these data, so inspecting these data is a quick method for determining whether a sudden drop or rise in the assay endpoint for a well might be the result of a technical error. A heat map for a 384-well plate is shown in Fig. 4. Usually, data from the primary endpoint is plotted in such maps to detect the presence of edge effects and outliers, particularly during the assay validation phase of assay development [62]. However, metadata such as exposure intensity, exposure time, focal distance, and other information recorded for each exposure can also be plotted in a heat map to alert an experimenter to a problem during the screen itself.

5   Conclusion

RNAi screening approaches have established an essential role in biological discovery and therapeutic target validation because they proved to be a definitive approach when used in the right context. The incorporation of RNAi screening with HCS was driven by more than as a natural fit, but because each method materially expands the range of studies for the other.

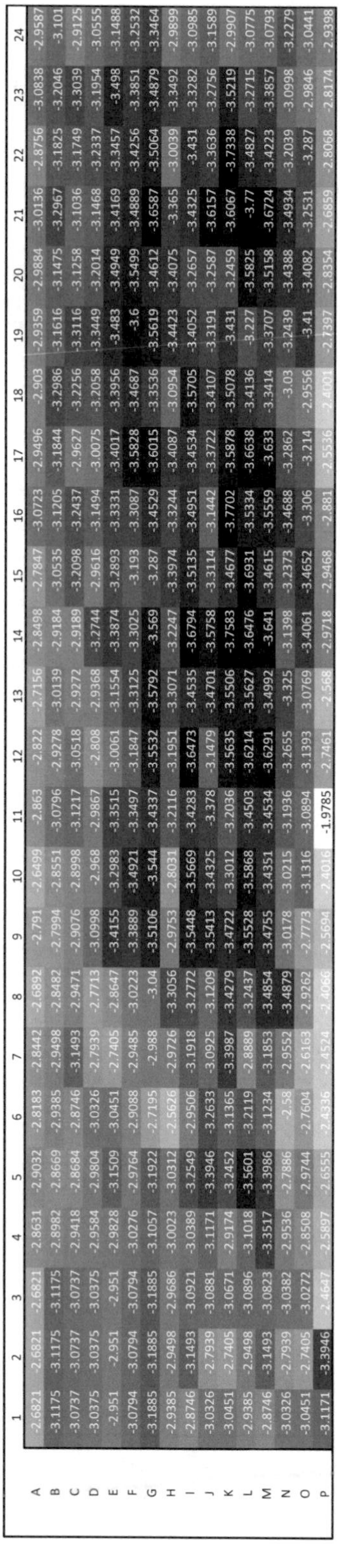

Fig. 4 Plate heat maps of primary and metadata can indicate image acquisition anomalies. Whereas heat maps of the primary endpoint are common to all HTS data analysis, metadata associated with image collection can also be used to identify problematic wells, dispensation problems or edge effects

References

1. Elbashir SM, Harborth J, Lendeckel W, Yalcin A, Weber K, Tuschl T (2001) Duplexes of 21-nucleotide RNAs mediate RNA interference in cultured mammalian cells. Nature 411:494–498

2. Lee SS, Lee RY, Fraser AG, Kamath RS, Ahringer J, Ruvkun G (2003) A systematic RNAi screen identifies a critical role for mitochondria in C. elegans longevity. Nat Genet 33:40–48

3. Pothof J, van Haaften G, Thijssen K, Kamath RS, Fraser AG, Ahringer J et al (2003) Identification of genes that protect the C. elegans genome against mutations by genome-wide RNAi. Genes Dev 17:443–448

4. Berns K, Hijmans EM, Mullenders J, Brummelkamp TR, Velds A, Heimerikx M et al (2004) A large-scale RNAi screen in human cells identifies new components of the p53 pathway. Nature 428:431–437

5. Willingham AT, Deveraux QL, Hampton GM, Aza-Blanc P (2004) RNAi and HTS: exploring cancer by systematic loss-of-function. Oncogene 23:8392–8400

6. Schulte J, Sepp KJ, Wu C, Hong P, Littleton JT (2011) High-content chemical and RNAi screens for suppressors of neurotoxicity in a Huntington's disease model. PLoS One 6: e23841

7. Agrotis A, Ketteler R (2015) A new age in functional genomics using CRISPR/Cas9 in arrayed library screening. Front Genet 6: e300. doi:10.3389/fgene.2015.00300

8. Munoz DM, Cassiani PJ, Li L, Billy E, Korn JM, Jones MD et al (2016) CRISPR screens provide a comprehensive assessment of cancer vulnerabilities but generate false-positive hits for highly amplified genomic regions. Cancer Discov 6:900–913

9. Li CX, Parker A, Menocal E, Xiang S, Borodyansky L, Fruehauf JH (2006) Delivery of RNAi interference. Cell Cycle 5:2103–2109

10. Echeverri CJ, Perrimon N (2006) High-throughput RNAi screening in cultured cells: a user's guide. Nat Rev Genet 7:373–384

11. Hannon GJ (ed) (2003) RNAi: a guide to gene silencing, 1st edn. Cold Spring Harbor Press, Cold Spring Harbor, NY

12. Rodriguez-Barrueco R, Marshall N, Silva JM (2013) Pooled shRNA screenings: experimental approach. Methods Mol Biol 980:353–370

13. Blakely K, Ketela T, Moffat J (2011) Pooled lentiviral shRNA screening for functional genomics in mammalian cells. Methods Mol Biol 781:161–182

14. Jackson AL, Linsley PS (2004) Noise amidst the silence: off-target effects of siRNAs? Trends Genet 20:521–524

15. Lin X, Ruan X, Anderson MG, McDowell JA, Kroeger PE, Fesik SW et al (2005) siRNA-mediated off-target gene silencing triggered by a 7 nt complementation. Nucleic Acids Res 33:4527–4535

16. Birmingham A, Anderson EM, Reynolds A, Ilsley-Tyree D, Leake D, Fedorov Y et al (2006) 3' UTR seed matches, but not overall identity, are associated with RNAi off-targets. Nat Methods 3:199–204

17. Mohr S, Bakal C, Perrimon N (2010) Genomic screening with RNAi: results and challenges. Annu Rev Biochem 79:37–64

18. Hornung V, Guenthner-Biller M, Bourquin C, Ablasser A, Schlee M, Uematsu S et al (2005) Sequence-specific potent induction of IFN-alpha by short interfering RNA in plasmacytoid dendritic cells through TLR7. Nat Med 11:263–270

19. LaPan P, Zhang J, Pan J, Haney S (2008) Quantitative optimization of reverse transfection conditions for 384-well siRNA library screening. Assay Drug Dev Technol 6:683–691

20. Sontheimer EJ, Carthew RW (2005) Silence from within: endogenous siRNAs and miRNAs. Cell 122:9–12

21. Tang G (2005) siRNA and miRNA: an insight into RISCs. Trends Biochem Sci 30:106–114

22. Jackson AL, Burchard J, Schelter J, Chau BN, Cleary M, Lim L et al (2006) Widespread siRNA "off-target" transcript silencing mediated by seed region sequence complementarity. RNA 12:1179–1187

23. Jackson AL, Burchard J, Schelter J, Chau BN, Cleary M, Lim L et al (2006) Widespread siRNA "off-target" transcript silencing mediated by seed region sequence complementarity. RNA 12:1–9

24. Birmingham A, Selfors LM, Forster T, Wrobel D, Kennedy CJ, Shanks E et al (2009) Statistical methods for analysis of high-throughput RNA interference screens. Nat Methods 6:569–575

25. Peters EC, Gray NS (2007) Chemical proteomics identifies unanticipated targets of clinical kinase inhibitors. ACS Chem Biol 2:661–664

26. Baell J, Walters MA (2014) Chemical con artists foil drug discovery. Nature 513:481–483

27. Barrangou R, Birmingham A, Wiemann S, Beijersbergen RL, Hornung V, Smith AM (2015) Advances in CRISPR-Cas9 genome engineering: lessons learned from RNA interference. Nucleic Acids Res 43:3407–3419

28. Mali P, Yang L, Xia Y (2013) RNA-guided human gene engineering vial Cas9. Science 339:823–826

29. Tan J, Martin SE (2016) Validation of synthetic CRISPR reagents as a tool for arrayedfunctional genomic screening. PLoS One 11: e0168968

30. Consortium TE (2012) An integrated encyclopedia of DNA elements in the human genome. Nature 489:57–74

31. Haney SA (ed) (2008) High content screening: science, techniques and applications. Wiley, Hoboken, NJ

32. Inglese J (ed) (2006) Measuring biological responses with automated microscopy. Academic, New York, NY

33. Taylor DL, Haskins JR, Giuliano K (2006) High content screening: a powerful approach to systems cell biology and drug discovery. Humana Press, New York, NY

34. Singh S, Carpenter AE, Genovesio A (2014) Increasing the content of high-content screening: an overview. J Biomol Screen 19:640–650

35. Boland MV, Markey MK, Murphy RF (1998) Automated recognition of patterns characteristic of subcellular structures in fluorescence microscopy images. Cytometry 33:366–375

36. Lapan P, Zhang J, Pan J, Hill A, Haney SA (2008) Single cell cytometry of protein function in RNAi treated cells and in native populations. BMC Cell Biol 9:43

37. Kiger AA, Baum B, Jones S, Jones MR, Coulson A, Echeverri C et al (2003) A functional genomic analysis of cell morphology using RNA interference. J Biol 2:27

38. Truong TV, Supatto W (2011) Toward high-content/high-throughput imaging and analysis of embryonic morphogenesis. Genesis 49:555–569

39. Chua KN, Sim WJ, Racine V, Lee SY, Goh BC, Thiery JP (2012) A cell-based small molecule screening method for identifying inhibitors of epithelial-mesenchymal transition in carcinoma. PLoS One 7:e33183

40. Azorsa DO, Robeson RH, Frost D, Meec hoovet B, Brautigam GR, Dickey C et al (2010) High-content siRNA screening of the kinome identifies kinases involved in Alzheimer's disease-related tau hyperphosphorylation. BMC Genomics 11:25

41. Arora S, Beaudry C, Bisanz KM, Sima C, Kiefer JA, Azorsa DO (2010) A high-content RNAi screening assay to identify modulators of cholesterol accumulation in Niemann-Pick type C cells. Assay Drug Dev Technol 8:295–320

42. Haney SA, Zhao J, Tiwari S, Eng K, Guey LT, Tien ES (2013) RNAi screening in primary human hepatocytes of genes implicated in genome-wide association studies for roles in type 2 diabetes identifies roles for CAMK1D and CDKAL1, among others, in hepatic glucose regulation. PLoS One 8(6):e64946

43. Giuliano K (2006) Optimizing the integration of immunoreagents and fluorecent probes for multiplexed high content screening assays. Methods Mol Biol 356:189–195

44. Howell BJ, Lee S, Sepp-Lorenzino L (2006) Development and implementation of multiplexed cell-based imaging assays. Methods Enzymol 414:284–300

45. Perlman ZE, Slack MD, Feng Y, Mitchison TJ, Wu LF, Altschuler SJ (2004) Multidimensional drug profiling by automated microscopy. Science 306:1194–1198

46. Adams CL, Kutsyy V, Coleman DA, Cong G, Crompton AM, Elias KA et al (2006) Compound classification using image-based cellular phenotypes. In: Inglese J (ed) Methods in enzymology: measuring biological responses with automated microscopy, vol 414. Academic, New York, pp 440–468

47. Tencza SB, Sipe MA (2004) Detection and classification of threat agents via high-content assays of mammalian cells. J Appl Toxicol 24:371–377

48. Tolosa L, Pinto S, Donato MT, Lahoz A, Castell JV, O'Connor JE et al (2012) Development of a multiparametric cell-based protocol to screen and classify the hepatotoxicity potential of drugs. Toxicol Sci 127:187–198

49. Abraham VC, Towne DL, Waring JF, Warrior U, Burns DJ (2008) Application of a high-content multiparameter cytotoxicity assay to prioritize compounds based on toxicity potential in humans. J Biomol Screen 13:527–537

50. Erdmann G, Volz C, Boutros M (2012) Systematic approaches to dissect biological processes in stem cells by image-based screening. Biotechnol J 7:768–778

51. Wang J, Zhou X, Li F, Bradley PL, Chang SF, Perrimon N et al (2009) An image score inference system for RNAi genome-wide screening based on fuzzy mixture regression modeling. J Biomed Inform 42(1):32–40

52. Misselwitz B, Strittmatter G, Periaswamy B, Schlumberger MC, Rout S, Horvath P et al (2010) Enhanced CellClassifier: a multi-class classification tool for microscopy images. BMC Bioinformatics 11:30

53. Jones TR, Kang IH, Wheeler DB, Lindquist RA, Papallo A, Sabatini DM et al (2008) Cell-Profiler analyst: data exploration and analysis software for complex image-based screens. BMC Bioinformatics 9:482

54. Tallarida RJ (2010) Combination analysis. Adv Exp Med Biol 678:133–137

55. Berns K, Bernards R (2012) Understanding resistance to targeted cancer drugs through loss of function genetic screens. Drug Resist Updat 15:268–275

56. Whitehurst AW, Bodemann BO, Cardenas J, Ferguson D, Girard L, Peyton M et al (2007) Synthetic lethal screen identification of chemosensitizer loci in cancer cells. Nature 446:815–819

57. Sachse C, Krausz E, Kronke A, Hannus M, Walsh A, Grabner A et al (2005) High-throughput RNA interference strategies for target discovery and validation by using synthetic short interfering RNAs: functional genomics investigations of biological pathways. Methods Enzymol 392:242–277

58. Beske O, Guo J, Li J, Bassoni D, Bland K, Marciniak H et al (2004) A novel encoded particle technology that enables simultaneous interrogation of multiple cell types. J Biomol Screen 9:173–185

59. Zhang J-H, Chung TDY, Oldenberg KR (1999) A simple statistical parameter for use in evaluation and validation of high throughput screening assays. J Biomol Screen 4:67–73

60. Kozak K, Csucs G (2010) Kernalized Z' factor in multiparametric screening technology. RNA Biol 7:615–620

61. Snijder B, Sacher R, Rämö P, Liberali P, Mench K, Wolfrum N et al (2011) Single-cell analysis of population context advances RNAi screening at multiple levels. Mol Syst Biol 8:579

62. Sittampalam GS, Coussens NP, Brimacombe K, Grossman A, Arkin M, Auld D, Austin C, Baell J, Bejcek B, Chung TDY, Dahlin JL, Devanaryan V, Foley TL, Glicksman M, Hall MD, Hass JV, Inglese J, Iversen PW, Kahl SD, Kales SC, Lal-Nag M, Li Z, McGee J, McManus O, Riss T, Trask OJ Jr., Weidner JR, Xia M, Xu X (eds) (2004–2012) Assay Guidance Manual. Eli Lilly & Company and the National Center for Advancing Translational Sciences, Bethesda, MD. PMID: 23469374

Chapter 9

Strategies and Solutions to Maintain and Retain Data from High Content Imaging, Analysis, and Screening Assays

K. Kozak, B. Rinn, O. Leven, and M. Emmenlauer

Abstract

Data analysis and management in high content screening (HCS) has progressed significantly in the past 10 years. The analysis of the large volume of data generated in HCS experiments represents a significant challenge and is currently a bottleneck in many screening projects. In most screening laboratories, HCS has become a standard technology applied routinely to various applications from target identification to hit identification to lead optimization. An HCS data management and analysis infrastructure shared by several research groups can allow efficient use of existing IT resources and ensures company-wide standards for data quality and result generation. This chapter outlines typical HCS workflows and presents IT infrastructure requirements for multi-well plate-based HCS.

Key words High content screening, Databases, Image visualization, Assay development, IT infrastructure

1 Introduction

"High-throughput" in High Content Screening (HCS) is relative: although instruments that acquire in the range of 100,000 images per day are commercially available, this is still not comparable to the throughput of classical high-throughput screening (HTS) notorious by drug discovery screening centers [1]. As assays get more and more complex, consequently assay development times become prolonged. Further, standardization of cell culture conditions is a major challenge. Informatics technologies are required to transform HCS data and images into useful information and then into knowledge to drive decision making in an efficient and cost-effective manner. Major investments must be made to gather a critical mass of instrumentation, image analysis tools, and information technology (IT) infrastructure. The data storage load per screening campaign easily goes beyond the one terabyte border, and the processing of the hundreds of thousands of images applying

Paul A. Johnston and Oscar J. Trask (eds.), *High Content Screening: A Powerful Approach to Systems Cell Biology and Phenotypic Drug Discovery*, Methods in Molecular Biology, vol. 1683, DOI 10.1007/978-1-4939-7357-6_9,
© Springer Science+Business Media LLC 2018

complex image analysis software and algorithms requires an extraordinarily powerful IT infrastructure to manage it. This chapter will give an overview of the considerations and guidelines that should be mindful when setting up the informatics infrastructure to implement and successfully run large-scale high-content experiments.

2   Assay Development—Short Description

Assays developed for HTS can be roughly characterized without cell line or cell-based in nature. The choice of either biochemical or cell-based assay design and the particular assay format is a balancing act between two broad areas. Successful assays must ensure that the measured signal provides relevant data on the biological process under investigation. For HCS assays this must be balanced with the availability of reagents that yield robust data in microtiter plate formats. The investigator must validate the assay methodology by proceeding through a series of steps which produce structured data.

Successful completion of validation at an earlier stage increases the likelihood of success at later stages and provides insight into data formats planned to produce. During method development, assay conditions and procedures are selected that minimize the impact of potential sources of invalidity on the measurement of the biological end-point in targeted sample matrices.

2.1 Local IT Infrastructure

To manage a tremendous amount of HCS data collected throughout the course of the experiment, an effective and automated data-flow must be developed. For an experiment with one single plate only, the local storage space on the microscope and an external hard drive are likely sufficient [2]. For larger more extensive HCS experiments with numerous images and multi-parameter data, the balance between network bandwidth, server, and storage system configurations, and each organization's unique model which determines how information will be accessed and shared, and these will need to be considered to optimize overall system performance. During the data-flow setup, the following questions may arise: who will be allowed to view and manage the data, how will the data be backed up, and what is the strategy for data retention, archiving, and/or deletion? Rules or procedures for storing HCS data need to be determined by each organization. Application of policies and communication between the IT department and HCS users must also be formulated. In many situations, organizations take a conservative approach and decide to store everything; in part, due to the low cost for data storage or not having a strategic plan before generating HCS data. The number of HCS instruments, number of users, the number of sites, and the network bandwidth within a site (i.e., Local Area Network) are a few of the key factors impacting the hardware requirements for an informatics solution. Sizing and scoping the optimal hardware for an informatics solution is an area where professional IT support is critical. It is also important to decide which parts of these systems will be operated by the informatics department and HCS unit within organization. There are several advantages of placing the HCS data management system in a central IT department. These include:

- Physical data center.
- Facility management (electricity, air conditioning, fire protection, physical security).
- Networking infrastructure (rules for sockets, switches).
- Security infrastructure.
- Existing backup and recovery mechanism.
- Standards for PCs and peripherals, servers, desktop applications, middleware applications, web standards.
- Investment in informatics infrastructure elements.

Managing permissions and/or access is another key point that underscores why user management (UM) applications are so important. The most powerful and commonly used application is a UNIX user management system including groups and users where read/write permissions are set at the folder level. The UM architecture is advantageous and allows users to assign permission to access the file storage. UM can also be organized to access relational databases that are used as storage for database and LIMS applications. In large organizations a dedicated IT department usually helps with the archiving of data. The key feature of a successful HCS backup strategy is to prevent the volume of data that needs to be backed up from growing beyond the manageable range of the backup solution. One of the best approaches to achieve this is to store the HCS data in different locations based on time. Regardless of who actually performs the archiving, coordination among users, knowledge of IT department rules, understanding communication between different IT experts and IT staff is vitally important to effectively manage HCS data.

2.2 Software Workflow Solutions

HCS requires software solutions which interact with the IT infrastructure. Workflow systems are recently beginning to emerge in image-based screening that gives users more flexibility on several fronts. These include applications such as image analysis and machine learning tools that are components of an analysis pipeline. Workflow systems can be used to build virtual systems for image acquisition and can perform feature extraction and high-level data analysis without writing complex scripts [3]. With the increasing need for sophisticated processing, image analysis, and high-level data interpretation, open-source workflow systems are gaining popularity. KNIME [4] is an open-source workflow system with a very broad set of domains that connects image analysis tools and other bioinformatics tools to create complex image processing and analysis workflows.

Some of the open-source image analysis tools can also be combined without using a workflow system. For example, CellProfiler, with its empowering integrative ability, can run an ImageJ macro or Ilastic machine-learning algorithm within the context of an automated image analysis pipeline. In the context of image-based systems biology, the main advantage of workflow systems is that it can construct workflows beyond the direct interoperability of available image analysis tools. For example, workflow systems can integrate the library ImgLib for n-dimensional image analysis from Fiji [5] into a workflow that was missing this functionality. A combination of databases with workflow systems such as KNIME can enable the integration of functionalities beyond the scope of classical image databases. For example, the KNIME node "1Click1-View" (1C1V) was developed to facilitate a link between large-scale image data sets from HTS and numeric data [6].

2.3 Data Handling Software: File System Based

The continued evolution of HCS will require the adoption of a generic and standardized image format suitable for long-term storage and capable of handling a very large collection of images on the file system. Currently, most automated HCS platforms save images in unique proprietary and incompatible image file formats, most of which are already obsolete. Several factors have contributed to the problem: rapid increases in the number of pixels per image; acceleration in the rate at which images are produced; changes in image designs to cope with new automated HCS microscope platforms, and research metadata dictionaries that must support frequent and rapid extensions. The existing solutions are suitable for archiving to maintain images far into the future. Some frameworks represent partial solutions: a few, such as XML, are primarily suited for interchanging metadata; others, such as CIF (Crystallographic Information Framework). The Open Microscopy Environment project (OME) has developed a standardized file format, OME-TIFF [7] which can store raw HCS images along with experimental meta-information. More recent approaches are based on the Hierarchical Data Format (HDF) (http://www.hdfgroup.org) which is optimized for efficient storage and rapid access of large-scale multi-dimensional data. A generic approach is described in [8], combining HDF5 and XML for metadata. The usage of HDF5 for image-analysis data, in particular for data on the level of cells and cell compartments, seems promising and is currently being investigated by projects like:

- CellH5Browserhttp://www.github.com/cellcognition.
- HCSCLD http://svncisd.ethz.ch/doc/hcscld.

- Image Cytometry Experiment (ICE) format https://www.denovosoftware.com/site/Image_Learn_more_link_ICE_Format.shtml.

3 Primary and Secondary Screening: Short Description

The possibility of simultaneous measurement of a multitude of cellular properties or features gives HCS tremendous power and challenging complexity [9]. Typical applications include primary and secondary screening to identify potential lead molecules that may be developed into drug candidates, or genetic screening. Primary and secondary screening assays typically involve the scanning of many multi-well microtiter plates containing cells or cellular components in each well, the acquisition of multiple images of cells, and the extraction of multiple features relevant to the designed assay, resulting in a large quantity of data and images. HCS assay "plates" or "microplates" can have 96-, 384-, or 1536-well densities, where each "well" is a container that contains an individual sample of cells. Each well can be made up of multiple fields. One image is taken from each field for each microscopy channel at a defined time point. The amount of HCS data and images generated from a single microtiter plate can range from hundreds of megabytes (MB) to multiple gigabytes (GB). Standardized and full automatic solutions to maintain, retain, and display data are needed to manage the large volume of HCS numerical data and images generated from primary and secondary screens to enable understanding of the samples under investigation. To fully exploit the potential of data and images from automated high content microscopes and robots, it is therefore crucial to understand the key factors in determining a suitable high content database solution.

The very large amounts of image data produced using modern automated HCS microscopic systems have exceeded the capacity of manual analysis; hence, the development of automated image analysis tools is essential. Due to the complexity and size of modern imaging data, the computational analysis of biological imaging data has already become a vital emerging sub-discipline of bioinformatics and computer vision [10]. Research using multiparametric imaging data relies heavily on computational approaches for image acquisition, data management, visualization, and correct data interpretation [11, 12]. The typical functions of dedicated computer vision systems are data preprocessing, image segmentation, feature extraction, and decision making [13, 14].

The high dimensionality of HCS data sets often makes it difficult to find interesting patterns. To cope with high dimensionality, low dimensional projections of the original data set must be generated; human perceptual skills are more effective in 2D or 3D displays. Software solutions have been developed to enable researchers to select low dimensional projections of HCS data, but most are static or do not allow users to view one frame and select what is of greatest interest to them. Software applications based on "Not interactive" mechanisms require users to make many independent clicks which in consequence changes content of the main view panel.

3.1 Databases: Data Provenance Tracking

Data provenance tracking describes all means used by scientific data management or workflow systems to track when (date) and how data have been created. It is the basis of reproducibility of any scientific data. Provenance tracking starts with the original raw data and extends to the published results (e.g., figures or tables), connecting the latter to the former. For that reason, proper tracking of data provenance is "good scientific practice." In classical empirical research, lab journals on paper have been used for this task.

When scaling up data measurement or analysis like it is done routinely in HCS, it becomes critical that data provenance tracking is done with computer databases and that the provenance data enters the database as soon as possible, immediately after the creation of those data and before they are processed any further. In computers, data provenance tracking metadata usually take two forms:

1. Properties of experiments, samples and data sets that are pairs of keys (like "date of measurement") and values (the date value).
2. Connections between experiments, samples, and data sets. For example, a sample can be derived from two other samples, e.g., the substance in a screening well can be derived from a clinical tumor sample and may be mixed with a chemical compound, making the tumor sample and the chemical compound "parent" samples of the sample in the screening well.

A capable data management system should be able to represent and make available for searches both types of provenance data. For HCS or similar high-throughput technologies, it is highly recommended to integrate management of data, workflows, and data provenance into a tightly integrated system, as provenance data entered manually tend to be incomplete and more error prone compared to automatically captured provenance data. Data management solutions such as openBIS have built-in support for ingestion and querying of data provenance tracking metadata.

3.2 Storage Solutions and Strategies: Data Lifecycle Management

The amount of data measured in one HCS experiment can easily extend into the two-digit terabyte range, in particular when the measurement is time resolved or 3-dimensional. It is thus important to carefully consider where to store what data and for how long before the experiment is started. The decision has to be guided by the plans for data analysis, the requirements of good scientific practice, and considerations of data safety and cost. A good data management system should be used to safeguard the data and help in managing its life-cycle. In large experiments, there are five phases of working with the data:

1. Data acquisition and analysis.
2. Re-analysis of the data.
3. Publication of the data.
4. Using the data as reference for later experiments.
5. Keeping the data available to fulfill good scientific practice.

In phase 1, new data are acquired and quality-controlled (missing values, type consistency, distribution check, file existence), and the initial image analysis is performed. During this phase, the data need to be available on "fast" storage. The driver for the storage used in this phase is the needs of the analysis. If the analysis is performed on a computer cluster or parallel supercomputer array, which is generally advisable for large experiments, then a storage device should be chosen that can cope with a large number of computer nodes accessing the same file system in parallel. "Parallel file systems" such as GPFS (General Parallel File System, [http://www-03.ibm.com/systems/software/gpfs]) or Lustre [http://www.lustre.org] are a common choice for this task. There are also "Network Attached Storage" (NAS) devices available which scale well in this respect. However, be aware that not all NAS devices scale well in this respect and that common NAS protocols such as NFS and CIFS do not ensure correct performance when multiple clients write to the same file in parallel. If you plan to use a regular NAS from multiple compute nodes, we would advise to check its parallel I/O performance and also check the I/O patterns of your analysis programs to see whether parallel write access to the same file can occur. Obviously, this needs to be done early enough to make it take effect on infrastructure and data analysis plans.

In phase 2, the findings of the experiment are consolidated and publication manuscripts are prepared. Statistical analysis and machine-learning algorithms such as computer-based classification are used on the image analysis results to distil knowledge from the HCS data. It is a common situation that findings on the level of statistical analysis reveal weaknesses of the original analysis which in turn trigger partial or full image reanalysis. Often disk space on the fast storage is reduced and image data can only be kept there while it is actively under real-time analysis and afterward it is moved to other (slower and less expensive) types of storage such as a regular NAS. The possibility of image reanalysis should be considered when choosing that storage: restaging the images to the analysis storage needs to be sufficiently speedy to effectively not prevent reanalysis. The network link between the storage systems is a critical piece of infrastructure which is often overlooked: a (dedicated) GBit ethernet line limits transfers to about 80 MB/s, leading to 3 days for a 20TB staging process. Another critical piece of infrastructure is a good data management software package that supports the researchers with metadata-based queries to find the right image set and can help with automating the staging process.

When the analysis is finished, in phase 3 of the life cycle a publication or report is prepared based on the data. Today, there are no public repositories for data from HCS experiments, so the authors of a publication need to decide on how to make available all or a subset of the published data. In the simplest case this may be a huge tar ball in the online materials; however, this is not a very manageable form for the reader. Better solutions will allow the user to browse the images and analysis in a web browser before downloading all or parts of them. In the simplest case, this may be done via a self-written web application running on a standard web server. More sophisticated systems such as *openBIS for HCS* "openBIS: a flexible framework for managing and analyzing complex data in biology research" [15], *HCS Road* [16], or *PE Columbus* [17] can ease that task considerably.

What data are made available in this way depends both on the specifics of the publication and on practicalities. A common choice is to make available all or a subset of the images with reduced resolution and image compression (e.g., JPEG 200) and color-depth (optimized for viewing) and selected feature values averaged over a well. This subset of data can reduce the total storage size compared to the original image and cell-level feature data by a factor of 100.

In phase 4 of the life cycle of a dataset, the focus of the work has moved on to new experiments. The dataset at hand is used as a reference to the datasets from these new experiments, e.g., a similar experiment on a different cell line, pathogen or screening library. How it will be used depends on the scientific question and difficulty to predict outcome up-front. The comparison may be on the level of well-averaged feature values and "typical" images.

In phase 5, data are not used any more to draw new conclusions from it, but have to be kept because "good scientific practice" requires published data to be kept for the purpose of validation of results; guidelines for how long data have to be kept differ, but vary usually between 7 and 30 years after publication.

When data volumes become very large and the cost of data storage becomes a relevant part of the cost of a project by factor 10%, there are confirmed habits from small-scale projects which should be reconsidered. In small-scale projects, it is common practice that all input data are copied as sub-directories into an analysis folder. This is a form of data provenance tracking (see above) which can turn out to be very expensive, particularly when the inputs are raw image or movie data. On a large scale, it is advisable to instead assign a data set used as input data a stable identifier (ID) and reference the ID in the data provenance graph. This can help in reducing the storage needs by a factor of 3 and still properly track data provenance.

3.3 System Workflow Solutions

A workflow management system (WMS) (KNIME, Labview) [18, 19] controls the flow of data through a series of processing steps. A typical HCS workflow may consist of at least the following steps: fetching the data from the microscope to the processing cluster or machine, detecting objects of interest (also called segmentation), extracting descriptive features, identifying phenotypes (possibly using machine learning), and finally aggregating high level features into a hit list or table for data mining. Flexible WMSs allow integrating different dedicated software tools for each of these tasks into one workflow. The WMS ensures that the data is passed between the tools, and does error checks and format conversion if required. This makes it possible to combine for example image deconvolution from ImageJ with object segmentation from CellProfiler [20].

Before taking the decision whether the installation of a full-blown WMS is reasonable for a lab or not, consider reviewing the key benefits of such a system. Users that need to analyze many image datasets in a very similar fashion will benefit from the automation capabilities of the WMS. If datasets are big, the WMS can distribute work-intense tasks on a computer cluster, grid, or cloud, and finally merge the results, which otherwise is tedious and error-prone manual work. Finally, a WMS can abstract the technical details when image analysts want to empower users to execute their own analysis. Labs that have neither many datasets, nor big datasets, might rather consider a desktop workflow manager, which typically is less effort to install and maintain.

iBRAIN2 is an example of a free and open source WMS for HCS [21]. iBRAIN2 integrates well with common cluster queuing systems, allows for execution of arbitrary non-interactive software, has robust data transfer, and a user-friendly web interface. Installation of iBRAIN2 should be done by an IT administrator to a machine (requires Windows 7, 8, 10) that has access to the cluster and to the labs file storage (or the openBIS database). Finally, the data analysis software (CellProfiler, ImageJ, ImageMagick) that should be used must be installed on the cluster, typically to the user's home directory.

Workflows in iBRAIN2 are represented as XML files. A template is supplied, which can be modified according to the user's needs. The modules-section of the XML contains the sequentially executed modules, where each module can be split into multiple tasks (task parallelism) to process a subset of the input data. Screening image data is typically analyzed site by site independently, so task parallelism can unfold its full potential and speed up execution. Software such as CellProfiler supports this natively by assigning a subset of images to each invocation of the non-interactive CellProfiler executable CPCluster. Afterward, a merge-step is required to unite the CPCluster results into a single output file or sqlite database.

iBRAIN2 monitors an incoming folder (dropbox) for new datasets. For MetaXpress (MDC File), iBRAIN2 provides journals that can be used to write a copy of the images into the dropbox. Other microscope vendors likely support similar means of image export during or after acquisition. On the Unregistered Datasets page, datasets are registered for storage, and a default workflow can be assigned. We recommend using the empty workflow as a default that will only trigger storage of the dataset. This allows for more flexibility when later assigning a processing workflow in the Assay Details page. The front page highlights successfully processed jobs in green, warnings in orange, and errors in red. Potential error messages and details of ongoing job processing can be seen on the Details-page for a process.

3.4 Data Visualization

Prior to all visual or statistical exploration of data stemming from HCS experiments, the basic experiment layout and the goal of the experiment have to be considered. Obviously, the HCS images play a key role for the scientist to access the phenotypes under examination; however, the large number of images render a systematic review of the images impossible (i.e., looking at each site of a 384-well plate for 1 min would require 2.5 days of continuous work). As the whole process—from imaging to data analysis—is automated, it is important to calibrate the whole process using controls of known biological effect. In contrast, typically compound libraries contain substances of unknown pharmacological effects.

While the controls are a necessity that can be used to judge experimental quality, normalize the data, and provide a solid reference for any result, the performance of the unknown substances or probes is the actual outcome of the experiment. There are at least two types of controls needed: one to resemble the unperturbed biological system (an experimental base line) and one for a known substance causing defined phenotypic changes (expected effect).

Each substance or probe can fall in one of the following three categories:

- No effect: the probe does not cause any change to the phenotypes as defined by the baseline controls—the probes behave like the no-effect control.

- Known effect: the probe induces a change of phenotype which is identical or like the one caused by the known positive control substance. The effect can be weaker, similar or (for stimulatory experiments) even be higher compared to the reference by the known-effect control.

- Unexpected effect: these are the promise of "High Content" experiments: probes can have an effect on the cells which was unforeseen (or foreseen but to be ruled out, like toxicity or autofluorescence).

The statistical and visual approaches to analyze HCS experiments must take these three categories into consideration:

- No effect: the features (and also derived signals from several features) have to behave or prove reproducibility across all control wells of that type. Usually, there are many such controls (typically 8 or 32 per 96- or 384-well plate respectively) sufficient to access the biological variance. Density plots and bar plots are the typical visualization tools used and standard deviation of replicates is used to address the question of reproducibility and variance.

- The same applies to the "expected effect" category. In addition, such data are combined with the "no effect" data to get an understanding of the signaling window (S/B Ratio, Z-factor coefficient, Normalization, T-Test).

- Most interesting is the "unknown effect"—for instance, a substance could have a similar expected effect as the reference controls, but shows in addition another effect (e.g., toxicity, slowed cell growth, etc.). Or a substance does not show a numeric relevant expected effect and is similar to the baseline, but induces a different phenotype.

In every visualization, outliers are candidates that deserve some interest. Depending on the context an outlier may point to: an unknown or unexpected effect of the substance under consideration, or a technical or biological artifact. While the former are interesting for follow-up, the latter may point to a potential shortcoming or required optimization of the experiment, and/or the data points may be excluded from further analysis.

A common problem in HCS experiments is the so-called plate effect or batch effect. Plate effects are systematic, and stem from changes in the conditions of a well depending on its x, y, and z-position on the plate, for example from higher or lower temperature or CO_2 in outer wells leading to changes in cell growth, or from pipetting errors with the robotic liquid handler. Positional effects or biases are best identified in plate heatmaps that can cover one or multiple plates from one batch.

Another common problem in HCS experiments is that certain effects are likely to be correlated, and in some cases this correlation is undesired and should be corrected. Scatterplots are a helpful tool for exploring correlation of up to four features (when including color and shape of points). A good starting point is correlation of cell number with other effects like infection (in infection assays). Strong correlation with cell number can also indicate population effects [22] that should be corrected by other means.

The key for a successful application of any visualization is the ability to change the context of visualization easily: from one plate to another, across groups of plates or the complete experiment. A very powerful commercial tool with good interactive visualization of very large data sets is Spotfire [23], which also can read data from a variety of formats and sources and has some easy-to-use computational methods. Not as powerful, but free and open source software is CellProfiler Analyst [20], which provides a good starting point for HCS image and feature visualization.

4   Case History

In this case history, a real-life solution for an advanced HCS group is described, including the requirements and the individual components of the solution. At the end, the benefits over the previous existing system are detailed.

4.1   Requirements

The HCS group had gathered experience with the setup, operation, and maintenance of several HCS platforms. These platforms range from a single vendor solution where the instrument is integrated out of the box, to more complex solutions involving multiple vendors and custom integrations, the specific experimental requirements may demand such complexity.

The following set of requirements was defined by the customer to build a new, complete, and standardized HCS workflow that could be transferred and established at different sites.

4.2   Imaging Infrastructure (Including Image Analysis)

- Support for multiple instrument platforms, image analysis software packages, and their combination, in particular the following:
    - GE Incell 2000, image analysis by GE investigator.
    - PE Opera, image analysis by Columbus.
    - Definiens as optional image analysis software.
- Workflow aspects.
    - Automated image storage workflow direct from the instruments without copying; maintenance of the original image file formats from the instrument.

- Creation of thumbnails.

- Possibility of adding results of image analysis (segmentation/overlays, cell level and well level) from instrument specific software and third party software.

4.3 Secondary Analysis

- For cell-level data: definition of cell populations per well; calculation of well results based on averaging of cellular features, e.g., by median or KS score [24].

- Plate data processing: normalization, calculation of quality scores (S/B Ratio, Z-factor coefficient, Normalization, *T*-Test) [24], invalidation of well results due to quality scores, e.g., by number of valid cells.

- Calculation of end results like average compound activity, IC50, or gene activity.

- Access to relevant and underlying images at all the stages of the data analysis workflow.

4.4 Results: Storage, Access, and Browsing

- Automated propagation of relevant results to a central result storage database, including references to relevant images.

- Browsing of experiments by type or material (cell lines, compounds, siRNAs, Genes), filtering by results or quality thresholds and further meta data.

- Display of graphical results (IC50 graphs) and HCS images in the context of results.

- Access to complete sets of raw data for reanalysis or to modify existing analysis parameters and review underlying details.

4.5 Solution

Solution is based on openBIS for HCS, Genedata Screener, and customizations via Application Programming Interfaces for data transfer and additional calculation methods.

openBIS [15] for HCS is a professional system for the automated ingestion and storage of HCS images and analysis results and has a wide range of options for retrieval and visualizing images and image analysis results. It has support for requirements that become important when "scaling up" quantitative imaging and screening approaches, like authorization management, independent user groups and data spaces, multi-share storage, data archiving, and customizable metadata for image search and retrieval. It is used by imaging facilities and research groups in academia and industry. Genedata Screener [25] is a software solution for all plate-based screening, in usage at more than half of the leading pharmaceutical organizations. It provides an efficient and standardized workflow for the analysis of plate-based data from all screening instruments. Genedata Screener for HCS [25] specifically supports cell-level data, provides immediate access to HCS images, and has the capability to process any number of features and plates, making it a good option for data analysis in HCS.

4.6 Solution Details: Imaging Workflow

openBIS runs on a Linux server and has access to a large shared drive that hosts all the images acquired by the HCS platforms. The HCS instrument control software pushes the images for each completely scanned plate onto this shared drive into a predefined directory structure (project → plate barcode → scanning time stamp). The instrument-specific image analysis software is configured to read the images directly from this directory structure and to write results into an additional results folder at the level of project (project → numeric results). Each result file contains the plate barcode and the scanning time stamp to be matched to the plates, in addition the files include a unique identifier for the individual analysis run. The latter is used in openBIS to identify the image analysis method used and to distinguish two consecutive runs of the same image analysis method with different parameterization.

On the shared drive there are additional directories, one for each different type of image analysis results. These directories are monitored by specific openBIS processes that automatically process any file moved into these directories. The combination of directory and process is called "a dropbox," and there can be many arbitrary dropboxes in a given openBIS installation. In this case, there are three dropboxes: one for results from Columbus/Acapella which also processes images acquired on the Opera HCS platform, one for the GE investigator result files that also processes images acquired on the GE Incell HCS platform, and one for image analysis results output by the Definiens software.

If the user drops now a set of Acapella alphanumeric image processing results into the appropriate dropbox, the process starts the parsing of the data files, one after the other. If the barcode is inked to an experiment already defined in openBIS and if the parsing of the data was successful, the processing of the image files starts next.

As a first step of the image registration, the shared drive with the images written by the HCS instrument is scanned to identify the directory matching the barcode of the plate being processed. If there are no images for this plate, the processing is stopped with an error notification to the user. If a folder with images is found, the processing starts by creating a hard-linked copy of the image containing directory and its contents into the image processing dropbox. In the transaction process, one image after the other is parsed for meta data (barcode, well, channel identifier plus specific conditions like wavelengths for excitation and detection). After the successful parsing of this information for the complete plate, all images are moved together into a data share under control of openBIS and are registered as an image dataset. As the images are processed as hard-links, no physical data moving happens and this process is fast. For a quick and efficient display of the images, there are also thumbnails created in custom-defined dimensions, these are stored as a separate dataset in openBIS. After this process is completed, the registration of well results is finalized by adding the references to the image datasets to the well result dataset, allowing now a programmatic identification of the images for which the results were calculated.

In summary, the only manual user intervention is to drop the result files into the dropbox—the complete processing and registration is done completely automatically.

The same principle is applied for all images and image analysis results from the different instruments and software packages listed in the requirements. The import of graphical results of the segmentation requires an additional functionality in the dropbox, the principle of function is identical, though. Similarly for cell-level data, the only difference here is that a specific dataset type has to be created and the dropbox has to be adapted.

5 Solution Details: Secondary Analysis

5.1 Data Import

The secondary analysis is run with Genedata Screener for HCS. To start with, the user gets a hierarchical view of the plate-data contents of openBIS and can navigate by space, project, and experiment. Selection of one or many entities in this tree results in the display of all the corresponding plates in a tabular view, from which the actual plates to be loaded are selected. After plates have been selected, the union of available features is displayed and the user can select which one(s) are to be loaded. These are automatically added to the experimental setup in Screener (which also encompasses process information like on normalization, correction methods, and further result generation). When the user starts the plate-loading process, all the information is automatically fetched from openBIS, imported into Screener and the data analysis session is prepared without any further action needed by the user.

5.2 Data Processing—Plate QC

Following the data loading, all features are automatically normalized according to the settings in the experiment definition, for signal features (as identified by a proper separation between the in-active "zero activity" and treated "full effect" control wells. The user can easily switch between overviews views on all plates for raw and such normalized data, and can also review quality parameters such as Z'-factor coefficient [24] or S/B that are calculated automatically. If needed, features can be combined (e.g., in a linear combination) to calculate results with a better separation of the control wells or higher reproducibility.

5.3 Result Calculation

To calculate compound-specific results like average activity or potency, the compound mapping information (compound ID and concentration) is imported automatically via a customer-specific API connection or other means to the customer's foundation system. With this information in place, simple condensing from well to compound activity is done by application of mean or median; a thorough replicate analysis is done utilizing auxiliary information on number of replicates, masking status, and errors. A further replicate block shows the activity of the active compound plotted against the deviation of the replicates and allows a graphical selection of wells to be excluded from the result calculation.

The more complex process of dose response curve fitting runs iteratively and quick: a heuristic identifies the best optimization of the fit to the measured points by automatic elimination of outliers and using additional information like expected plateaus. The fit results including the quality of the underlying fit are displayed to the user and allow focusing immediately on fits with poor quality or high potency. If needed, the user can also manually interfere with the fit process and manually provide optimal fit settings. As there are often multiple features showing a dose-dependent effect, also the simultaneous fitting of multiple features and the differential analysis of the results is easily supported.

Similar to the calculation of average compound activity, also siRNAs experiments are analyzed: the mapping contains the siRNAs per well plus the annotation of the gene silenced. For such experiments, the user applies the RSA [26] condensing method that averages first the technical replicates, the result of this is subject to the actual RSA algorithm. The user gets a table with the p-values and the gene activity—information relevant in siRNA screening experiments.

5.4 Image Viewing

The images showing the biological objects in the well are the only source for the scientist to judge on the accuracy of the whole image analysis process—the ability to see them is crucial to identify any processing artifact in the laboratory or later data analysis stage of the experiment. For all relevant result entities (like wells or IC50s) the user has immediate access to the underlying images. For instance, if the user has identified some 100 wells of suspicious quality due to too low cell count, he can display with a single mouse click the images for precisely those wells. Similarly, the DRC image viewer shows the complete set of images for all dilutions and replicates for the 25 most potent compounds in a DRC experiment.

5.5 Results: Storage, Access, and Browsing

The results from all the screening experiments should be stored in a centralized, vendor-independent database which allows also upload of results from other experimental sources. For screening, the following result types were defined: DRCs (fit results like IC50, hill coefficient and the plateaus, information derived from the fit like IC80 or efficacy, quality information like χ^2/f or number of fitted and masked data points), condensed results (activity average from replicates for compounds and siRNAs), gene results (gene activities plus p-value); single well results were not stored. Associated with these results are DRC plot Ids to have always fast access to a graphical representation of the DRC plots. Further, references to the experiments defined in openBIS and the data processing sessions in Screener allow the user to go back easily to every step of the data analysis pipeline to review the stored analysis or repeat it.

5.6 Unique Points of the Solution

- Independence from the HCS instrument vendors—any new instrument or image analysis software can be added to the workflow at any time.

- Built-in image life cycle support: image data sets can be flagged automatically or manually to become archived. Successfully archived datasets are deleted automatically if the remaining disk space is reduced below a predefined threshold (watermark). Important here is that the complete data analysis workflow is still functional and the users have full access to the images, as the thumbnails are stored in separate datasets and are persistent.

- Immediate access to the HCS images during the different stages of the data analysis pipeline.

- Full flexibility in automated or manual annotations (e.g., experimental parameters or conditions) relevant both for result calculation, storage, and propagation. This allows easily queries showing for example HCS images for a specific selection of compounds and cell lines.

References

1. Liu B, Li S, Hu J (2004) Technological advances in high-throughput screening. Am J Pharmacogenomics 4(4):263–276

2. Kozak K (2010) Large scale data handling in biology. Ventus Publishing ApS, Frederiksberg. ISBN: 978-87-7681-555-4

3. Michel P, Antony A, Trefois C, Stojanovic A, Baumuratov AS, Kozak K (2013) Light microscopy applications in systems biology: opportunities and challenges. Cell Commun Signal 11:24. doi:10.1186/1478-811X-11-24

4. Berthold MR, Cebron N, Dill F, Gabriel TR, Kötter T, Meinl T, Ohl P, Thiel K, Wiswedel B (2009) KNIME - the Konstanz information miner. ACM SIGKDD Explor Newsl 11:26

5. Schindelin J, Arganda-Carreras I, Frise E, Kaynig V, Longair M, Pietzsch T, Preibisch S, Rueden C, Saalfeld S, Schmid B, Tinevez JY, White DJ, Hartenstein V, Eliceiri K, Tomancak P, Cardona A (2012) Fiji: an open-source platform for biological-image analysis. Nat Methods 2012(9):676–682

6. Zwolinski L, Kozak M, Kozak K (2013) 1Click1View: interactive visualization methodology for RNAi cell-based microscopic screening. Biomed Res Int 2013:1–11

7. Linkert M et al (2010) Metadata matters: access to image data in the real world. J Cell Biol 189:777–782

8. Millard BL et al (2011) Adaptive informatics for multifactorial and high-content biological data. Nat Methods 8:487–493

9. Nichols A (2007) High content screening as a screening tool in drug discovery. Methods Mol Biol 356:379–387

10. Myers G (2012) Why bioimage informatics matters. Nat Methods 9:659–660

11. Swedlow JR, Eliceiri KW (2009) Open source bioimage informatics for cell biology. Trends Cell Biol 19:656–660

12. Peng H (2008) Bioimage informatics: a new area of engineering biology. Bioinformatics 24:1827–1836

13. Eliceiri KW, Berthold MR, Goldberg IG, Ibánez L, Manjunath BS, Martone ME, Murphy RF, Peng H, Plant AL, Roysam B, Stuurmann N, Swedlow JR, Tomancak P, Carpenter AE (2012) Biological imaging software tools. Nat Methods 9:697–710

14. Walter T, Shattuck DW, Baldock R, Bastin ME, Carpenter AE, Duce S, Ellenberg J, Fraser A, Hamilton N, Pieper S, Ragan M, Schneider JE, Tomancak P, Hériché JK (2010) Visualization of image data from cells to organisms. Nat Methods 7:S26–S41

15. Bauch A, Adamczyk I, Buczek P, Elmer FJ, Enimanev K, Glyzewski P, Kohler M, Pylak T, Quandt A, Ramakrishnan C, Beisel C, Malmstrom L, Aebersold R, Rinn B (2011) openBIS: a flexible framework for managing and analyzing complex data in biology research. BMC Bioinformatics 12:468. doi:10.1186/1471-2105-12-468

16. Jackson D, Lenard M, Zelensky A, Shaikh M, Scharpf JV, Shaginaw R, Nawade M, Agler M, Cloutier NJ, Fennell M, Guo Q, Wardwell-Swanson J, Zhao D, Zhu Y, Miller C, Gill J (2010) HCS road: an enterprise system for integrated HCS data management and analysis. J Biomol Screen 15(7):882–891

17. Columbus Platform. http://www.perkinelmer.com/pages/020/cellularimaging/products/columbus.xhtml

18. Berthold MR et al (2009) KNIME - the Konstanz information miner: version 2.0 and beyond. SIGKDD Explor. Newsletter 11:26–31

19. Peter A (2007) Blume: The LabVIEW Style Book Part of the National Instruments Virtual Instrumentation Series. Prentice Hall, Upper Saddle River

20. Carpenter AE, Jones TR, Lamprecht MR, Clarke C, Kang IH, Friman O, Guertin DA, Chang JH, Lindquist RA, Moffat J, Golland P, Sabatini DM (2006) CellProfiler: image analysis software for identifying and quantifying cell phenotypes. Genome Biol 7:R100. PMID: 17076895

21. Rouilly V, Pujadas E, Hullár B, Balázs C, Kunszt P, Podvinec M (2012) iBRAIN2: automated analysis and data handling for RNAi screens. Stud Health Technol Inform 175:205–213

22. Snijder B et al (2012) Single-cell analysis of population context advances RNAi screening at multiple levels. Mol Syst Biol 8:579

23. Datta SSR (2006) "Spotfire's predictive technology helps companies mine data". CNNMoney.com. Spotfire, a Business-intelligence software company based in Somerville

24. Zhang JH, Chung TD, Oldenburg KR (1999) A simple statistical parameter for use in evaluation and validation of high throughput screening assays. J Biomol Screen 4(2):67–73

25. Screener Platform: https://www.genedata.com/products/screener/

26. Konig R et al (2007) A probability-based approach for the analysis of large-scale RNAi screens. Nat Methods 4:847–849

Chapter 10

Live-Cell High Content Screening in Drug Development

Milan Esner, Felix Meyenhofer, and Marc Bickle

Abstract

In the past decade, automated microscopy has become an important tool for the drug discovery and development process. The establishment of imaging modalities as screening tools depended on technological breakthroughs in the domain of automated microscopy and automated image analysis. These types of assays are often referred to as high content screening or high content analysis (HCS/HCA). The driving force to adopt imaging for drug development is the quantity and quality of cellular information that can be collected and the enhanced physiological relevance of cellular screening compared to biochemical screening. Most imaging in drug development is performed on fixed cells as this allows uncoupling the preparation of the cells from the acquisition of the images. Live-cell imaging is technically challenging, but is very useful for many aspects of the drug development pipeline such as kinetic studies of compound mode of action or to analyze the motion of cellular components. Most vendors of HCS microscopy systems offer the option of environmental chambers and onboard pipetting on their platforms. This reflects the wish and desire of many customers to have the ability to perform live-cell assays on their HCS automated microscopes. This book chapter summarizes the challenges and advantages of live-cell imaging in drug discovery. Examples of applications are presented and the motivation to perform these assays in kinetic mode is discussed.

Key words Drug development, Imaging, Image analysis, Live cell, Kinetic, Environmental control

1 Introduction

Automated imaging and automated image analysis has been widely adopted by the drug discovery and development industry. It is expected that the increased sophistication of imaging assays will enhance the quality of compounds emerging from the drug discovery pipeline translating into higher success rates in clinical studies. Furthermore, it has emerged in recent years that the multiplexing capacity of imaging assays can be exploited to reduce the development time of new drugs reducing development costs. These advances come at a time when the industry is struggling with a loss of productivity concomitant with rising costs of developing new drugs, fueling a demand for improved screening technologies [1–3].

Paul A. Johnston and Oscar J. Trask (eds.), *High Content Screening: A Powerful Approach to Systems Cell Biology and Phenotypic Drug Discovery*, Methods in Molecular Biology, vol. 1683, DOI 10.1007/978-1-4939-7357-6_10,
© Springer Science+Business Media LLC 2018

The amount of information encoded in images is what lead Kenneth Giuliano and colleagues at Cellomics to coin the term "high content" for imaging screens [4]. The possibility of documenting several spatio-temporal cellular features quantitatively and simultaneously has been exploited to develop novel types of assays of hitherto unparalleled complexity. The localization of intensities in several fluorescence channels allows measurement of spatial relationships of cellular markers. Examples of assays quantifying translocation events are subcellular localization of proteins or organelles, changes in morphology (tube formation, neurite outgrowth), or growth properties of cellular populations (scatter assay, wound-healing models). Microscopy allows also measuring more traditional readouts such as transcriptional response, posttranslational modifications (phosphorylation, sumolyation). HCS is therefore a very flexible assay platform capable of carrying out many different types of screening assays.

Many parameters can be extracted from digital images such as localization, intensity, area, shape, and texture. The ability to measure several parameters for the event measured (for instance induction of a GFP fusion protein and its localization to the nucleus) has several beneficial consequences. First, compared to single parameter readouts, scoring a response with several orthogonal parameters increases the confidence in the readout, as the probability of having a false negative or positive signal for all parameters is reduced. Second, unexpected phenotypes can be discovered (for instance induction of a GFP fusion protein but no nuclear translocation), enriching the possible mode of actions. Third, several biological events can be scored simultaneously (i.e., nuclear translocation, morphology, and toxicity). Information about toxicity and mode of action can be obtained very early during the drug development process, leading to better informed hit selection decisions and reducing cycle times during lead optimization [5–9]. Fourth, by implementing cell type specific markers or morphological analysis permits the development of assays with mixed cell cultures for organotypic assays. These types of cultures increase the physiological relevance of the assays possibly improving the quality of compounds issuing from the drug discovery pipeline and thus reducing the attrition rate in later clinical stages [10, 11].

HCS/HCA has entered all the stages of the drug discovery pipeline and has helped bring about the paradigm shift away from target-driven in vitro screening to pathway-driven cellular screening. In pathway-driven screening, no or few assumptions about targets are made. Cells are screened for physiological changes toward a desired cellular state. The desired state is dependent upon the therapeutic application (i.e., apoptosis in oncology, insulin secretion in diabetes, neurite outgrowth in neuropathologies, ES cell differentiation into specific cell type, or toxicology, etc.). In some instances, the targets may not be identified until later stages of drug development, but once verified will help drive the lead optimization process. Various strategies can be applied to discover the target, several of which are again based on HCS. For instance, in functional genomics, the phenotypic signature resulting from gene silencing through RNAi is compared to the phenotypic signature of a hit compound [12]. Another strategy exploiting the richness of HCS is to generate reference phenotypic fingerprints with several assays representing various pathways of cell biology with a panel of known inhibitors at different concentrations. Hit compounds can be screened against the collection of assays and the resulting phenotypic fingerprint compared to the reference fingerprint to determine the affected pathway [13–15].

Most imaging in drug discovery is carried out using fixed cells, with few live-cell applications being reported to date. Live-cell imaging is thought to be more challenging than fixed-cell imaging for a number of reasons ranging from synchronization across plates, culture conditions, and imaging conditions (see below). Although live-cell imaging has its own set of challenges and requires careful characterization, it can in some cases be easier to perform than with fixed cells because it involves fewer liquid handling steps for fixing, staining, and other manipulations. Live cells expressing fluorescent proteins can be directly imaged and the fluorescence signals tend to be more consistent over time and in some instances the overall costs of the assay may be less expensive than antibody-based methods. Additionally, fixing samples for antibody staining can introduce artifacts [16]. Furthermore, live-cell imaging can yield information about the cell dynamics and kinetics of both biological processes and compound activity. Eadweard Muybridge, an early pioneer of time-lapse photography, showed the analytical power of kinetic imaging. Using time-lapse photography, he demonstrated that all four hooves of a horse leave the ground during gallop. Only time-lapse imaging offers the sufficient temporal resolution to measure such transient events [17].

2   Live-Cell Imaging Modalities

There are three main modalities of live-cell imaging: single image acquisition, tracking, and intermittent imaging. The single image modality does not capture any kinetic information, but takes advantage of the intactness of cells to measure membrane potentials or exclusion of dyes. These types of assays are mainly used in toxicology studies, where the permeability of the plasma membrane and mitochondrial potential are measured to assess viability and fitness of cells during drug treatment. Assaying live cells avoids the fixing and washing steps, thereby streamlining robotic processing and potentially reducing both variability and the costs of capital investment in equipment.

The tracking of objects live-cell imaging modality is the most challenging one in terms of cell handling and data processing. During this type of time-lapse microscopy, the frame rate and the length of acquisition are adjusted to be able to determine reliably the identity of objects from frame to frame and thus follow objects in space and time. To adjust the frame rate and duration of acquisition correctly, movies of the objects need to be acquired at different speeds (sampling rates) with varying lengths of recording time (sampling sizes) and the kinetic readouts analyzed. The sampling rate determines the fastest object that can be reliably tracked and the length of the recording determines the sampling size necessary to obtain statistically significant data. The sampling rate is dependent both on the speed of the objects being tracked and on their density (relative abundance) as object identification from frame to frame in a sparse population can be more easily determined than from objects in a dense population. The slowest frame-rate and shortest movie allowing unambiguous tracking of objects with stable readouts should be chosen to minimize photo damage, bleaching and reduce data volume. Using this modality, the movement of cells or subcellular structures can be analyzed [18]. Parameters that can be measured are speed, persistence of movement (processivity), direction, distance, length of pauses versus movement, changes of intensity, or shape over time. Several tracking algorithms have been developed for a variety of applications. The general approach is to segment objects in each frame and link the objects between frames based on their similarity using various statistical methods. Correlation of location, intensity, shape, or texture is determined and the most similar objects are deemed identical. This process can be repeated over several frames both forward and backward in time before a track is assigned [19–21]. Depending on the approach, tracking can be computationally very demanding. For screening purposes, few software solutions exist that can track objects unambiguously at reasonable computing speed. Examples of commercial software for tracking objects in HCS are Definiens, Matlab, Kalaimoscope and some HCS instruments have tracking software such as Thermofisher BioApplications. An open source solution for tracking is offered by CellProfiler [22]. Often, custom-made software solutions are developed and used, because the existing solutions are either too expensive or do not yield the desired results [18, 23–27]. The amount of data generated by tracking objects in live-cell imaging modalities is very large [28].

The third modality of time-lapse microscopy involves measuring the changes of a population over time without tracking individual objects. The frame rate is adjusted to obtain sufficient sampling of the process studied. For example, when analyzing the phases of the cell cycle, images might be acquired every 6 h to obtain a reasonable sample of all the phases of the cell cycle over time in each field of view. Short phases will be less often recorded in the population as the changes in events rarely occur and the probability of capturing them is lessened. The sampling of these events must be sufficient to score statistically significant changes in those rare populations. It follows that the fastest or rarest event (for stochastic events) will determine the acquisition rate in this type of live-cell imaging. The image analysis of the data does not require special tools, as each time point is analyzed individually and the evolution of the parameters is studied. Nevertheless, ready-made statistics tools to analyze the time series are lacking and custom software solutions have been developed to automatically extract kinetic information from time-lapse series [29]. The computational overhead is much smaller than for tracking, although the amount of data is still significantly higher compared to endpoint assays. Imaging populations over time measure the kinetics of the onset and duration of compound action, and the kinetics of slow biological processes. This imaging modality also allows distinguishing between primary and secondary phenotypes and allows sampling rare events over time.

Objects can also be tracked at different time points resulting in two temporal dimensions combining both time-lapse modalities. This leads of course to the generation of even larger data sets.

3 Light: The Enemy of Cells

The main challenge in live-cell imaging is that cells do not respond well to light. Exposure to light leads to photodamage or phototoxicity due to the absorption of photons by biological molecules and nonradiative release of the energy leading to the production of oxygen radicals [30, 31]. In consequence, all the settings for acquisition are a compromise between effective sampling and reducing damage. As a rule, long exposure times with low intensity of light energy are less toxic than short exposures with high light intensity [32, 33]. To achieve this, using a microscope with neutral density filters or adjustable power output for the illumination is advisable. To further reduce the energy of the light, long wavelengths are preferable, especially since near UV light damages DNA resulting in cell death. Excitation wavelengths between 560 and 670 nm are best suited for live-cell imaging as they damage cells the least. The sensitivity of cells to light also imposes the choice of laser-based autofocus systems for the automated microscope, because the multiple image acquisition steps of an image-based autofocus can be damaging to cells.

Few organic dyes are useful for live-cell imaging due to phototoxicity and bleaching problems. Most fluorescent dyes are somewhat toxic to cells and concentrations need to be titrated to the absolute minimum required to obtain sufficient signal at low excitation energy. This is particularly problematic for long-term imaging, as the dye is diluted at each cell division. Photo-bleaching of the dye will exacerbate the problem of decreasing signal over time. Repeated excitation of dyes leads to bleaching of fluorophores due to transition of absorbed photons to the triplet state that also results in photo damage through production of oxygen radicals. For many fluorescent dyes, scans of 10 repeated exposures will deteriorate the signal to the extent of rendering it too weak for quantitative image analysis. To evaluate the phototoxicity of the acquisition settings and the toxicity of the dyes, a plate can be prepared in which half the cells are stained with all the combinations of the desired stains at various concentrations and half remain unstained. Half of the stained cells and half of the unstained cells are then imaged with the single and combinations of wavelengths, at a desired frame rate and length of observation. At the end of the experiment, the cells are fixed, stained with DAPI and other cell health indicators such as apoptosis markers can be applied. In this setup, a quarter of the cells were unstained and not illuminated and constitute the negative control. One quarter of the cells were stained but not illuminated and control for the toxicity of the dye. The quarter of the unstained cells that were illuminated serve to assess photo damage. Last, the quarter of the stained cells that were illuminated serve to assess phototoxicity, photo damage, and bleaching. The least disturbing conditions can be used to fine-tune the frame rate and length of acquisition. The impact of dye loading and illumination needs to be reassessed after the final acquisition parameters have been determined to ensure that cells are not suffering too much under the acquisition regimen.

The implementation of recombinant green fluorescent proteins and the generation of mutants with different spectral properties have revolutionized live-cell imaging [34, 35]. Fluorescent proteins are less prone to bleaching than most organic dyes and, in living cells, are constantly replenished through de novo synthesis. Furthermore, when a fluorescent protein bleaches, it is less likely to cause photo damage, as the chromophore produces the radical oxygen within the ß barrel of the fluorescent protein where the oxygen radical is more likely to react with the fluorescent protein rather than damaging other molecules. The disadvantage of using fluorescent proteins is that recombinant proteins need to be constructed and stably transfected; preferably without significantly altering either the physiology or biology of interest. Usually, the cDNA of the protein of interest is fused to fluorescent cDNA in plasmids containing artificial introns and viral promotors leading to non-physiologically high expression levels. Careful characterization of the clones is required to rule out physiological disturbances or mis-localization due to high fluorescent fusion protein expression levels. The use of cDNA limits the analysis to a single splice variant.

Two alternative methodologies are currently emerging to reduce the effects of over expression and to provide all splice variants. One technology uses bacterial artificial chromosomes (BACs) and recombinant engineering to fuse fluorescent proteins either at the amino- or at the carboxy-terminal end of a protein of interest. Due to the large insert size of BACs, large pieces of chromosomes bear the gene of interest maintaining the exon/intron structure and most cis-regulatory elements [36]. The second technology uses various genome-editing technologies with endonucleases such as Zinc finger nucleases, TALENS, and recently CRIPSR to trigger sequence-specific cleavage of the genome [37]. Homologous recombination of fluorescent protein sequences with flanking sequences homologous to the cleavage site repairs the cleavage of the chromosome [38]. Thus, the fluorescent protein is integrated directly into the genome. Both technologies offer the advantage of expressing tagged proteins from the endogenous promoter with all the exons and introns present ensuring in most cases physiological expression levels of the various splice variants. When using any of the above-mentioned fluorescent construct technologies, several clones need to be isolated and compared to untransformed cells. The expression level needs to be quantified with Western blots and by microscopy to ensure that expression levels are not much above two-fold. Immunostaining of the fixed cells with antibodies against the tagged proteins needs to be carried out to quantify the colocalization of the fluorescent protein with the endogenous protein and to control for normal morphology. A new promising technology that combines the advantages of antibodies and fluorescent proteins are chromobodies [39]. Chromobodies are constructed from the heavy chain of *Camelidae* antibodies fused to fluorescent proteins. Due to their small size and monomeric structure, chromobodies sequences can be coded on plasmids and either transfected into cells or produced in vitro and transduced. The advantage is that, in addition to recognizing proteins, they can also recognize posttranscriptional modifications, adding a very important tool for live-cell imaging. Chromobodies need to be carefully characterized for specificity and special care must be taken to ensure that binding to their target protein does not inhibit its function.

4   Assays that Require or Benefit from Live-Cell Imaging

Generally, live-cell assays are developed only when absolutely required, given the challenges. The necessity can arise either: (1) because an intact cell with working membrane potentials is required, (2) because the events under study are rare and need to be sampled over time, (3) because the biological process involves multiple stages and the various stages need to be characterized over time, or (4) kinetic information needs to be gathered. Examples for each requirement are given below.

G protein coupled receptor (GPCR) signaling is highly complex both at the level of the receptors and at the signaling level. Drugs can modulate GPCR signaling via many different modes and are a very important drug target class [40]. Consequently, determining the kinetics of GPCR signaling is important for candidate drugs to characterize their mode of action and to infer their therapeutic indications [41]. To obtain kinetic data, GPCR second messenger probes for live-cell imaging have been developed. For instance, an assay requiring intact cells with a functioning membrane potential is the cAMP assay developed by Tang and coworkers [42]. The assay is designed to capture the production of cAMP in live cells measuring membrane depolarization due to the opening of mutated cyclic nucleotide binding ion channel (CNGC). In resting cells, the dye HLB 021-152 accumulates in the plasma membrane and does not enter the cytoplasm due to the membrane potential. Agonist binding to Gαs coupled GPCRs stimulates activation of adenylate cyclase by the Gαs subunit of the heterotrimeric small GTPase, and intracellular cAMP concentrations rise. cAMP then binds to the mutated CNGC leading to plasma membrane depolarization permitting the dye to enter the cell. Using this assay it is possible to study the kinetics of GPCR agonists and antagonists and thereby characterize the mode of action of these drugs on cAMP. Other probes monitor a second important GPCR second messenger: calcium. Calcium transients in the cytoplasm occur either through entry across the plasma membrane or through release from internal stores. Which mechanism of calcium entry into the cytoplasm is triggered depends on the type of heterotrimeric G protein activated by the GPCR. Therefore, analyzing spatio-temporal calcium changes in the cell can reveal the nature of the signaling pathway activated by a GPCR. To detect these calcium changes, dyes such as Fura-2 or calcium binding proteins such as aequorin or Cameleon are used [43, 44]. Live-cell imaging in conjunction with onboard pipetting is required to analyze fast calcium concentration changes in response to drugs. Calcium is an important second messenger and recording calcium responses is important for many fields in drug discovery besides GPCR signaling [15, 45]. For instance, increased intracellular calcium has been implicated in PDGF-mediated neurite outgrowth [46]. Speed of propagation, synchronization, and amplitude of calcium waves are important markers of cellular activity in various biological processes and can only be analyzed with live cells [47, 48]. Other typical assays requiring intact cells are toxicology studies where cells are imaged live to maintain the cellular integrity and biological gradients. Fixing cells dissipates cellular gradients across membranes and can compromise membrane integrity. Generally, cells are imaged only once to avoid light-induced stress and bleaching. Examples of measured toxicity parameters are the permeability of the plasma membrane, calcium concentrations, or the potential of mitochondrial membranes [49–51].

Migration assays are another type of assay often carried out with live cells with multiple acquisitions. Cell migration is a relevant therapeutic target in development, angiogenesis, inflammation, cancer metastasis, and wound closure. In wound closure assays, a monolayer of cells is scratched using physical devices such as a pipet tip and closure of the wound is dynamically monitored. Only with live cells is it possible to measure the kinetics of wound closure and to distinguish between migration and division [52–54]. Other assays monitor the migration of cells over time in response to external stimuli [18, 55]. Such assays require tracking the movement of cells in three-dimensional matrices and can therefore only be carried out with live cells.

Assays monitoring multistage biological processes with transient events require also live-cell imaging for optimal resolution of the events [27, 56, 57]. For instance, Cervantes and coworkers have developed RNA dyes for studying the infection of *Plasmodium* parasites in erythrocytes [58]. The parasite undergoes several stages within red blood cells and these can be monitored over time with the RNA dyes they developed in conjunction with DAPI staining. Using this assay, compounds can be screened and characterized that affect various stages of the *Plasmodium* life cycle. Screening of fixed cells could only target a single stage of the life cycle of *Plasmodium*. By monitoring the parasite over time, all the stages can be targeted, increasing the number of potential targets and mode of actions. Other multistage biological processes that benefit from live-cell imaging are cell cycle assays, as S, G2, and M phases are rare in a population. By sampling the populations at different time points, more events can be scored and the length of the different phases can be measured [25, 59, 60]. HCS allows classifying cells into specific cell phases by analyzing DNA content and localization of different proteins. In this manner, a dynamic description of the entire cell population in respect to the cell cycle is possible, describing the length of each phase. Inhibition of the cell cycle often leads to apoptosis and this secondary phenotype is not informative [61]. To observe the primary phenotype and determine the phase of the cell cycle and the morphology before apoptosis, it is necessary to monitor live cells.

A popular HCS assay is nuclear translocation and these assays are typically carried out with fixed cells. Often nuclear import is followed by nuclear export due to negative feedback loops. Cells undergo dynamic equilibrium shifts between states where the reporter protein accumulates in the nucleus and then in the cytoplasm. The frequency and the length of the two states cannot be captured by fixed point assays [62, 63]. Many other cellular events are controlled by dynamic equilibrium between different locations

[64] and screening for changes of period of these fluctuations might offer the possibility of designing drugs that modulate cellular events with completely novel modes of action.

Last, two powerful imaging technologies require live-cell imaging: Förster resonance energy transfer (FRET) and fluorescence life time imaging (FLIM). FRET relies on the transfer of energy from one fluorophore to another and is widely used to monitor protein conformation changes and protein-protein interactions with spatio-temporal resolution. FLIM measures the time elapsed between the absorption of a photon and the emission of a photon and can be used as a sensor for the molecular environment of a fluorophore. This technology is also widely used in conjunction with FRET (FRET-FLIM) as a more reliable method than FRET alone for detecting protein-protein interactions. The discussion of these technologies goes beyond the scope of this article and the reader is referred to other reviews [65–68]. Particularly, FLIM will probably play an important role in drug discovery in the future [69, 70].

5   Live-Cell Imaging in Drug Screening and Development

Live-cell imaging is currently not a major technique used in drug development, but because it has the capacity to provide more insights into drug actions we anticipate that it will gain in importance in the future. This is reflected by the fact that nearly all HCS instruments can be equipped for live-cell imaging and many have on board pipetting options.

When developing novel assays, the kinetics of the cellular processes need to be determined. Since the discovery and application of fluorescent proteins, live-cell imaging has been extensively used to analyze the spatio-temporal characteristics of signaling pathways and cellular processes [23, 35, 71, 72]. Some of the parameters determined in assay development are the kinetics of the process under study and the optimal compound incubation times with available controls. Time-lapse microscopy might be useful for determining incubation times to optimize cellular responses. As live-cell imaging generally involves generating fluorescent protein constructs, serial fixation and staining of cells over a time course is often preferred. If a fluorescent protein construct is to be used for the screen, for instance due to the lack of a suitable antibody or due to cost and automation considerations, incubation times are most easily determined using live-cell imaging instead of fixing the cells at various time points [73]. After the determination of optimal incubation times, the assay can be further optimized and the screen carried out with fixed cells.

In primary screening, cell treatment is generally uncoupled from data acquisition by fixing the cells and therefore live-cell primary screening is rare but may include special cases where the kinetics of the target is the desired measurement. Uncoupling cell treatment from acquisition is preferred because both processes have very different throughputs. Furthermore, with fixed cells it is possible to reacquire plates in case of automation problems during the first acquisition, whereas live cells leave little room for errors or faulty equipment. A further reason for preferring endpoint primary screens is the scale of the data that needs to be acquired, as often more than 100,000 compounds are screened. With multiple acquisitions per compound, this leads to the generation of enormous data sets that are very difficult to transfer, store, and analyze. To complicate matters further, the incubation time of each compound needs to be the same for each well and this can be challenging in live-cell applications.

Pipetting schemes can be complex for primary screening depending on the kinetics of the biology. If the kinetics of the biological process to be screened is slow compared to the time required to acquire a whole plate, compounds can be added to the whole plate in one pipetting step. Under these conditions, the difference of time between compound addition and imaging is deemed negligible compared to the time scale of the biological process screened. These types of live-cell assays are easily amenable to primary screening and are less laborious than endpoint assays, as cells do not need to be fixed and no antibody staining is performed. Several time points can be acquired in this screening mode, allowing documentation of the changes in phenotype over time. Since every compound acts with different kinetics, a live-cell assay enables the investigator to analyze optimal time points for each compound, avoiding missing late appearing phenotypes or recording secondary phenotypes for fast acting compounds. To be able to achieve respectable throughput, the HCS microscope needs to be connected to an incubator with a robotic arm, to allow sequential loading and scanning of several plates repeatedly.

For other applications, where either the biology studied is fast or the time of image acquisition of a plate is long, compound addition has to be synchronized with the imaging. We have performed a late endosomes tracking screen in live cells (manuscript in preparation). Four movies of 80 frames were acquired per well resulting in 10 h of acquisitions per 384-well plate. Adding compounds simultaneously to the entire plate would have resulted in incubation time differences of 10 h between the first and the last movie. We therefore dispensed compounds over a 10 h period, pipetting column by column with pauses of 24 min in between compound additions. Between pipetting steps, the plates were placed automatically back into an incubator. In this manner, all the wells within a row had the same incubation time before acquisition and wells within columns had a maximum incubation time difference of 24 min. Another possibility is to use onboard pipetting, but problems with evaporation can occur and should be carefully controlled. Several HCS microscope platforms with live-cell chambers have the ability to pipette compounds onboard.

After a successful primary screen and hit verification, many compounds comprising several scaffolds are selected for lead optimization. Several criteria are applied to the lead candidates, such as physicochemical properties of the compounds, cellular toxicity, mode of action, specificity, efficacy, and pharmacokinetic properties. Live-cell imaging is very useful for answering several of these questions with high accuracy. To study the kinetics of compounds in cells, it is necessary to record the cellular response over time. Live-cell imaging allows measuring the onset of a biological response and its duration. These data can help in deciding about the frequency of administration of drugs in animal studies. These data can also be used to determine the cellular response time to compounds to measure IC50 values at steady state. Measuring IC50 values before equilibrium is attained can lead to false conclusions, as the maximum response is not measured. Cellular toxicity should be assayed at several time points and repeated imaging of live cells is the most convenient way to obtain such data. As mentioned above, in toxicology studies, membrane potentials and permeability and calcium concentrations are often analyzed and these can only be documented in living intact cells.

6 Conclusions

Live-cell imaging remains a niche application in drug discovery and development. The reasons are (1) it is challenging to design an experimental setup that does not damage the cells, (2) pipetting schemes allowing synchronization with the acquisition rate are complex, (3) often large amounts of data are produced, (4) multiple image acquisitions lead to lower throughput, and (5) few off-the-shelf analysis tools are available. Given the challenges, a careful cost-benefit analysis needs to be conducted before embarking on a live-cell imaging project, especially for large-scale primary screening projects. Assays requiring intact membranes and ion gradients, sampling rare events, or tracking objects in time and space clearly require live cells. Assays monitoring the dynamics of biological events, such as tracking of organelles movements or recording calcium waves, also need live-cell imaging. For other assays, fixed endpoint assays are currently more appropriate. With increased computing power and better analysis tools in the future, more assays might be carried out with live cells to allow capturing transient events and avoid scoring secondary phenotypes. Table 1 summarizes the strengths and weaknesses of live-cell imaging in drug discovery.

Table 1
Summary of strengths and weaknesses of live-cell imaging for drug discovery

Strengths and weaknesses of live-cell imaging for drug discovery	
Strengths	**Weaknesses**
Identification of optimal incubation time for biological process	Challenging experimental setup to ensure cell fitness
Ability to score both fast and slow acting compounds	Challenging tracking problems
Possibility to measure biological phenomena dependent on membrane integrity (cell viability, mitochondrial potential)	Mostly restricted to transgenic fluorescent proteins
Possibility to measure kinetic events (vesicular transport, muscle contraction, neuronal signaling, etc.)	Challenging synchronization of data acquisition and compound addition for all wells
	Very large data volume
	Challenging analytical workflow, especially for tracking problems
	Less robust workflow, as any interruption modifies the assay

References

1. Adams CP, Brantner VV (2010) Spending on new drug development. Health Econ 19:130–141

2. Hughes B (2009) 2008 FDA drug approvals. Nat Rev Drug Discov 8:93–96

3. Pearson H (2006) The bitterest pill. Nature 444:532–533

4. Giuliano KA, DeBiasio RL, Dunlay RT et al (1997) High-content screening: a new approach to easing key bottlenecks in the drug discovery process. J Biomol Screen 2:249

5. Denner P, Schmalowsky J, Prechtl S (2008) High-content analysis in preclinical drug discovery. Comb Chem High Throughput Screen 11:216–230

6. Drake PJM, Griffiths GJ, Shaw L et al (2008) Application of high-content analysis to the study of post-translational modifications of the cytoskeleton. J Proteome Res 8:28–34

7. Gabriel D, Simonen M (2008) High content screening as improved lead finding strategy. Eur Pharm Rev 2:46–52

8. Granas C, Lundholt BK, Loechel F et al (2006) Identification of RAS-mitogen-activated protein kinase signaling pathway modulators in an ERF1 redistribution(R) screen. J Biomol Screen 11:423–434

9. Chang YC, Antani S, Lee DJ et al (2008) CBIR of spine X-ray images on inter-vertebral disc space and shape profiles. 21st IEEE international symposium on computer-based medical systems, 17–19 June 2008, Jyvaskyla, Finland, pp 224–229

10. Oellers P, Schallenberg M, Stupp T et al (2009) A coculture assay to visualize and monitor interactions between migrating glioma cells and nerve fibers. Nat Protoc 4:923–927

11. Evensen L, Micklem DR, Link W et al (2010) A novel imaging-based high-throughput screening approach to anti-angiogenic drug discovery. Cytometry A 77A:41–51

12. Eggert US, Kiger AA, Richter C et al (2004) Parallel chemical genetic and genome-wide RNAi screens identify cytokinesis inhibitors and targets. PLoS Biol 2:e379

13. MacDonald ML, Lamerdin J, Owens S et al (2006) Identifying off-target effects and hidden phenotypes of drugs in human cells. Nat Chem Biol 2:329

14. Young DW, Bender A, Hoyt J et al (2008) Integrating high-content screening and ligand-target prediction to identify mechanism of action. Nat Chem Biol 4:59–68

15. Richards GR, Smith AJ, Parry F et al (2006) A morphology- and kinetics-based Cascade for human neural cell high content screening. Assay Drug Dev Technol 4:143–152

16. Schnell U, Dijk F, Sjollema KA et al (2012) Immunolabeling artifacts and the need for live-cell imaging. Nat Methods 9:152–158

17. Wing A, Beek P (2004) Motion analysis: a joint centenary. Hum Mov Sci 23(5):iii–iiv. https://doi.org/10.1016/j.humov.2004.11.001

18. Bahnson A, Athanassiou C, Koebler D et al (2005) Automated measurement of cell motility and proliferation. BMC Cell Biol 6:19

19. Kalaidzidis Y (2007) Intracellular objects tracking. Eur J Cell Biol 86:569

20. Morimoto T, Kiriyama O, Harada Y, Adachi H, Koide T, Mattausch HJ (2005) Object tracking in video pictures based on image segmentation and pattern matching. IEEE 4:3215–3218

21. Sacan A, Ferhatosmanoglu H, Coskun H (2008) CellTrack: an open-source software for cell tracking and motility analysis. Bioinformatics 24:1647–1649

22. Carpenter A, Jones T, Lamprecht M et al (2006) CellProfiler: image analysis software for identifying and quantifying cell phenotypes. Genome Biol 7:R100

23. Rink J, Ghigo E, Kalaidzidis Y et al (2005) Rab conversion as a mechanism of progression from early to late endosomes. Cell 122:735

24. Bacher C, Reichenzeller M, Athale C et al (2004) 4-D single particle tracking of synthetic and proteinaceous microspheres reveals preferential movement of nuclear particles along chromatin. BMC Cell Biol 5:45

25. Harder N, Mora-Bermúdez F, Godinez WJ et al (2009) Automatic analysis of dividing cells in live cell movies to detect mitotic delays and correlate phenotypes in time. Genome Res 19:2113–2124

26. Li F, Zhou X, Ma J et al (2010) Multiple nuclei tracking using integer programming for quantitative cancer cell cycle analysis. IEEE Trans Med Imaging 29:96–105

27. Wenus J, Düssmann H, Paul P et al (2009) ALISSA: an automated live-cell imaging system for signal transduction analyses. BioTechniques 47:1033–1040

28. Jameson D, Turner D, Ankers J et al (2009) Information management for high content live cell imaging. BMC Bioinformatics 10:226

29. Walter T, Held M, Neumann B et al (2010) Automatic identification and clustering of chromosome phenotypes in a genome wide RNAi screen by time-lapse imaging. J Struct Biol 170:1–9

30. Dixit R, Cyr R (2003) Cell damage and reactive oxygen species production induced by fluorescence microscopy: effect on mitosis and guidelines for non-invasive fluorescence microscopy. Plant J 36:280–290

31. Hoebe RA, Van Oven CH, Gadella TWJ et al (2007) Controlled light-exposure microscopy reduces photobleaching and phototoxicity in fluorescence live-cell imaging. Nat Biotechnol 25:249–253

32. Bernas T, ZarĘBski M, Cook RR et al (2004) Minimizing photobleaching during confocal microscopy of fluorescent probes bound to chromatin: role of anoxia and photon flux. J Microsc 215:281–296

33. Chen T-S, Zeng S-Q, Luo Q-M et al (2002) High-order Photobleaching of green fluorescent protein inside live cells in two-photon excitation microscopy. Biochem Biophys Res Commun 291:1272–1275

34. Chalfie MTY, Euskirchen G, Ward WW, Prasher DC (1994) Green fluorescent protein as a marker for gene expression. Science 263:802–805

35. Wang Y, Shyy JYJ, Chien S (2008) Fluorescence proteins, live-cell imaging, and mechanobiology: seeing is believing. Annu Rev Biomed Eng 10:1–38

36. Poser I, Sarov M, Hutchins JRA et al (2008) BAC TransgeneOmics: a high-throughput method for exploration of protein function in mammals. Nat Methods 5:409–415

37. Marx V (2012) Genome-editing tools storm ahead. Nat Methods 9:1055–1059

38. Maeder ML, Thibodeau-Beganny S, Osiak A et al (2008) Rapid "open-source" engineering of customized zinc-finger nucleases for highly efficient gene modification. Mol Cell 31:294–301

39. Rothbauer U, Zolghadr K, Tillib S et al (2006) Targeting and tracing antigens in live cells with fluorescent nanobodies. Nat Methods 3:887–889

40. Bosier B, Hermans E (2007) Versatility of GPCR recognition by drugs: from biological implications to therapeutic relevance. Trends Pharmacol Sci 28:438–446

41. Nikolaev VO, Hoffmann C, Bünemann M et al (2006) Molecular basis of partial Agonism at the neurotransmitter alpha2A-adrenergic receptor and Gi-protein heterotrimer. J Biol Chem 281:24506–24511

42. Tang Y, Li X, He J et al (2006) Real-time and high throughput monitoring of cAMP in live cells using a fluorescent membrane potential-sensitive dye. Assay Drug Dev Technol 4:461–471

43. Ng S-W, Nelson C, Parekh AB (2009) Coupling of Ca2+ microdomains to spatially and temporally distinct cellular responses by the tyrosine kinase Syk. J Biol Chem 284:24767–24772

44. Eglen RM, Reisine T (2008) Photoproteins: important new tools in drug discovery. Assay Drug Dev Technol 6:659–672

45. Floto RA, MacAry PA, Boname JM et al (2006) Dendritic cell stimulation by mycobacterial Hsp70 is mediated through CCR5. Science 314:454–458

46. Black MJ, Woo Y, Rane SG (2003) Calcium channel upregulation in response to activation of neurotrophin and surrogate neurotrophin receptor tyrosine kinases. J Neurosci Res 74:23–36

47. Buibas M, Yu D, Nizar K et al (2010) Mapping the spatiotemporal dynamics of calcium signaling in cellular neural networks using optical flow. Ann Biomed Eng 38(8):2520–2531

48. Jaffe L (2008) Calcium waves. Philos Trans R Soc B Biol Sci 363:1311–1317

49. O'Brien P, Irwin W, Diaz D et al (2006) High concordance of drug-induced human hepatotoxicity with in vitro cytotoxicity measured in a novel cell-based model using high content screening. Arch Toxicol 80:580–604

50. Jan E, Byrne SJ, Cuddihy M et al (2008) High-content screening as a universal tool for fingerprinting of cytotoxicity of nanoparticles. ACS Nano 2:928–938

51. Abraham VC, Towne DL, Waring JF et al (2008) Application of a high-content multiparameter cytotoxicity assay to prioritize compounds based on toxicity potential in humans. J Biomol Screen 13:527–537

52. Liang C-C, Park AY, Guan J-L (2007) In vitro scratch assay: a convenient and inexpensive method for analysis of cell migration in vitro. Nat Protoc 2:329–333

53. Menon MB, Ronkina N, Schwermann J et al (2009) Fluorescence-based quantitative scratch wound healing assay demonstrating the role of MAPKAPK-2/3 in fibroblast migration. Cell Motil Cytoskeleton 66:1041–1047

54. De Rycker M, Rigoreau L, Dowding S et al (2009) A high-content, cell-based screen identifies micropolyin, a new inhibitor of microtubule dynamics. Chem Biol Drug Des 73:599–610

55. Kumar N, Zaman MH, Kim H-D et al (2006) A high-throughput migration assay reveals HER2-mediated cell migration arising from increased directional persistence. Biophys J 91:L32

56. Brand P, Lenser T, Hemmerich P (2010) Assembly dynamics of PML nuclear bodies in living cells. PMC Biophys 3:3

57. Antczak C, Takagi T, Ramirez CN et al (2009) Live-cell imaging of Caspase activation for high-content screening. J Biomol Screen 14:956–969

58. Cervantes S, Prudhomme J, Carter D et al (2009) High-content live cell imaging with RNA probes: advancements in high-throughput antimalarial drug discovery. BMC Cell Biol 10:45

59. Burney RO, Lee AI, Leong DE et al (2007) A transgenic mouse model for high content, cell cycle phenotype screening in live primary cells. Cell Cycle 15:2276–2283

60. Tsui M, Xie T, Orth JD et al (2009) An intermittent live cell imaging screen for siRNA enhancers and suppressors of a kinesin-5 inhibitor. PLoS One 4:e7339

61. Bitomsky N, Hofmann TG (2009) Apoptosis and autophagy: regulation of apoptosis by DNA damage signalling - roles of p53, p73 and HIPK2. FEBS J 276:6074–6083

62. Nelson DE, Ihekwaba AEC, Elliott M et al (2004) Oscillations in NF-kappa B signaling control the dynamics of gene expression. Science 306:704–708

63. Szymanski J, Mayer C, Hoffmann-Rohrer U et al (2009) Dynamic subcellular partitioning of the nucleolar transcription factor TIF-IA under ribotoxic stress. Biochim Biophys Acta 1793:1191–1198

64. Saxena G, Chen J, Shalev A (2009) Intracellular shuttling and mitochondrial function of thioredoxin-interacting protein. J Biol Chem 285(6):3997–4005

65. Wallrabe H, Periasamy A (2005) Imaging protein molecules using FRET and FLIM microscopy. Curr Opin Biotechnol 16:19–27

66. Levitt JA, Matthews DR, Ameer-Beg SM et al (2009) Fluorescence lifetime and polarization-resolved imaging in cell biology. Curr Opin Biotechnol 20:28–36

67. Lohse MJ, Hoffmann C, Nikolaev VO et al (2007) Kinetic analysis of G protein-coupled receptor signaling using fluorescence resonance energy transfer in living cells, Advances in protein chemistry, vol 74. Academic, New York, pp 167–188

68. Herbst KJ, Ni Q, Zhang J (2009) Dynamic visualization of signal transduction in living cells: from second messengers to kinases. IUBMB Life 61:902–908

69. Errington RJ, Ameer-beg SM, Vojnovic B et al (2005) Advanced microscopy solutions for

monitoring the kinetics and dynamics of drug-DNA targeting in living cells. Adv Drug Deliv Rev 57:153–167

70. Talbot CB, McGinty J, Grant DM et al (2008) High speed unsupervised fluorescence lifetime imaging confocal multiwell plate reader for high content analysis. J Biophotonics 1:514–521

71. Wolter KG, Hsu Y-T, Smith CL et al (1997) Movement of Bax from the cytosol to mitochondria during apoptosis. J Cell Biol 139:1281

72. Yu H, West M, Keon BH et al (2003) Measuring drug action in the cellular context using protein-fragment complementation assays. Assay Drug Dev Technol 1:811

73. Bandara S, Schlöder JP, Eils R et al (2009) Optimal experimental design for parameter estimation of a cell signaling model. PLoS Comput Biol 5:e1000558

Chapter 11

Challenges and Opportunities in Enabling High-Throughput, Miniaturized High Content Screening

Debra Nickischer, Lisa Elkin, Normand Cloutier, Jonathan O'Connell, Martyn Banks, and Andrea Weston

Abstract

Within the Drug Discovery industry, there is a growing recognition of the value of high content screening (HCS), particularly as researchers aim to screen compounds and identify hits using more physiologically relevant in vitro cell-based assays. Image-based high content screening, with its combined ability to yield multiparametric data, provide subcellular resolution, and enable cell population analysis, is well suited to this challenge. While HCS has been in routine use for over a decade, a number of hurdles have historically prohibited very large, miniaturized high-throughput screening efforts with this platform. Suitable hardware and consumables for conducting 1536-well HCS have only recently become available, and developing a reliable informatics framework to accommodate the scale of high-throughput HCS data remains a considerable challenge. Additionally, innovative approaches are needed to interpret the large volumes of content-rich information generated. Despite these hurdles, there has been a growing interest in screening large compound inventories using this platform. Here, we outline the infrastructure developed and applied at Bristol-Myers Squibb for 1536-well high content screening and discuss key lessons learned.

Key words Phenotypic screening, High content screening, HCS, High-throughput screening (HTS), Cell-based assays, Image analysis, Mulitparametric data, Hierarchical clustering, Data analysis

1 Introduction

Since its inception in the mid-1990s, high content screening has seen steady growth in its application throughout drug discovery [1]. The value of high content screening (HCS) relative to more traditional in vitro assays includes the ability to capture multiple cellular endpoints simultaneously, to monitor spatial distribution of molecules within the cell, and to derive cell population information from each sample. Combined, these capabilities provide more richly descriptive information regarding the phenotypic state of the cell relative to biochemical assays and other cell-based platforms that typically monitor a single endpoint with well-level resolution [2, 3]. While HCS has been in routine use for over a decade, most labs

Paul A. Johnston and Oscar J. Trask (eds.), *High Content Screening: A Powerful Approach to Systems Cell Biology and Phenotypic Drug Discovery*, Methods in Molecular Biology, vol. 1683, DOI 10.1007/978-1-4939-7357-6_11,
© Springer Science+Business Media LLC 2018

within the biopharmaceutical industry have only integrated this platform using 96- or 384-well plate formats. HCS has historically been a tool primarily used for hit assessment or small focused screens, as the cost, throughput, and capabilities of this platform have prohibited large screening efforts. Recently, however, the technology (including the peripheral instrumentation and consumables) has matured to enable HCS in a miniaturized (1536-well) format, opening the door to enable large-scale primary screening efforts using high content imaging.

In recent years, there has been a growing interest in applying HCS to lead identification efforts that interrogate large compound screening collections of over one million compounds. This increased adoption of high-throughput HCS can be attributed to multiple factors, most of which relate to the alarming decrease in efficiency of drug discovery efforts across the pharmaceutical industry [4, 5]. The steady decline in new molecules entering clinical development is ascribed to attrition due to a lack of in vivo efficacy, safety issues, or both. This attrition problem has driven two interwoven trends: (1) the use of better cell models in early drug discovery efforts and (2) a resurgence of phenotypic screening. With respect to the first trend, many throughout the biopharmaceutical industry and certainly within Bristol-Myers Squibb are looking to expand the use of in vitro assay models that more closely reflect human physiology and disease to identify, characterize, and assess lead compound drug candidates. The expectation is that assays carried out using this more context-appropriate approach will produce hits with improved translation of efficacy to in vivo animal models and, ultimately, humans. To meet this goal, efficiency gains in the cost and throughput of HCS assays are critical.

Where lack of human efficacy is concerned, there has been much literature and discussion recently surrounding the paradigm of hit-identification efforts, acknowledging that target-centric screens have predominated over phenotypic-based approaches for the past 25 years, and this singular focus may have contributed to high attrition rates [6, 7]. The decreased productivity of early drug discovery efforts combined with recent studies analyzing target selection models has revived the phenotypic screening approach, whereby hit identification efforts seek to monitor one or more desirable phenotypic endpoints irrespective of the molecular target(s) or mechanism(s) of action (MOA). These latter approaches necessitate the ability to prosecute large compound inventories such that novel targets or MOAs are not excluded by limiting the chemotypes evaluated. Given that high content imaging is capable of capturing a broad range of phenotypic profiles, HCS is well suited to these efforts and can provide significant value to phenotypic drug discovery efforts, as outlined here and in a companion chapter (Wardwell and Hu).

Like many in the industry, a goal within the Lead Discovery group at Bristol-Myers Squibb has been to enhance our capabilities for high-throughput phenotypic screening. It is important to emphasize, however, that target-based screening still remains an appropriate choice for many screening efforts. Depending on the nature of program goals, target-based screening efforts may be preferred, given the efficiency and relative simplicity of developing tractable structure-activity relationships with target-based screens. To this end, multiple non-HCS-based assay formats, both cellular and biochemical, are leveraged wherever possible. Given the challenges inherent in conducting large high-throughput HCS screening campaigns, we reserve this approach only for those instances where it is critical for program support. Indeed, there are many targets for which biochemical or conventional cellular assays are not suitable. These include assays that seek to monitor endpoints with subcellular resolution, including translocation assays and those measuring cell morphology changes. Additionally, cell-based phenotypic assays offer the ability to interrogate targets such as enzymes in their native environments, accounting for endogenous full-length conformations and cofactors that cannot be readily recapitulated with in vitro *assays* [8–10]. Finally, there are many instances in which the assay signal window achieved from monitoring average well-level fluorescence is too low, but subpopulation analysis and/or measurements restricted to a specific cellular compartment overcome the challenge of a low signal, thereby enabling a robust cellular assay. In our experience, these examples are growing, as program targets become more sophisticated and as different modalities are sought out. For instance, the potentiation of an allosteric modulator is typically lower in amplitude relative to activation of the same target by an orthosteric agonist, thereby requiring more sensitive assays for robust hit identification [11].

Fortunately, in parallel to the growth in the underlying scientific and strategic objectives driving interest in high-throughput HCS, there have been tremendous improvements in the development of hardware and software platforms, as well as informatics solutions, that collectively have made higher throughput HCS feasible. With advances in 1536-well microtiter imaging plates, liquid handling instrumentation, HCS readers, and informatics solutions, we have developed and applied a robust 1536-well high content screening platform within the Lead Discovery group at Bristol-Myers Squibb. The key challenges and our lessons learned through this initiative are outlined here.

1.1 Why Miniaturize High Content Assays?

Cost has historically been one of the most significant hurdles in applying HCS to large hit-identification efforts. For those assays that utilize cell lines with built-in biosensors, this is less of a concern as these engineered systems can circumvent the requirement for expensive detection reagents. However, many assays leverage antibodies or other expensive reagents, and this expense can limit the ability to conduct very large primary screening campaigns with a 384-well format. In these lower-density formats, large screens of one million or more compounds are simply not cost effective relative to the traditional HTS methods. Miniaturizing assays from 384-well to 1536-well format significantly reduces the reagent costs and has enabled HCS screening campaigns that are several fold larger than would be feasible with 384-well assays. In addition to reagent costs, a significant efficiency savings is achieved in terms of the cell requirements for executing a large screen. In our experience, for those assays that can be transferred to a 1536-well format while retaining assay robustness and appropriate pharmacological characteristics, a three- to fourfold reduction in cell number per well is typical when moving from a 384- to 1536-well format. This reduction in cell number is especially impactful as efforts to utilize more physiologically relevant cells for HTS continue to increase. The use of more relevant models includes primary cells from animals or humans and stem cell-derived models including embryonic stem cells and induced pluripotent stem cells (iPSCs). Primary cells are difficult to obtain in large quantities, particularly where the goal is to minimize batch-to-batch variability or to leverage a single donor. While stem cells have long-held promise for circumventing this availability challenge, the reality is that these models are expensive to obtain commercially and still hold challenges for developing large batches of consistent, quality cells; the protocols for pushing stem cells along specific cell differentiation pathways are long and complex and often utilize expensive cell growth factors. In short, obtaining large quantities of more physiologically relevant cells is not as straightforward relative to immortalized cell lines in terms of cost, consistency, and quality [8]. For these reasons, reduction in cell requirements for a given high-throughput screen offers significant benefits toward enabling more physiologically relevant assays with HCS. Perhaps the most notable benefit is the reduction in the use of animals for those assays that utilize primary cells from animal models.

In addition to the cost, the lower throughput of HCS relative to more traditional HTS platforms has also limited the use of HCS for screening large libraries. One distinguishing factor of an imaging approach is the comparatively long plate read times, especially for assays that require the acquisition of multiple images in different fluorescent channels on a well-by-well basis and to apply analysis algorithms to those images. The image acquisition and analysis times associated with HCS almost always represent the rate limiting step for each plate in an automated protocol. As a result, although transitioning to 1536-well plates reduces the overall number of plate cycles for a given automated run or screening campaign, the time to complete the rate-limiting step may not be significantly reduced, and thus the throughput gains that were achieved by miniaturization for other platforms are not realized with HCS to the same extent. This latter fact has led many to question the value of miniaturizing HCS. The answer not only lies in the cost reduction and efficiencies described earlier, but also includes the reduction in consumption of test compounds and the decreased complexity associated with handling four-fold fewer assay plates. At Bristol-Myers Squibb, our entire collection of source compound plates for HTS exists in a 1536-well format, and while it is possible to dispense from these sources to 384-well plates, it introduces complexity both in terms of assay timing and informatics tracking. Additionally, 384-well assays require more compound and overall, a screening campaign that requires more assay plates has increased potential for automation failures which can also be costly.

2 The Challenges of Enabling Miniaturized, High-Throughput High Content Screening

Given the efficiencies outlined, we believe miniaturizing HCS assays to accommodate a 1536-well format has clear value; in our experience, overcoming various hurdles for enabling miniaturization of HCS has been well worth the effort. The challenges of this transition, however, have been significant and, in some cases, unique relative to the challenges of miniaturizing other assay formats. Miniaturizing HCS requires special considerations related to consumables, instrumentation, and data acquisition and analysis: HCS has more stringent requirements than many assay formats for the specifications of microtiter plates; most HCS assays are non-homogeneous requiring the use of high-density plate washers; the scale of HCS data far exceeds that of most other platforms presenting a data management challenge; and finally, more sophisticated approaches to data analysis are needed to truly capitalize on the content-rich information that HCS provides. Combined, these challenges have set a high hurdle for miniaturized, high-throughput HCS. Here, we describe our application of new hardware and software solutions to make 1536-well HCS a reality.

2.1 Considerations for 1536-Well HCS Microplates

Identifying suitable microplates for automated, 1536-well HCS has been a significant challenge that was not initially anticipated. The major geometry considerations for HCS plates are displayed in Fig. 1. Due to the mechanics of imaging via a microscope objective located below the plate, not all clear-bottom plates are suitable for HCS. For example, it can be particularly challenging to access all wells within the plate using high magnification, high numeric aperture (NA) or water objectives, due to the dimensions of the objectives and the lack of space between the outer wells and the rim of the plate. For most typical 1536-well plates, the rim of the plate is

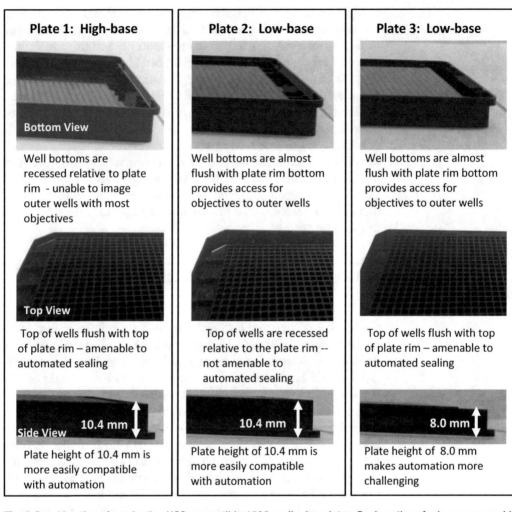

Fig. 1 Considerations for selecting HCS-compatible 1536-well microplates. Basic options for image-amenable 1536-well microplates are shown, including images of a typical high-base plate (plate 1) and two variations of a low-base plate (plates 2 and 3). Plate 3 is most suited for automated HCS. The low-base nature of plate 3 enables imaging of all wells because of the non-recessed wells relative to the plate rim. Plate 3 also poses no challenges for automated sealing; however, specific considerations need to be addressed for robotic handling of these shorter plates

lower than the well bottoms, precluding large objectives from accessing the outer wells as they would collide with the rim (Fig. 1, Plate 1: high-base plate). This has been addressed by many vendors through the development of a "low-base" plate in which the bottom of the wells is either flush with the rim bottom (ultra low-base), or only slightly recessed (Fig. 1, Plates 2 and 3: low-base plates). However, these alterations can pose additional challenges for robust and accurate plate movement and manipulation in automated protocols. For instance, some companies have raised the entire rim of the plate such that the top of the wells (instead of the bottom) is recessed (Fig. 1, Plate 2: low-base), which allows more real estate for robotic plate grippers. While these plates have a height that is consistent with that of the high-base plate, thereby avoiding the challenge of inadequate robot gripping space, they are not amenable to automated sealing given the raised lip around the entire top of the plate. Other vendors have addressed this by generating low-base plates with a rim that is flush with the top of the wells (Fig. 1, Plate 3: low-base). These plates enable high-quality, consistent sealing with automated sealers, but the significant height reduction (8.0 mm relative to 10.4 mm) makes robust movement of these plates with robotic grippers very challenging. While the shorter plates ("Plate 3") are most suitable for our automated HCS work, we have had to modify our platform to leverage either overhead grips or alligator tooth grippers for side-gripping of this plate. These plates can also offer challenges with other accessory instrumentation on robotic platforms. For example, when this microplate resides within the nest of the microplate washer, there is insufficient access for the robot gripper to place and remove plates from the washer tray. To address this, we developed a custom insert that fits within the nest of the plate washer and serves to elevate the plate, thereby enabling access for gripping. Finally, because of the reduced height of "Plate 3," only a subset of available plate lids are compatible with automated handling of the plates, specifically, those that do not completely cover the entire sides of the plate to further reduce real estate available for robotic grippers.

In addition to the frame dimensions, an important consideration in choosing microplates for HCS is the flatness across the bottom of the plate; we have found this to vary substantially not only among manufacturers, but even among different lots from the same source. Specifically, we have observed a range of topographies across the bottom of the plate with deviations as little as 50 μm (top to bottom) or as great as 400 μm. Examples of two 1536-well plates that vary significantly in flatness and topology are shown in Fig. 2a. Z-positions for all rows across 48 columns were measured using our LED platform as an approach to evaluate the evenness across the plate bottom. Plate 1 has a concave pattern that is typical, with a range of ~150 μm; plate 2 has a pattern from left to right which spans 400 μm. The ability to accurately identify objects when there are large variations across the plate will depend largely on the focusing method used in image acquisition [12]. Image-based autofocusing can better handle variations in plate flatness given that with this approach, the focus is based on a specific fluorescent target within each well. In contrast, while laser-based autofocus is much faster, this method relies on the feature of interest being consistent relative to the defined focal point. If large changes in the focal plane are introduced, the laser-based method is unable to correct for this, resulting in out-of-focus images. The differential

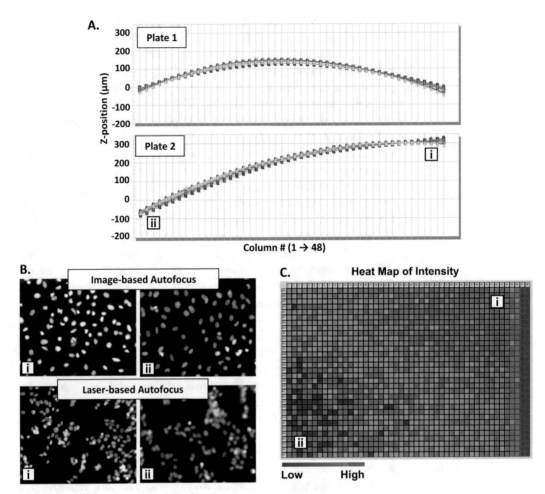

Fig. 2 Plate flatness and focusing methods. (**a**) The z-positions are plotted for all wells of two 1536-well plates, one that has ~150 μm top-to-bottom deviation with a concave topology (plate 1; *top panel*); and another that has ~400 μm deviation that is graded from left to right (plate 2; *bottom panel*). The effects of the variation from plate 2 differ based on the focusing method used (**b**). Images of Hoescht-stained nuclei were acquired using the LED instrument (**b**; *top panel*) which relies on image-based autofocusing and the confocal instrument (**b**; *bottom panel*) which relies on laser-based autofocus. Despite the unevenness of this plate, the nuclei obtained from opposite regions of the plate are accurately identified using image-based focusing (*top panel*). In contrast, when laser-based autofocusing is used, nuclei from several wells are out-of-focus as shown in a representative image (*bottom right panel*) which was acquired from a well on the left side of the plate (region ii). A heat map of the feature intensity obtained from the confocal instrument is shown in (**c**), demonstrating the effect of an uneven plate on images acquired by laser-based

effects of image- and laser-based focusing with plate 2 (bottom panel of Fig. 2a) are shown in Fig. 2b. Specifically, nuclei are accurately identified by image autofocus (top panel), but are out of focus in some wells of the plate when laser-based autofocus is used (bottom panel). In the latter case, an uneven plate results in an inconsistent pattern of feature intensity that mirrors the z-position differences across the plate (Fig. 2c). While the issue of plate flatness is not specific to the higher-density, 1536-well plates, it can be extra challenging given the limited imaging area per well.

2.2 Liquid Handling, Immunostaining, and General Protocol Considerations

One of the greatest advantages of miniaturized HCS has been the cost reduction associated with screening of large compound libraries using assays that require immunostaining. The significant reduction in antibody costs afforded by moving from 384-well to 1536-well assays has opened the door for ultra-high-throughput screening with immunofluorescent detection of cellular proteins. Immunostaining protocols, however, are not trivial to automate given the numerous washing and liquid exchange steps involved. Figure 3 outlines a typical procedure for immuno-detection of a cellular protein by HCS. To complete these multi-step protocols while avoiding cell loss and minimizing background staining, the precision of these liquid-handling steps is essential. A plate washer with tips that are angled to reduce the mechanical force of liquid addition is essential, particularly for live-cell washes. In addition, leaving small volumes in the well during liquid removal is preferred over completely emptying the well during each exchange,

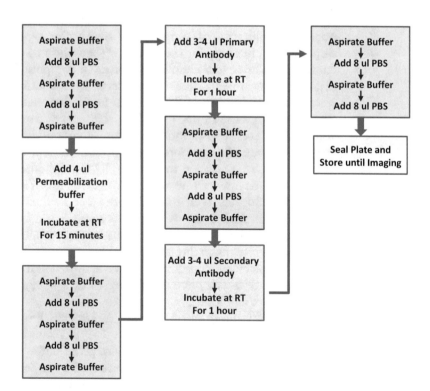

Fig. 3 Flow chart of a typical immunostaining protocol. The multi-step process for immunostaining cells with a specific antibody is outlined. The number of liquid-handling steps exceeds that of most HTS assays and requires iterative washing steps (*blue*) and antibody incubations (*green*) which can introduce bottlenecks into the overall HCS process

particularly for wash steps. Finally, it is essential to utilize a washer that is not prone to tip clogging to ensure the appropriate exchanges in all wells. Fortunately, some high-density plate washers currently on the market have a built-in sonicating function to minimize crystal formation that can occur with various buffers and often leads to pervasive tip clogging. Since most HCS assays are non-homogeneous requiring one or more wash steps and/or liquid-handling exchanges, the advent of high-density microplate washers is an innovation that has been critical to enabling the execution of these challenging protocols in 1536-well plates. For these reasons, this capability is a critical component that is used routinely in our automated HCS protocols.

Aside from the integration of plate-washing capabilities, automation considerations for the liquid-handling steps of HCS assays are not unlike those for other high-throughput 1536-based platforms. Where compound addition is concerned, the approach of choice at Bristol-Myers Squibb is to apply "just-in-time" addition of compound to bioassays through use of acoustic dispensing instruments integrated into robotic assay platforms, such that compound delivery occurs in-line with the assay protocol. This avoids use of pre-stamped compounds stored in assay plates, which can significantly reduce the accuracy of results due to drying and inadequate resolubilization of compound in assay buffer. We use acoustic dispensing technology for compound dispensation to enable non-contact transfer of low nanoliter volumes of compound. Acoustic dispensing instruments have transducers which use sound energy to eject nanoliter droplets of fluid into a destination plate accurately and without the use of tips [13]. The ability to dispense such low volumes reduces consumption of our compound library and enables screening at appropriate compound concentrations while keeping the DMSO load in the assay well at a minimum. Additionally, the acoustic low volume printing avoids the use of intermediate compound dilutions, which can alter the soluble fraction of compound in the final assay well. To accomplish just-in-time, low volume compound delivery, our automated systems all have acoustic dispensers integrated directly on the platform. In addition to the washer and acoustic dispensers, other standard liquid handlers suitable for 1536-well plate dispensing, with a working range of 0.5–10 μL, are employed.

The volume used when initially plating cells into 1536-well plates for HCS is often approximately 5–7 μL. This low volume can lead to evaporation, particularly along the outer wells of the plate; the severity of this issue, often referred to as "edge effect," varies with the duration of culture. We have successfully cultured cells in 1536-well HCS plates for up to 21 days without an edge effect, but this is only achievable by periodic addition of media throughout the culture period. If media additions are not an option, alternative approaches for addressing evaporation include increasing the humidity within the incubator and utilizing alternative lids on the plates. Multiple lid options have become available with the promise to reduce evaporation from 1536-well plates. In our experience, however, none of these options has adequately circumvented edge effect. Evaporation for 1536-well assays remains a significant hurdle not just for HCS assays, but for high-throughput screening in general; solving this will greatly enhance efforts to enable miniaturized cell-based assays.

2.3 Choosing the Best HCS Instrument(s)

As outlined, dedicated instruments for performing high content screening have been in routine use for over a decade, but this use has historically been limited to 96- and 384-well screening. In recent years, vendors have increasingly adapted their hardware and software solutions to accommodate high-throughput screening with lower density, 1536-well plates. In addition to hardware and software adaptations that enable imaging of all wells of a 1536-well plate, instrument cameras have become more sensitive and improved light sources, such as light-emitting diodes (LEDs), have been applied to HCS instruments. There are multiple instrument options now available for 1536-well high content screening and these fall along a broad spectrum in terms of cost, throughput, and image resolution. Aside from ensuring that the instrument is compatible with the microplate of choice (recall that for 1536-well imaging, this is not trivial and currently options are limited), consideration needs to be given to the typical resolution needs and throughput requirements. Additionally, the capacity for the instrument to store large amounts of data is important; otherwise, the investment necessary to enable automated data transfer to a core database is a key factor to address. At Bristol-Myers Squibb, our high-throughput infrastructure centers around two platforms that differ primarily in resolution and ease of use. For high-resolution assays, we use a laser-based spinning disk confocal system equipped with a water immersion objective option. This instrument offers superior resolution that, in our experience, has afforded more robust assays for select applications. The image analysis software is also flexible, which allows the end user to tailor algorithms very specifically to the assay needs. Relative to many systems, however, this instrument is very complicated to use, requiring dedicated operators with in-depth training and dedicated support in-house for algorithm development. Importantly, this latter support requires individuals with unique skill requirements, namely solid script writing expertise and a firm knowledge of cell-based assays. If interested in very high-resolution image capabilities and algorithm flexibility, the level of support to efficiently leverage these sophisticated instruments can be significant and is thus an important consideration in building an HCS infrastructure.

In addition to our confocal instrument for 1536-well HCS, we also have much simpler instruments with LED-based, air-cooled

CCD cameras, which provides the resolution that is adequate for most HTS applications. The two instrument platforms complement one another well and give us broad capabilities for miniaturized assays. Importantly, the LED-based instruments are affordable and, when two LED systems are used in parallel, we reach a much higher throughput relative to the confocal instrument alone, particularly when using a low $2.5\times$ or $5\times$ objective. For simple assays, the LED-based instruments are the top choice given the speed and ease of use; where improvements in resolution are needed to improve an assay window, we leverage the confocal instrument. When investing in multiple distinct platforms within a company, as we have done, the informatics solutions to support the handling and storage of data become increasingly complex. Therefore, where multiple groups are leveraging HCS, the choice of whether or not to standardize to one instrument across multiple groups is an important one. Each of the major HCS instruments has unique advantages and disadvantages with regard to both hardware and integrated image analysis software, and having more than one can provide a broader range of capabilities, as outlined here. However, if centrally managing data from these instruments is required, the challenge of incorporating data into a standardized format is not to be underestimated.

2.4 Putting the Hardware Components Together: A Modular Approach to High-Throughput HCS

As outlined earlier, image acquisition represents a main bottleneck in automated HCS protocols. Leveraging faster acquisition instruments (LED, or laser-based) and using lower magnification objectives wherever possible can significantly reduce this bottleneck; this has certainly not gone unrecognized by vendors, as evidenced by the evolution of instruments on the market. In addition, adding one or more imagers to your HCS platform and using these in parallel has become much less cost-prohibitive in recent years. We have employed this strategy in our labs by integrating two LED-based instruments and found that this can reduce and even eliminate the imaging bottleneck for many screens. Further, to truly maximize the throughput of our automated HCS platform, we have also taken a modular approach to integrating HCS stations into our automated systems. Initially, our HCS capabilities were consolidated into a single robotic platform that included reagent addition, immunostaining, and imaging capabilities. As demand for larger capacity screens increased, however, we chose to take advantage of the fact that a majority of our assays are fixed endpoint and therefore various steps within an average HCS method can be uncoupled from one another. The result is our current modular infrastructure (Fig. 4). A similar approach to HCS has been outlined by others whereby instrumentation used to perform discrete and common tasks are grouped together in units or "modules" that can operate in parallel for improved efficiency [14, 15]. This modular approach not only addresses the image analysis bottleneck, but

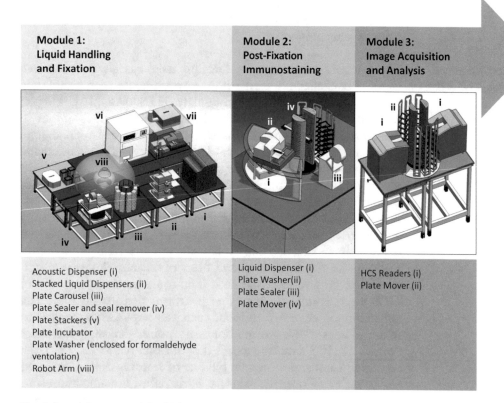

Module 1:
Liquid Handling
and Fixation

Module 2:
Post-Fixation
Immunostaining

Module 3:
Image Acquisition
and Analysis

Acoustic Dispenser (i)
Stacked Liquid Dispensers (ii)
Plate Carousel (iii)
Plate Sealer and seal remover (iv)
Plate Stackers (v)
Plate Incubator
Plate Washer (enclosed for formaldehyde ventolation)
Robot Arm (viii)

Liquid Dispenser (i)
Plate Washer(ii)
Plate Sealer (iii)
Plate Mover (iv)

HCS Readers (i)
Plate Mover (ii)

Fig. 4 A modular approach for high content screening. Computer drawings for three modules that make up our HCS infrastructure are shown. Module 1 consists of a track-based platform for handling the live-cell biology portions of the protocol including on-line acoustic dispensing capabilities, incubator storage, tip-based liquid handlers, and a formaldehyde fixation component (vii). Assay plates are prepared and fixed prior to imaging or post-fixation staining steps. Module 2 contains the essential components for post-fixation staining including a liquid dispenser (*i*), a microplate washer (*ii*), a sealer (*iii*) and a plate mover (*iv*). Module 3 enables off-line image acquisition and analysis with two LED-based HCS instruments (*i*) and a plate mover (*iii*). Our confocal HCS reader is not shown, but is predominantly used off-line, decoupled from the liquid-handling steps

also improves the efficiency of immunostaining protocols, which can also create bottlenecks given the length and complexity of these multi-step protocols. Figure 4 outlines the specific modules developed in-house for HCS. With this infrastructure, we frequently decouple the live-cell portion of the protocol ("module 1"; Fig. 4) from the post-fixation section ("module 2"; Fig. 4), and handle image acquisition and analysis separately from these plate preparation steps ("module 3"; Fig. 4). Decoupling the immunostaining and imaging steps of HCS protocols enables these processes to proceed in parallel on modules 2 and 3, while initiating the next batch of screening plates on module 1. This approach maximizes efficiencies and shortens overall screening efforts. When used in conjunction with multiple HCS readers which have low magnification objectives, this modularity affords a throughput average of

30–40 1536-well plates per day with a typical two-color assay and acquisition of the two to three features often captured at the primary screening stage.

2.5   Considerations for Managing HCS Data

Managing high-throughput imaging data has always presented significant challenges even within environments with established expertise and tools to accommodate HTS data [16]. Despite many parallels with other HTS platforms, HCS data is unique in many aspects. First, unlike most other platforms, end-users require access not only to the numerical results, but also to the underlying images that are invaluable for monitoring assay performance, for identifying artifacts and other false-positives, and for understanding any alternative phenotypes that arise from test compound treatments. To readily access images requires significant storage capacity and an effective means to link images with the algorithm-derived data and metadata. Second, HCS assays provide multiple biological readouts from a single well and users often seek to leverage cell-level data or well-level statistics. The multi-parametric nature of HCS assays is unlike many other platforms that usually limit data collection to one or two endpoints at a well-level. Given the need to retain images and store multiple data points per well, the sheer volume of HCS data far exceeds that of other platforms. For large screening campaigns, literally billions of features can be collected along with millions of images; it is therefore not uncommon to produce multiple Gigabytes of data from a single 1536-well assay plate, and up to Terabytes within a single screening campaign [17, 18]. These key challenges are intensified by the need for efficiency in high-throughput environments and, where multiple platforms are in use, by the diversity of informatics solutions provided by various vendors of HCS platforms. Finally, in addition to the data handling and processing challenges outlined, to truly derive maximal benefits from the content-rich data generated requires novel approaches to analysis and interpretation of the data. This latter challenge is addressed in a later section in this chapter and further elaborated on in a companion chapter (Wardwell and Hu).

While HCS enables one to collect images initially and apply analysis algorithms later, we prefer to conduct "on-the-fly" analysis whereby the algorithm is applied immediately following image acquisition. This in-line analysis allows the end-user to immediately evaluate the assay performance and reduces the overall time and complexity of the screening effort. To enable this efficiently however, a critical requirement is to keep image analysis in-step with image acquisition. One of the early challenges we faced with our confocal imaging platform was insufficient storage capacity and memory on the instrument computer such that when complex algorithms were in use, analysis lagged behind image acquisition and this effect was cumulative. This was addressed by increasing the RAM availability within the instrument computer and by dedicating a Gigabit local network specifically to this platform to enable fast consistent removal of data from the local computer. Our LED-based platforms offer a scalable server architecture to manage imaging protocols and automatic data file backup and retrieval from the laboratory to a managed storage location. In establishing high-throughput capabilities for HCS, a detailed understanding of the vendor-provided solutions and an awareness in any informatics gaps that need to be addressed for adequate efficiency is critical. Contingency plans for network or database outages are also important (*see* **Note 1**).

Despite the significant database storage needs for saving high content images, at Bristol-Myers Squibb our HCS group has opted to retain all images, even at the primary screening stage of single-concentration endpoint data acquisition. While it is not feasible to review one million images or more, having these images available for future use has proven to offer multiple benefits. We have often found it useful to retrieve images selectively, especially to monitor assay performance, for troubleshooting, and to quickly eliminate false positive hits that are the result of obvious observable artifacts. In addition, review of screening images has allowed us to identify alternative phenotypes that were not anticipated during assay development and validation. Finally, saving images has enabled us to initiate new analyses to collect additional features without the need to repeat the entire wet biology portion of the screen. The benefits of retaining images outweigh the cost, namely the need for servers that are scalable and a system whereby the end-user can readily access images that are associated with data values of interest.

In order to maximize the use of collected and stored images, we leverage in-house data management tools to access HCS data and accompanying images, and to perform preliminary data analysis including quality control assessment and data normalization. A system, HCS Road, was recently developed as an enterprise solution for the storage and processing of HCS images and results [19]. HCS Road supports all imaging platforms used at Bristol-Myers Squibb, and enables the linkage of experimental meta data, a relational database for results storage, data normalization, assay performance analysis, and preliminary data analysis. Additionally, this system includes a reagent registry and a reagent inventory database and enables facile data export to third-party applications for further analysis and exploration [19]. The decision to invest in a custom, in-house-developed informatics framework results from our recognition that high content screening offers significant potential in our efforts to improve drug discovery efforts, and the unique nature of HCS data necessitated new solutions for data management and evaluation. However, the resources for this endeavor are significant and, fortunately, commercial solutions for improved handling of HCS data are evolving to keep up with the increased demand and use of HCS.

3 Case Study: Applying High-Throughput HCS to Identify Leads for a Cardiovascular Target

One of the first high-throughput HCS efforts at Bristol-Myers Squibb was in support of a target for the cardiovascular program. A phenotypic assay was developed to identify small molecules that would reduce or modify the uptake of a fluorescently labeled ligand, ligand X. When ligand X is added to the wells, in the absence of an inhibitor compound it binds to a plasma membrane-associated receptor and causes the rapid internalization and lysosomal degradation of the ligand-receptor complex. The phenotypic screen described here was established initially with the goal of identifying compounds that reduce levels of labeled ligand X within the cell (Fig. 5a). This case study is of particular interest in that despite the early goal to find hits reducing cell entry of the ligand, images from this assay revealed alternative phenotypes that were of subsequent interest to the program and redirected the approach for data analysis. The assay protocol and lessons learned regarding feature capture and analysis are outlined.

3.1 Image Analysis

Cellular uptake of fluorescent ligand X-receptor complexes resulted in punctate perinuclear spots that were best resolved by imaging a narrow focal plane. For this reason the screen was conducted using a confocal instrument with laser autofocus. In this instrument, the focal plane is determined by a Pifco laser scan that focuses on the bottom of the plate. Laser autofocus is quick and finds the bottom of the plate by identifying the peak voltage versus the height of the plate (*see* **Note 8**). The determined focal plane does not automatically achieve in-focus images or focal positions where the target of interest is found. The internalized ligand X was visualized at 4.0 μm above the focal plane; the best position for in-focus nuclear staining was 10 μm. Images were analyzed using an analysis script to quantify the average spot area per cell in a defined region of the cytoplasm. The script first identified nuclei using the Hoechst 33342 staining (Fig. 5b, top right), and then a 10 pixel ring was extended from the nuclear boundary to designate the cytoplasmic area in which spots would be identified (Fig. 5b, bottom left). Finally, size and intensity thresholds were applied in the ligand X channel to maximize detection of spots in the control condition while minimizing detection of artifacts, such as ligand aggregates (Fig. 5b bottom right). After thresholding based on spot size/intensity, the average spot area and intensity per cell was determined for each well. Each test compound was evaluated for percent inhibition of uptake, relative to a nontreated control (maximal uptake) and control inhibitor (minimal uptake). Images from negative control positive controls are shown in Fig. 5a. In the absence of an inhibitor, the uptake of the labeled ligand X is easily observed as

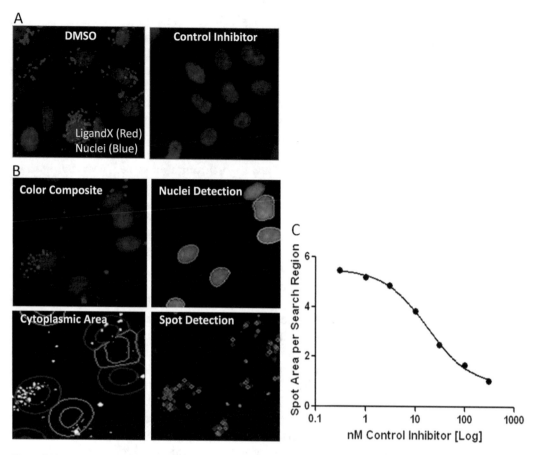

Fig. 5 A high content assay to monitor cellular uptake of fluorescently labeled ligand X. Composite images are shown for untreated (DMSO) and treated (control inhibitor) cells are shown in (**a**). Hoescht-stained nuclei are *blue* and the AlexaFluor-labeled ligand is *red*. In DMSO-treated wells, cellular uptake of ligand X is observed as spots surrounding the nucleus whereas (*left image*) there is little to no detectable uptake of ligand X in response to a known inhibitor molecule (*right image*). Images of a field from a nontreated well are shown in (**b**) with an algorithm overlay. Hoescht-stained nuclei are first identified as individual objects (*top right*) and a ring surrounding the nuclei was chosen as the cytoplasmic area (*bottom left*) within which spots were identified and their area and intensity quantified (*bottom right*). A concentration-response curve for a known inhibitor of ligand X – receptor Y interactions was generated by quantifying the spot area within the defined cytoplasmic search area (**c**)

punctuate, peri-nuclear fluorescent spots characteristic of endosomal localization (Fig. 5a left). A molecule known to physically disrupt binding of ligand X to its cognate membrane receptor results in the complete elimination of detectable perinuclear spots (Fig. 5a right). Images with the algorithm overlay are shown in Fig. 5b, and a concentration response curve for the control inhibitor was generated by quantifying spot area per cytoplasmic area (Fig. 5c).

3.2 Validation and Assessment of High Content Assays for High-Throughput Screening

For the most part, high-throughput, high content screening should be conducted in a manner very similar to that of other high-throughput screening campaigns, with robust validation studies, a well mapped-out screening strategy, and appropriate quality control measures implemented during screening. Not surprisingly, because of the complexity and throughput issues associated with high content imaging, high-throughput screens using high content approaches also face additional challenges above and beyond those experienced with traditional high-throughput screens.

At the outset of any high-throughput screen, a screening strategy is established to maximize the potential for identifying hits. This includes selecting the compound libraries to screen, establishing screen validation studies, developing a plan to confirm hits identified, and devising a suitable strategy for prioritizing hit follow-up in selectivity and mechanism of action studies. This is particularly critical in a phenotypic screen, where multiple hit phenotypes may be observed.

In our screening facility, high-throughput primary screens are typically conducted at a single concentration without replicates. Primary hits are then followed up at the hit confirmation or retest stage at the same single concentration, in replicates to confirm activity and to eliminate false positives. Following hit confirmation, compounds that are consistently active at retest are then prioritized for assessment in a concentration series. Our high-throughput screening efforts typically consist of screening one million or more compounds, selected from the larger screening collection on the basis of diversity and/or chemical modeling. Use of modeling to bias compound selection at the primary stage is usually only a minor component of the compound selection process, since serendipity and breadth of chemical space is still highly valuable at this early screening stage. A hit cutoff for the primary screen is typically established at the median of all values plus three times the median absolute deviation (Median + 3*MAD). All compounds with primary results greater than the hit cutoff are then requested for hit confirmation. Hit rates from primary screens may range from 0.01% to 10%, and may result in over 10,000–60,000 compounds to be selected for hit confirmation. Hit rates can be estimated during screen validation and primary screening concentration can be reduced if hit rates are too high. However, it is particularly important during phenotypic screening to understand the mechanisms behind high hit rates seen in validation. A large number of potent hits with a less-than-desirable mechanism of action may not be as valuable as a number of weaker hits with novel and equally efficacious mechanisms. In a typical high-throughput screening campaign, 2000–4000 hits may be screened for concentration response following hit confirmation. These hits may be prioritized for concentration response using both chemical and biological triage; in the case of phenotypic screening, biological triage should address both potency and a sampling of various phenotypes observed during the screen. Chemical triage typically focuses on chemical diversity as well as tractability.

For all high-throughput screening cascades a rigorous validation is conducted to verify appropriate primary screening concentrations, to assess run-to-run variability, and to estimate false positive and false negative rates. In the absence of a number of known reference compounds, validation can also provide additional references to include as quality control during the full screen. For this screen, ten 1536-well compound source plates (~15,000 compounds) were randomly selected from the screening set to use in validation studies. These compound source plates, a "blank" DMSO control plate, and a reference plate containing one known reference randomly spiked throughput the plate at different initial starting concentrations, were tested on three separate test occasions in our screening assay over multiple days at two final test concentrations (10 μM and 20 μM). This effort allowed us to observe day-to-day variability, consistency, and frequency of artifacts. This exercise was also valuable in providing general false positive and false negative estimates for each respective screening concentration (Fig. 6a–e). From these results it was determined that the screen would be conducted with a 10 μM test compound concentration, since ligand aggregation was seen more often with some compounds at the 20 μM test compound concentration (likely caused by compound microaggregates at 20 μM facilitating ligand precipitation) (Fig. 6b). At a 10 μM screening concentration, the day-to-day variability was deemed acceptable (<20% CV for >90% of replicates) (Fig. 6e), the DMSO plates consistently showed few to no "hits" more than three median absolute deviations from the median of all samples tested, and the reference compound was routinely detected (>95% of all test occasions down to 10 nM test concentration). The hit rate at 10 μM was estimated to be approximately 2%, which was as expected for this difficult target and was well within our capacity for hit follow-up.

Based on the results from the validation studies, a single final concentration of 10 μM was chosen to screen the larger compound library. Although both spot area and intensity per cell were measured, we initially focused our hit criteria on the average spot area per cell and chose 50% or greater inhibition of this feature as the primary selection criteria. As a control for cytotoxicity or reduction in cell adherence, cell counts of less than 40% of the control were excluded from further analysis. After chemical triage and removal of blatant false positives (fluorescent compounds and nonspecific aggregation of the ligand X (Fig. 7c)), approximately 1500 compounds were ordered for evaluation with a 10-point concentration series of compounds carried out in triplicate. In addition to the phenotype of primary interest, namely a reduction in the number of spots similar to the control molecule (Fig. 7a), alternative

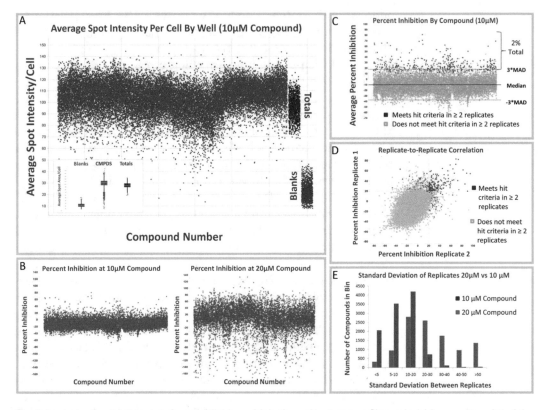

Fig. 6 Analysis of validation data for establishing a high-throughput screen Shown in (**a**) is a scatter plot of the average spot area per cell for test compounds (*green*), maximal controls/total (*red*), and minimal controls/blanks (*blue*) for a validation test at 10 μM compound. Scatter plots of percent inhibition by test compound at 10 μM compound and 20 μM compound are shown F (**b**). A scatter plot is displayed (**c**) of the average percent inhibition of test compounds at 10 μM, highlighting hits that were consistently greater than the hit cutoff (3*MAD + Median) in at least two out of three replicates (*red*). Data from two replicate runs completed on two separate days are compared (**d**). A histogram (**e**) in which the standard deviations of replicates in the 10 μM and 20 μM validation test sets are binned, demonstrating much larger variability in the 20 μM replicates (likely due to a significant increase in the number of artifacts observed in the 20 μM validation set; an example of such artifacts can be seen in Fig. 7c)

phenotypes were observed in response to some test compounds (Fig. 7b), and multiple artifacts could be readily eliminated following visual inspection of images corresponding to some hits (Fig. 7c). The identification of these alternate phenotypes and the obvious artifacts highlight a key advantage of retaining the images from a high content screen. As a first pass, compounds showing a reduction in the number of spots were prioritized and, although a handful of these test compounds showed promise in downstream studies, very few yielded the potency or chemical tractability to continue with lead generation. Given the different phenotypes observed, coupled with new knowledge of potential alternative mechanisms of inhibiting ligand-receptor interaction, it was decided to re-analyze the screen to broaden our classification of hits.

A   Controls   B   Distinct Phenotypes   C   Obvious Artifacts

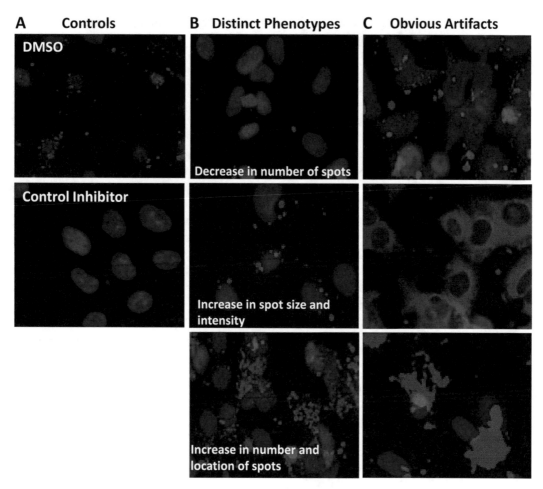

Fig. 7 Identification of multiple phenotypes by high content imaging. Images from control-treated cells are shown (**a**) alongside representative images in which three distinct phenotypes are observed (**b**). In addition to the identification of alternative phenotypes, images can be used to identify obvious assay artifacts caused by factors such as compound aggregation and fluorescent compounds (**c**)

To reanalyze the screen, we made use of the raw images collected during primary screening and re-applied a broader algorithm for spot analysis. The analysis parameters in the new script were expanded to include all spot sizes and intensities. The goal of this analysis was not only to locate and identify additional inhibitors of lysosomal internalization of the ligand X-receptor complex, but to also identify larger vesicles located in the perinuclear region of the cell which are most likely representative of localization to the endosomal recycling compartment . For reanalysis, three intracellular regions were defined including the nucleus, the perinuclear region, and the outer cytoplasm area. Within the disparate cellular regions, several features were analyzed and collected ranging from spot intensity, area, and number, to several additional attributes of the ligand X "spots" and the nuclei. In total, over 20 features across the two channels were collected for use in a phenotypic clustering approach to more thoroughly examine the screening data. This in-depth approach to data analysis was chosen given the clear variety in observed phenotypes combined with the potential value of these alternative phenotypes for the program. Additionally, the concept of clustering multidimensional imaging data has been shown by others and through previous in-house efforts to be an effective means of differentiating compounds that may be acting through distinct mechanisms [20–22]. A more extensive image reanalysis was expected to broaden the capture of compounds that could be of value to the program, but were missed in the initial effort limited to minimal features. Moreover, identifying phenotypic clusters of these compounds was viewed as a valuable approach for evaluating the diversity in MOAs for the compounds tested and for providing a means for prioritizing hits for further analysis. A detailed outline of the methods used for this phenotypic profiling effort as well as the outcomes is presented in the companion chapter from Wardwell and Hu.

4   Materials

1. Microtiter plates:1536-well low based COP, PDL coated mir-cotiter plate.

2. Cells: HepG2 cells.

3. Assay Reagents.

    (a) HepG2 Growth Medium: RPMI 1640 Phenol Red Free, 10% FBS, 1%
    PennStrept.

    (b) Assay medium: RPMI 1640, 2.5% LPDS (Lipid Protein Deficient Serum).

    (c) Formaldehyde.

    (d) Phosphate-Buffered Saline (No Calcium or Magnesium), (PBS).

    (e) Hoechst 33342 (Nuclear dye).

4. Equipment.

    (a) Plate Incubator regulating CO_2 (5%), humidity (95%) and temperature (37 °C).

    (b) Multidrop Combi Dispensers.

    (c) Plate sealer.

    (d) Plate seal remover.

    (e) Acoustic compound delivery system.

    (f) Plate Washer/Dispenser with 1536-well function.

5. High content Imager.

(a) High Content Imager.

(b) Image Analysis Software.

5 Methods

1. Harvest the HepG2 cells with 0.25% Trypsin (*see* **Note 2**).

2. Count and dilute the cells in assay medium.

3. Plate the cells into the 1536-well PDL-coated microtiter plates (*see* **Note 3**) at 1000 cells per well in 5 µL of assay medium.

4. Incubate the plates overnight in a humidified incubator set to 37 °C with 5% CO_2.

5. 18–24 h later, add 1 µL of the controls to columns 45–48.

(a) 1 µL of assay medium to columns 45–46 as a nontreated control.

(b) 1 µL of a control inhibitor compound [300 nM final concentration] to columns 47–48.

6. Add 1 µL of assay medium to wells 1–44, to maintain consistent assay volumes across all wells.

7. Add the compounds [2 mM] to the plates at a volume of 35 nL per well by acoustic dispense for a final screening concentration of 10 µM and 0.5% DMSO.

8. Return the plates to the incubator for 30 min.

9. Dispense 1 µL of AlexaFluor 647 labeled ligand X to all wells.

10. Return the plates to the incubator for 4 h.

11. Wash the plates once with 8 µL assay medium (*see* **Note 4**), leaving 5 µL in the wells.

12. Fix the plates by adding 1 µL of Fixative Solution to yield 2% final formaldehyde in PBS containing a 1:500 dilution of Hoechst 33342 stain.

13. Incubate the plates for 10 min at room temperature.

14. Wash the plates twice with 10 µL of PBS, leaving 10 µL in the well (*see* **Note 5**).

15. Seal the plates (*see* **Note 6**).

16. Acquire one image field per well with a 20× objective using a high content imager capable of confocal or pseudo-confocal imaging (*see* **Note 7**) and apply in-line image analysis.

(a) To image the Hoescht stain, acquire images with an XE bulb or equivalent. If necessary, choose a 425 nm excitation filter, along with a 450/50 nm band pass and 475 nm emission filter. The exposure time used to capture the Hoechst stained image was 10 ms.

(b) Acquire images of the Alexa647-labeled ligand X, using a 640 nm laser. For detection use a 405/488/640 nm primary dichroic filter set or equivalent with a 568 emission dichroic and a 690/50 nm emission filter. The exposure time used to capture the Alexa647-labeled ligand was 2400 ms.

(c) Apply an image analysis script to identify cells and to quantify intracellular levels of labeled Ligand X (outlined in detail below).

5.1   Image Analysis

1. Image Acquisition.

   (a) Measure the cellular uptake of fluorescent ligand X-receptor complexes which are visualized as punctate perinuclear spots that are best resolved by imaging a narrow focal plane with a confocal imager (*see* **Notes 8** and **9**).

   (b) Determine focal z position where the image or target of interest is in focus for each fluorescent channel used (*see* **Note 10**).

2. Image Analysis.

   (a) Analyze the images using an analysis script designed to quantify the average spot area per cell in a defined region of the cytoplasm.

   (b) Identify the nuclei using the Hoechst 33342 staining (Fig. 5b, top right)

   (c) Create a 10 pixel ring extended from the nuclear boundary to designate the cytoplasmic area in which spots will be identified (Fig. 5b, bottom left).

   (d) Define the parameters to identify the perinuclear spots if the ligand X channel.

   - Define the spot size (2 pixel spot signal per area) and spot intensity thresholds (average intensity between 100 and 500) (Fig. 5c, bottom right) (*see* **Note 11**).

   (e) Determine the average spot area and intensity per cell was determined for each well.

   (f) Evaluate each compound for percent inhibition of uptake, relative to a non-treated control (maximal uptake) and control inhibitor (minimal uptake). Images from negative control positive controls are shown in Fig. 5a.

6 Notes

1. Contingency plans for network or database outages include having the ability to allow acquisition and analysis of images when not connected to the established managed database. The images and data should be retrospectively, and ideally effortlessly, backed up to the managed database when the network or database connections are reestablished.

2. To reduce clumping of the HeG2 cells after harvesting the cells, they were titrated through a 200 μL pipette tip at the end of a 10 mL serological pipette five to six times.

3. The HepG2 cells were plated on PDL microtiter plates which allowed the cells to attach to the plate in a single cell morphology enabling the cytoplasmic area to be more visible for imaging.

4. The plates were washed to help reduce the presence of unassociated sticky LigandX from the wells in order to reduce unwanted background fluorescence.

5. The total volume of PBS used to wash the wells of the 1536-well plate will depend on the plate itself. Well volumes can range from 8 to 12 μL capacity.

6. Sealing the plates will reduce well evaporation of the PBS. If a foil seal is used, light exposure and potential signal quenching is reduced, although lids can still be used.

7. We determined that for this assay, confocality was necessary for the resolution and separation of the spots within the cytoplasm.

8. The focal plane is determined by a Pifco laser scan that focuses on the bottom of the plate. Laser autofocus is quick and finds the bottom of the plate by identifying the peak voltage versus the height of the plate.

9. The flatness of the plates is essential when identifying objects in a defined region or z section within the cell. We found that not only maintaining the same lot number of assay plates is critical for consistent results across the screen, but established vendor quality control of plates lot to lot is essential for confidence in a product.

10. The focal plane determined by the laser autofocus does not automatically achieve in-focus images or focal positions where the target of interest is found. The internalized ligand X was visualized at 4.0 μm above the focal plane; the best position for in-focus nuclear staining was 10 μm.

11. Gating and intensity thresholds were applied in the ligand X channel to maximize detection of spots in the control condition while minimizing detection of artifacts, such as ligand aggregates (Fig. 5b bottom right).

Acknowledgments

The authors wish to thank members of the Core Automation team including Jennifer Zewinski, Jeffrey Cheicko, Lisa Simoni, David Connors, Christian Ferrante and Jessica Devito for providing critical support in enabling high-throughput HCS assays. Additionally, we acknowledge the efforts of Michael Lenard and Donald Jackson in the development and implementation of HCS Road, as well as informatics support to enable phenotypic clustering. Judi Wardwell-Swanson and Yanhua Hu were instrumental in the development and execution of the phenotypic clustering methods described. Finally, thanks are due to Edward Pineda for providing us with the CADD drawings in Fig. 3.

References

1. Zock JM (2009) Applications of high content screening in life science research. Comb Chem High Throughput Screen 12:870–876

2. Boutros M, Heigwer F, Laufer C (2015) Microscopy-based high-content screening. Cell 163:1314–1325

3. Nierode G, Kwon PS, Dordick JS et al (2016) Cell-based assay design for high-content screening of drug candidates. J Microbiol Biotechnol 26(2):213–225

4. Munos B (2009) Lessons from 60 years of pharmaceutical innovation. Nat Rev Drug Discov 8:959–968. doi:10.1038/nrd2961

5. Pammolli F, Magazzini L, Riccaboni M (2011) The productivity crisis in pharmaceutical R&D. Nat Rev Drug Discov 10:428–438. doi:10.1038/nrd3405

6. Swinney DC (2013) Phenotypic vs. target-based drug discovery for first-in-class medicines. Clin Pharmacol Ther 93:299–301. doi:10.1038/clpt.2012.1236

7. Swinney DC, Anthony J (2011) How were new medicines discovered? Nat Rev Drug Discov 10:507–519. doi:10.1038/nrd3480

8. Eglen R, Reisine T (2011) Primary cells and stem cells in drug discovery: emerging tools for high-throughput screening. Assay Drug Dev Technol 9:108–124

9. Eglen RM, Reisine T (2010) Human kinome drug discovery and the emerging importance of atypical allosteric inhibitors. Expert Opin Drug Discovery 5:277–290

10. Gasparri F, Sola F, Bandiera T et al (2008) High-content analysis of kinase activity in cells. Comb Chem High Throughput Screen 11:523–536

11. Bertekap RL Jr, Burford NT, Li Z et al (2015) High-throughput screening for allosteric modulators of GPCRs. Methods Mol Biol 1335:223–240

12. Johnston P (2008) Automated high content screening. In: Haney S (ed) High content screening: science, techniques and applications. Wiley, Hoboken

13. Ellson R, Mutz M, Browning B et al (2003) Transfer of low Nanoliter volumes between Microplates using focused acoustics – automation considerations. J Assoc Lab Autom 8:29–34

14. Bruner J, Liacos J, Siu A et al (2003) Work-cell approach to automating high content screening. J Assoc Lab Autom 8:66–67

15. Unterreiner V, Gabriel D (2011) When high content screening meets high throughput. Drug Discov World 13:19

16. Collins MA (2009) Generating 'omic knowledge': the role of informatics in high content screening. Comb Chem High Throughput Screen 12:917–925

17. Buchser W, Collins M, Garyantes T, et al. Assay Development Guidelines for Image-Based High Content Screening, High Content Analysis and High Content Imaging. 2012 Oct 1 [Updated 2014 Sep 22]. In: Sittampalam GS, Coussens NP, Brimacombe K, et al., editors. Assay Guidance Manual [Internet]. Bethesda (MD): Eli Lilly & Company and the National Center for Advancing Translational Sciences; 2004-.Available from: https://www.ncbi.nlm.nih.gov/books/NBK100913/

18. Young DW, Bender A, Hoyt J et al (2008) Integrating high-content screening and

ligand-target prediction to identify mechanism of action. Nat Chem Biol 4:59–68

19. Jackson D, Lenard M, Zelensky A et al (2010) HCS road: an enterprise system for integrated HCS data management and analysis. J Biomol Screen 15:882–891. doi:10.1177/1087057110374233. Epub 1087057110372010 Jul 1087057110374216

20. Feng Y, Mitchison TJ, Bender A et al (2009) Multi-parameter phenotypic profiling: using cellular effects to characterize small-molecule compounds. Nat Rev Drug Discov 8:567–578. doi:10.1038/nrd2876

21. Low J, Chakravartty A, Blosser W et al (2009) Phenotypic fingerprinting of small molecule cell cycle kinase inhibitors for drug discovery. Curr Chem Genomics 3:13–21. doi:10.2174/1875397300903010013

22. Low J, Stancato L, Lee J et al (2008) Prioritizing hits from phenotypic high-content screens. Curr Opin Drug Discov Devel 11:338–345

Part III

Harnessing Translocation, Signaling Pathways, Single Cell Analysis, and Multidimensional Data for Phenotypic Analysis

Chapter 12

Translocation Biosensors—Versatile Tools to Probe Protein Functions in Living Cells

Verena Fetz, Roland H. Stauber, and Shirley K. Knauer

Abstract

In this chapter, you will learn how to use translocation biosensors to investigate protein functions in living cells. We here present three classes of modular protein translocation biosensors tailored to investigate: (1) signal-mediated nucleo-cytoplasmic transport, (2) protease activity, and (3) protein-protein interactions. Besides the mapping of protein function, the biosensors are also applicable to identify chemicals and/or (nano) materials modulating the respective protein activities and can also be exploited for RNAi-mediated genetic screens.

Key words Chemical biology, Export, Fluorescence microscopy, High content screening, Import, Protein-protein interaction, Protease, Translocation

1 Introduction

Fluorescent protein biosensors are powerful tools for dissecting the complexity of cellular processes. As cellular biosensors have the advantage of acting in a physiological and/or pathophysiological environment, such as cancer cells, these are used to define the dynamics of cellular regulation, especially when combined with automated multi-parameter imaging technologies [1]. The increase in the use of cell-based assays during all major steps of drug discovery and development has increased the demand for cellular biosensors. Such biosensors are expected to allow the detection of a wide variety of signaling molecules and bear the potential for novel assay applications. Intensifying the use of kinetic, in contrast to snapshot drug screening assays, is expected to reveal subtle, but discrete effects of compounds, aiding the interpretation of their mode of action and leading to an improved understanding of key regulatory cellular pathways, based on "functional cellular responses" [2].

Redistribution approaches, a cell-based assay technology that uses protein translocation as the primary readout, have great

Paul A. Johnston and Oscar J. Trask (eds.), *High Content Screening: A Powerful Approach to Systems Cell Biology and Phenotypic Drug Discovery*, Methods in Molecular Biology, vol. 1683, DOI 10.1007/978-1-4939-7357-6_12,

potential to study the activity of cellular signaling pathways and other intracellular events [3–5]. Protein targets are labeled with a variety of autofluorescent proteins (e.g., the green fluorescent protein – GFP), and the assays are read using high content or high-throughput microscope-based instruments [2]. Protein translocation assays have the potential for the profiling of lead series, primary screening of compound libraries, or even as readouts for gene-silencing studies [6–8]. However, any realistic applications of high-content and high-throughput cell-based assays critically depend on robust and reliable biological readout systems with a high signal-to-noise ratio [9, 10]. Hence, the spatial and functional division into the nucleus and the cytoplasm marks not only two dynamic intracellular compartments vital for the cell but that can also be easily distinguished by microscopy and thus, be exploited for translocation biosensor assays.

We here present three cell-based translocation-based biosensor-systems to investigate nucleo-cytoplasmic transport, protease activity, and protein-protein interaction networks (Fig. 1, Table 1) in living cells.

The biosensor "prototype" is composed of functional modules, tailored to meet the specific biosensors' applications (Fig. 1, Table 1). These modules are autofluorescent proteins, subcellular

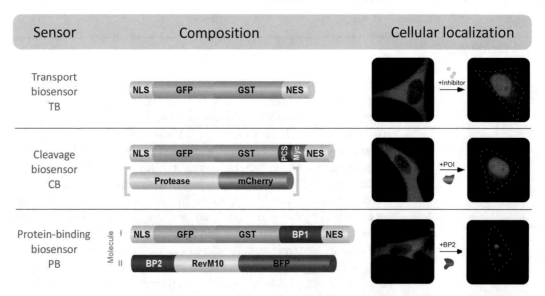

Fig. 1 Modular composition of the biosensors. The biosensor proteins consist of functional modules, tailored to meet the biosensors' specific applications. The transporter biosensor (TB) is composed of a nuclear localization signal (NLS), GFP, GST and nuclear export signal (NES). The protease cleavage biosensor (CB) and the protein-binding biosensor (PB) are derived from the TB. For the CB, a protease cleavage site (PCS) and a Myc-epitope is integrated between GST and the NES. If desired, the protease of interest (POI, shown in red) can be ectopically co-expressed as a fusion with mCherry. The PB system consists of two components. Molecule I is based on the TB, containing in addition binding partner (BP) 1. BP2 is expressed as a nucleolar fusion protein containing the RevM10 protein and BFP (shown in *blue*)

Table 1
Overview of the different biosensors

Biosensor	Application
Transport biosensor (TB)	Nuclear import and export
Cleavage biosensor (CB)	Protease activity
Protein-binding biosensor (PB)	Protein-protein interactions

targeting signals (STS), and gluthatione S-transferase (GST). As such, all biosensors can be visualized by fluorescence microscopy in living or fixed cells. Their intracellular localization is controlled by the rational combination of opposing STS, namely nuclear localization signals (NLS) and nuclear export signals (NES), facilitating nuclear import and export, respectively. Integration of Glutathione S-transferase (GST) increases the molecular weight up to >55 kDa, preventing passive diffusion through the nuclear pore complex. In the following paragraphs, we will describe the composition and applications of the different biosensors.

1.1 Transport Biosensor (TB)

The eukaryotic cell is divided into distinct subcellular functional compartments, which enables the fine-tuning of fundamental biochemical processes. In particular, the modulation of genetic programs critically depends on regulated nucleo-cytoplasmic transport through the nuclear pore complex (NPC), which is controlled by specific signals and transport factors [11]. Those include proteins of the NPC (nucleoporins), the RanGTPase, and transport receptors. Active nuclear import requires energy and is mediated by short stretches of basic amino acids, termed nuclear localization signals (NLS). The best characterized nuclear export signals (NESs) consist of a short leucine-rich stretch of amino acids and interact with the export receptor CRM1/Exportin1 [11]. NESs have been identified in an increasing number of disease-relevant cellular and viral proteins. As regulated subcellular localization provides also an attractive way to control the activity of regulatory proteins and RNAs, targeting nucleo-cytoplasmic transport as a novel therapeutic principle has attracted major interest by academia and industry [12].

To investigate nuclear import and export, our transport biosensors (TB) are applicable to map and quantify the activity of transport signals (NLS/NES) [13, 14], test transport (specific) cofactors, and identify compounds interfering with nucleo-cytoplasmic transport [2]. The TB fusion protein is composed of a NLS, GFP, GST and NES (Fig. 1). Upon ectopic expression in eukaryotic cells, it is continuously shuttling between the nucleus and the cytoplasm but localizes predominantly to the cytoplasm due to the different strengths of the transport signals (NES > NLS).

Upon inhibition of nuclear export, the NLS mediates nuclear import, resulting in the nuclear accumulation of the TB (nuclear translocation). NLS-/NES-sequences are interchangeable in the expression constructs and allow for the easy testing of putative transport signals [2, 13, 15].

1.2 Protease Cleavage Biosensor (CB)

In any physiological or patho-physiological state, numerous proteins are being processed or degraded in a highly controlled fashion [16]. Intrinsic hydrolytic cleavage is performed by proteases, playing critical roles in innumerable biological processes. As protease signaling is mostly irreversible, all proteases are strictly regulated. Consequently, protease de-regulation often leads to patho-physiological states that in principle could be medicated by specific protease inhibitors or activators [16, 17]. Proteases are therefore important drug targets in the pharmaceutical industry as well as potential disease markers.

Hence, our translocation-based protease cleavage biosensor-systems (CB) (Fig. 1, Table 1) can be used to identify chemicals that modulate the proteolytic activity of a protease of interest (POI) [1], map the POI's cleavage-recognition site, and test potential POIs' substrates and cofactors [18]. The CB is based on the transport biosensor, but is additionally equipped with a POI-recognizable cleavage site (PCS) recognized by the POI preceding the NES. Upon processing of the biosensor by the POI, the NES is cleaved-off and thus, the CB accumulates in the nucleus due to the remaining NLS sequence. Hence, cytoplasmic-nuclear translocation is indicative of protease activity in living cells [18]. An example for such an application is represented by the combination of the protease Taspase1 with its cleavage site derived from its target, the Mixed Lineage Leukemia (MLL) protein [1] (Fig. 1). Integration of an additional Myc-epitope tag allows the optional detection of the cleaved CB by immunofluorescence staining using a tag-specific antibody. The CB can be used to study both endogenous or ectopically expressed POI. For detecting endogenous POI activity only the CB construct has to be transfected into eukaryotic cells. If a non-endogenous POI should be investigated, we recommend ectopical expression as a fusion with the red fluorescent protein mCherry. In this case the CB and the POI-mCherry encoding plasmid have to be cotransfected. The mCherry fusion will enable the selection of cells expressing both, the CB and the POI, which is specifically important for object selection during screening.

1.3 Protein-Binding Biosensor (PB)

Protein-protein interaction networks are critical for the majority if not for all cellular events, and require distinct contact areas of both binding partners. Interfering with protein-protein interactions (PPIs) via enforced expression of dominant-negative mutants and/or application of small molecules has emerged as a promising, though challenging strategy for human therapeutics [19–21]. For a long time it has even been thought that small molecules are not feasible to target PPIs. However, this view changed over the last few years, and interest in pharmaceutical applications rose [22]. A prominent example is the interaction of p53 and mdm2 [23, 24]. If their binding is inhibited, activated p53 can accumulate in the nucleus, triggering apoptosis-mediated (tumor) cell death. Albeit most efforts focused on the inhibition of protein interactions, currently the stabilization of PPIs is considered an alternative but promising approach to target disease-relevant signaling pathways [21, 25, 26]. Numerous biochemical in vitro and several cell-based methods have been developed for detecting and studying PPIs [25, 27]. As most of those are laborious and time consuming, the screening of large compound collections for PPI modulators is still an exception.

We here present a two-component translocation-based biosensor systems (PB) (Fig. 1, Table 1) to test PPI and/or to identify compounds interfering with PPI. Molecule I is based on the green fluorescent biosensor, allowing the integration of binding partner 1 (BP1) N-terminal of the NES. The "binding partner" may represent a complete protein or distinct domains thereof. The second molecule (II) consists of binding partner 2 (BP2), fused to a nucleolar, export-deficient HIV-1 Rev protein (RevM10) [28], and the blue fluorescent protein (BFP). Although molecule I is continuously shuttling between the nucleus and the cytoplasm, it predominantly localizes to the cytoplasm due to the unequal strength of the transport signals (NES > NLS). In contrast, molecule II is anchored at the nucleolus, due to the nucleolar localization signal in the HIV-1 RevM10 protein [28]. Upon specific interaction between both binding partners in living cells, the shuttling molecule I will be entrapped in the nucleus, and thus, accumulate at the nucleolus ("nucleolar trapping").

For the construction of biosensor derivatives for your demands, modules (NLS, NES, cleavage sites, BP) can be exchanged in the respective mammalian expression vectors (Fig. 2). These plasmids can be distributed upon request.

For high-throughput analysis, the generation of cell lines stably expressing the respective biosensors is of great advantage. For one, this reduces the need for biosensor plasmid transfection prior to perform screens. Additionally, stably expressing cell populations often display more homogenous biosensor expression levels, facilitating microscopic image capture and subsequent analysis. If available, positive (e.g., leptomycin B/LMB for TB) and negative controls should be included to verify translocation biosensor-based results. For the CB system, biosensors with a mutated or unrelated protease cleavage site, and/or co-expression of a non-functional protease mutant are helpful negative controls [18].

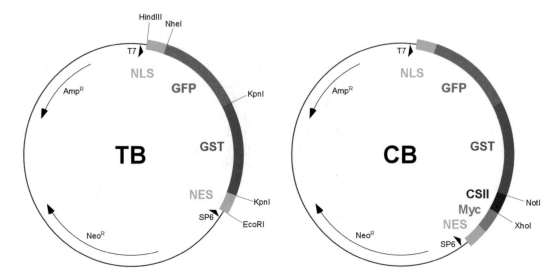

Fig. 2 Vector maps of the transporter biosensor and protease cleavage biosensor expression plasmids. Schematics depicting the composition of the respective biosensors as well as relevant restriction sites. Transporter biosensor (TB)—*left*, protease cleavage biosensor CB—*right*. Transcription in eukaryotic cells is under the control of the CMV-promoter. Amp[R], ampicillin resistance; Neo[R], neomycin resistance; SP6 and T7, respective promoter sequencing primer sites

Table 2
Biosensor - backbone vectors

Biosensor	Vector name
TB	pC3_NLS_GFP_GST_RevNES
CB	pC3_NLS_GFP_GST_CSII_Myc_NES
PB- I	pC3_NLS_GFP_GST_bZipJun_NES
PB- II	pC3_Fos_RevM10_BFP

Employing PPI-deficient mutant proteins and/or empty PB biosensor backbones, lacking the respective protein interaction partners will uncover unspecific protein-protein interaction.

2   Materials

2.1   Generation of Expression Constructs

1. Vector backbone (can be requested, *see* Table 2).
2. Oligonucleotides
   (a) For complementary annealing and subsequent cloning (NES, NLS, or PCS).
   (b) For PCR and subsequent cloning (POI or BP2).
   (c) For sequencing: T7 5′-TAATACGACTCACTATAGGG-3′

SP6 5′-ATTTAGGTGACACTATAG-3′.

3. Restriction enzymes: HindIII, NheI, KpnI, EcoRI, NotI, XhoI.

4. Other enzymes: T4 DNA Ligase (e.g., M0202S, New England BioLabs), T4 Polynucleotide Kinase (PNK, e.g., M0201S, New England BioLabs), Calf Intestinal Phosphatase (CIP, e.g., M0290S, New England BioLabs).

5. Reagents: deoxyribonucleotides (dNTPs, 10 mM), aqua dest., LB-agar-plates containing 100 µg/mL ampicillin, LB-medium containing 100 µg/mL ampicillin, molecular weight standard, agarose, TAE buffer (40 mM Tris, 20 mM acetic acid, 1 mM EDTA), DNA gel-loading dye.

6. Kits: plasmid purification (e.g., Qiagen plasmid mini kit, 12123), gel purification kit (e.g., QIAquick gel purification kit, 12123), PCR purification kit (e.g., QIAquick PCR purification kit, 28104).

7. Equipment: 37° heater and rotary shaker, chamber for agarose-gel electrophoresis, scalpel, plastic tubes, cup, microwave, NanoDrop or photometer, water bath.

8. Bacteria: chemically competent *E. coli* Top10 (C3019I, New England BioLabs) or similar.

2.2 Transfection and Cell Sorting

1. Reagents: Lipofectamine2000 (11668030, Invitrogen, or similar), OptiMEM, DMEM, DMEM 1.6% G418 (480 U/mL), PBS, FCS, DMSO.

2. Equipment: flow cytometer for fluorescence activated cell sorting, fluorescence microscope, laminar flow, sterile pipets, sterile cell culture flasks and plates, 70 µm cell strainer (e.g., Corning Life Sciences, 352350).

2.3 High Content Screening (HCS)

1. Reagents: PFA (4% in PBS), Hoechst 33342 (25 µg/mL in PBS), Triton X-100 (0.1% in PBS).

2. Equipment: High content imager (ArrayScanVTI, Thermo Scientific, or similar), plate washer/dispenser (EL406, BioTek, or similar), electronic multichannel pipets, pipetting robot (Biomek Nxp, Beckman Coulter, or similar), sterile pipets and cell culture flasks, black cell culture plates with clear thin bottom for microscopic imaging (e.g., Greiner µclear, 781092), clear adhesive plate seal.

3 Methods

3.1 Preparation of Vector Backbones

Depending on your intended application of the biosensors the vector backbones have to be prepared for inserting NLS, NES, CB, or BP.

Table 3
Restriction enzymes for vector preparation

Vector	Biosensor	Enzymes
pC3_NLS_GFP_GST_RevNES	TB	For NES: KpnI, EcoRI For NLS: HindIII, NheI
pC3_NLS_GFP_GST_CSII_Myc_NES	CB	NotI, XhoI
pC3_NLS_GFP_GST_Jun_NES	PB (I)	NotI, XhoI
pC3_Fos_RevM10_BFP	PB (II)	KpnI

1. Digest vector backbones with the restriction enzymes as depicted in Table 3 for 1 h at 37 °C.

2. Add calf intestinal phosphatase CIP (*see* **Note 1**) and incubate for another 30 min at 37 °C.

3. Add 1/6 Vol 6× DNA-loading dye.

4. Separate the restriction reaction by agarose gel electrophoresis (1% agarose gel).

5. Slice and collect the vector band out of the gel.

6. After KpnI digestion of the TB vector, the GST fragment will also be released. Collect and elute this fragment for re-integration later.

7. Purify the vector DNA from agarose gel using a plasmid-purification kit.

8. Determine concentration of purified vector using a NanoDrop or equivalent photometric method.

3.2 Annealing of Complementary Oligonucleotides for Inserting NES/NLS/ PCS

The NES and NLS of the TB (*see* **Note 2**) and the PCS of the CB can be replaced by complementary DNA oligonucleotides containing the respective restriction sites (*see* Fig. 2). For the CB, subsequently the Myc-epitope can be inserted into the XhoI restriction site using complementary oligonucleotides, if desired.

1. Order the oligonucleotides with the restriction sites as depicted in Table 4.

2. For phosphorylation of 5'ends, prepare a mixture of the ingredients listed in Table 5 for each single oligonucleotide.

3. Incubate the mixtures for 15 min at 37 °C.

4. Combine both phosphorylated complementary oligonucleotides in an aerosol-tight reaction tube.

5. Heat water in a cup in the microwave and directly insert the tube containing both oligonucleotides. Leave it on the bench until the water temperature decreased to room temperature (*see* **Note 3**).

Table 4
Restriction sites for oligonucleotides to insert NES/NLS/PCS via oligonucleotide annealing

Function	5′ end (enzyme)	3′ end (enzyme)
NES	KpnI	EcoRI
NLS	HindIII	NheI
PCS	NotI	XhoI

Table 5
Oligonucleotide phosphorylation

Reagent	Concentration	Amount (μL)
Oligonucleotide	100 μM	1
dNTP	10 mM	0.5
PNK buffer (supplier)	10×	1
Aqua dest.	–	7
PNK	10.000 U/mL	0.5

6. Use 1 μL of the annealing product for ligation into 100 ng linear vector backbone using T4 DNA-Ligase (*see* **Note 4**).

7. Mix ligation product or equivalent of sterile water with 50 μL chemically competent *E. coli* Top10 respectively and incubate on ice for 30 min.

8. Incubate in water bath at 42 °C for 30 s followed by 2 min on ice.

9. Add 1 mL LB-medium and incubate for 1 h at 37 °C on a rotary shaker.

10. Spin down transformed bacteria at 10,000 rpm for 1 min.

11. Discard supernatant LB medium and resuspend bacteria in 100 μL LB-medium.

12. Plate transformed bacteria on ampicillin-containing agar plates and incubate overnight at 37 °C.

13. Seed clones into 5 mL LB-medium containing 100 μg/mL ampicillin and incubate at 37 °C overnight on a rotary shaker.

14. Isolate plasmids using a plasmid purification kit.

15. Check proper ligation by digesting plasmids with matching restriction enzymes and by subjecting plasmids to sequencing using T7 and/or SP6 primer.

3.3 Integrating PCR-Amplified Sequences of BP1 and BP2

BP1 and BP2 can be replaced by inserting PCR-amplified sequences containing the restriction sites, as mentioned in Table 6 (*see* Fig. 3).

1. Order oligonucleotides for PCR amplification of your genes of interest fused to the restriction sites depicted in Table 6.

2. PCR amplify your gene of interest.

3. Check proper amplification by agarose gel electrophoresis and elute the product out of the gel using a gel purification kit.

4. Digest the PCR product with the restriction enzymes mentioned in Table 6.

5. Purify restricted PCR product using a PCR purification kit.

6. Ligate your gene of interest into the respective backbones, prepared in Subheading 3.1 and follow the instruction from Subheading 3.2.15 for verification of your plasmid.

Table 6
Restriction sites flanking amplification products for BP1 and BP2

Product	5′ end (enzyme)	3′ end (enzyme)
BP1	NotI	XhoI
BP2	KpnI	KpnI

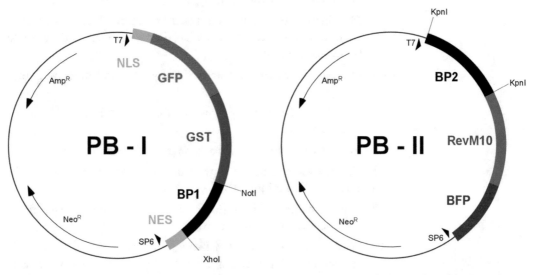

Fig. 3 Vector maps of the protein-binding biosensors PB I and PB II. Schematics depicting the composition of the respective protein-binding biosensors as well as relevant restriction sites. Binding partner I (BP I)—*left*, binding partner II (BP II)—*right*. Transcription in eukaryotic cells is under the control of the CMV-promoter. Amp^R, ampicillin resistance; Neo^R, neomycin resistance; SP6 and T7, respective promoter sequencing primer sites

3.4 Generation of Cells Stably Expressing the Biosensors

For high content screening, the generation of cells stably expressing the biosensors is of great advantage (*see* **Note 5**). This will suspend the use of freshly transfected cells with high differential expression intensity and also shorten the procedure of screening. Additionally, the time required for imaging will decrease due to the equalized expression level of stable cell lines. Cells stably expressing any biosensor will be named "biosensor cells" from here on.

1. Transfect the biosensor constructs into adherent eukaryotic cells using Lipofectamine2000 or similar transfection reagent. For PB, both molecules have to be co-transfected (*see* **Notes 6** and **7**).

2. 16 h post-transfection, add G418 (0.4 mg/mL) to the culture medium to select for cells expressing the biosensor constructs.

3. Culture cells continuously in G418 containing medium until enough cells are available for sorting (*see* **Note 8**).

4. Preparation for cell sorting: Detach cells using trypsin and wash once in PBS. Filter cells using a cell strainer and store on ice until sorting. Sort cells into pure FCS. If possible, sort into certain subpopulations with different expression levels of GFP (and BFP/mCherry for co-transfected cells). After sorting centrifuge cells and resuspend in medium containing G418.

5. Repeat **step 4** when enough cells recovered from first sorting.

6. Observe cellular fluorescence microscopically for proper localization of the biosensor (*see* **Note 9**).

7. Preservation: Freeze cells in FCS containing 10% DMSO until use (*see* **Note 10**).

3.5 Establish Screening Conditions (Prescreen)

For successful high content screening the cell seeding and treatment time course have to be carefully optimized.

1. Seed the biosensor expressing cells into 384-well plates (black with clear bottom, *see* **Notes 11** and **12**) at different densities (HeLa: approx. 9500 cells/384-well in 50 µL medium).

2. Visually monitor growth over the desired time frame of the planned treatment (*see* **Note 13**) and select optimal seeding density for further proceeding.

3. Seed biosensor expressing cells at optimized density and, if available, treat half of the wells with positive control, i.e., export inhibitor for TB, cleavage inhibitor for CB (*see* **Note 14**), PPI-inhibitor for PB (*see* **Note 15**).

4. Fixation: remove the medium and add 50 µL of 4%PFA to each well (*see* **Note 16**). Incubate at room temperature for 20 min (*see* **Note 17**).

5. Permeabilization: remove PFA with a washer/dispenser, wash thrice with 50 µL PBS, and add 50 µL PBS 0.1% TritonX to each well. Incubate at room temperature for 5 min (*see* **Note 17**).

6. Nuclear staining: remove PBS 0.1% TritonX and add PBS 25 μg/mL Hoechst 33342. Incubate at room temperature for 30 min (*see* **Note 17**).

7. Finalization: remove Hoechst and perform 3 washing steps with a washer/dispenser, leaving 50 μL PBS in the wells. Seal plates with clear adhesive seal.

8. High content imaging: capture plates with your high content imager.

9. Analyze the cellular localization of the biosensors. Use the respective software of the high content imager for evaluating the intracellular localization of the biosensor. For ArrayScan VTI, use "*Molecular Translocation*" or "*Cytoplasm to Nucleus Translocation*" (*see* **Note 18**).

10. If positive controls were measured, calculate Z'-value using the following formula [10]:

$$Z = 1 - \left(\frac{3^* [SD^+ + SD^-]}{R} \right)$$

For screening, the Z'-value should be 0.3 or higher.

3.6 High Content Screening Assays Using Biosensors

After optimizing cell density and treatment schedule, the biosensor cells can be used for screening compound libraries.

Procedure for screening:

1. Generate an appropriate amount of biosensor cells for seeding the required number of plates. Routinely check biosensor fluorescence of your cell population.

2. Seed cells at optimized cell density into 384-well plates using electronic multichannel pipets in dispensing mode (*see* **Note 11**). Leave in an incubator overnight.

3. Treatment: add compounds to the cells using a pipetting robot (*see* **Note 19**).

4. Incubation: leave cells in an incubator for the desired incubation period.

5. Follow **steps 4–10** from Subheading 3.5.

If desired, use antibody stains to detect additional targets in the cells. Thereby, it is possible to, e.g., simultaneously detect apoptotic effects. Calculate Z'-factor coefficients for every plate. Identify hit compounds and retest them. If possible, test at varying concentrations for obtaining dose response curves. Evaluate validity of your novel inhibitors in further experiments.

4   Notes

1. Preparation of vector backbone:
   It is of advantage to dephosphorylate the vector ends to prevent self-ligation of the vector.

2. TB:
   Select the combination of NLS and NES carefully. Highly active NLS will cause the biosensor to accumulate in the nucleus without any inhibition of export. Cells expressing TB with a highly active NES will show a prominent GFP signal on the nuclear membrane. This does not interfere with functionality of the TB. The nuclear membrane signal will vanish upon inhibition of export.

3. Annealing of complementary oligonucleotides:
   For proper annealing, insert the mix of both phosphorylated oligonucleotides directly into the container (cup) containing boiling water. Only use aerosol tight tubes. Store annealed oligonucleotides at 4 °C.

4. Ligation:
   Usually ligase reaction is most efficient if incubated at 4 °C overnight.

5. Cell type:
   Use adherent flat growing cells for high-quality microscopic images and optimal image analysis. They are easier to evaluate due to their more prominent cytoplasmic region.

6. PB:
   As internal transport signals of the BP may interfere with proper localization of the PB molecules, these have to be checked prior to cloning. Experience shows that it is advantageous to integrate both BPs in both PB (molecule I/II) backbones and check for their intracellular localization [13].

7. Expression-vector for POI: Cloning of the protease into a different vector backbone could be of advantage by allowing double-selection together with another antibiotic (e.g., puromycin, blasticidin).

8. Transfection:
   Directly add G418 the day following transfection and perform first round of cell sorting as soon as possible. Do not cultivate cells too long without using them for the assay. Freeze cells as early as possible for preserving biosensor expression.

9. Biosensor cells: Validate the localization of the biosensor for cells with different expression levels. It may be necessary to choose the cells with a medium expression level, as too strong overexpression may cause mis-localization of the biosensor. In this case, use biosensor cells with lower expression level.

10. Cultivation of biosensor cells:
Do not cultivate stable cell lines over a long time. Expression may decrease with time. Only defreeze when planning to use them.

11. Cell seeding for HCS:
Establish cell seeding well. Do not shake freshly seeded plates (cells will not adhere in the middle of the well or be equally distributed). If possible leave on the bench for 20 min after seeding. Use plate lids with condensation rings. If possible spare outer wells and fill these with PBS. Plates should be optimized for microscopic imaging of cells. Test them before by staining cells with PBS/Hoechst 33342 (25 μg/mL).

12. Plastic ware:
Do not use polystyrene plates or pipette tips because of their hydrophobic surface. Charged compounds might adhere to the polystyrene and thus, may not be available for the assay.

13. Cell density:
Cell density should be optimal for image analysis. Too few cells will require capturing many more images and cells growing too dense may lead to problems during image analysis.

14. CB:
Using cells with intrinsic POI activity might result in nuclear accumulation of the CB, dependent on the protease activity. Once processed, the cytoplasmic localization is solely generated by the expression of new CB. Thus, the incubation time with the compounds has to be properly established to generate a significant CB localization difference. The CB can also be used to test cell lines for the expression of POI.

15. HCS control:
Whenever possible, include positive and negative controls on every assay plate! Carefully establish the optimal incubation time for the control compound to obtain a maximum signal difference.

16. Fixation:
If using loosely attaching cells, 8%PFA may be added to the growth medium directly for fixation.

17. Handling of assay plates:
Adjust washer and dispenser to slowly remove/add solutions to an outer well area to protect integrity of the cell layer.

18. Quantify intracellular localization of the biosensor:
It is of advantage to use the ratio or difference of cytoplasmic to nuclear intensity for evaluating intracellular biosensor localization. This will yield a clearer and more stable discrimination between positive and negative controls. As the biosensor may completely translocate from cytoplasm to nucleus, you should

not apply any detection threshold for the cytoplasmic area. In case of complete translocation no cytoplasmic area would be detected, which may obscure your results.

19. Compound transfer:
Take care of properly distributing the compounds in the medium and not to detach the cells during this process.

References

1. Knauer SK et al (2011) Bioassays to monitor Taspase1 function for the identification of pharmacogenetic inhibitors. PLoS One 6(5): e18253

2. Fetz V, Knauer SK, Bier C, von Kries JP, Stauber RH (2009) Translocation biosensors - cellular system integrators to dissect CRM1-dependent nuclear export by Chemicogenomics. Sensors 9:5423–5445

3. Hua Y et al (2014) High-content positional biosensor screening assay for compounds to prevent or disrupt androgen receptor and transcriptional intermediary factor 2 protein-protein interactions. Assay Drug Dev Technol 12(7):395–418

4. Korn K, Krausz E (2007) Cell-based high-content screening of small-molecule libraries. Curr Opin Chem Biol 11(5):503–510

5. Lundholt BK et al (2006) A simple cell-based HTS assay system to screen for inhibitors of p53-Hdm2 protein-protein interactions. Assay Drug Dev Technol 4(6):679–688

6. Heydorn A et al (2006) Protein translocation assays: key tools for accessing new biological information with high-throughput microscopy. Methods Enzymol 414:513–530

7. Loechel F et al (2007) High content translocation assays for pathway profiling. Methods Mol Biol 356:401–414

8. Zanella F et al (2007) An HTS approach to screen for antagonists of the nuclear export machinery using high content cell-based assays. Assay Drug Dev Technol 5(3):333–341

9. Fuller CJ, Straight AF (2010) Image analysis benchmarking methods for high-content screen design. J Microsc 238(2):145–161

10. Zhang JH, Chung TD, Oldenburg KR (1999) A simple statistical parameter for use in evaluation and validation of high throughput screening assays. J Biomol Screen 4(2):67–73

11. Turner JG, Dawson J, Sullivan DM (2012) Nuclear export of proteins and drug resistance in cancer. Biochem Pharmacol 83 (8):1021–1032

12. Hutten S, Kehlenbach RH (2007) CRM1-mediated nuclear export: to the pore and beyond. Trends Cell Biol 17(4):193–201

13. Bier C et al (2011) The Importin-alpha/Nucleophosmin switch controls Taspase1 protease function. Traffic 12(6):703–714

14. Knauer SK et al (2006) The Survivin-Crm1 interaction is essential for chromosomal passenger complex localization and function. EMBO Rep 7(12):1259–1265

15. Knauer SK et al (2005) Translocation biosensors to study signal-specific nucleo-cytoplasmic transport, protease activity and protein-protein interactions. Traffic 6(7):594–606

16. Turk B (2006) Targeting proteases: successes, failures and future prospects. Nature reviews. Drug Discov 5(9):785–799

17. Clausen T et al (2011) HTRA proteases: regulated proteolysis in protein quality control. Nature reviews. Mol Cell Biol 12 (3):152–162

18. Bier C et al (2011) Cell-based analysis of structure-function activity of threonine aspartase 1. J Biol Chem 286(4):3007–3017

19. Kar G, Gursoy A, Keskin O (2009) Human cancer protein-protein interaction network: a structural perspective. PLoS Comput Biol 5 (12):e1000601

20. Arkin MR, Whitty A (2009) The road less traveled: modulating signal transduction enzymes by inhibiting their protein-protein interactions. Curr Opin Chem Biol 13 (3):284–290

21. Bier C et al (2012) Allosteric inhibition of Taspase1's pathobiological activity by enforced dimerization in vivo. FASEB J 26 (8):3421–3429

22. Mullard A (2012) Protein-protein interaction inhibitors get into the groove. Nat Rev Drug Discov 11(3):173–175

23. Dudgeon DD et al (2010) Characterization and optimization of a novel protein-protein interaction biosensor high-content screening assay to identify disruptors of the interactions between p53 and hDM2. Assay Drug Dev Technol 8(4):437–458

24. Carry JC, Garcia-Echeverria C (2013) Inhibitors of the p53/hdm2 protein-protein interaction-path to the clinic. Bioorg Med Chem Lett 23(9):2480–2485

25. Rose R et al (2011) Identification and structure of small-molecule stabilizers of 14-3-3 protein-protein interactions. Angew Chem Int Ed Engl 49(24):4129–4132

26. Ottmann C et al (2009) A structural rationale for selective stabilization of anti-tumor interactions of 14-3-3 proteins by cotylenin A. J Mol Biol 386(4):913–919

27. Berggard T, Linse S, James P (2007) Methods for the detection and analysis of protein-protein interactions. Proteomics 7 (16):2833–2842

28. Stauber RH et al (1998) Analysis of intracellular trafficking and interactions of cytoplasmic HIV-1 rev mutants in living cells. Virology 251 (1):38–48

Chapter 13

High Content Positional Biosensor Assay to Screen for Compounds that Prevent or Disrupt Androgen Receptor and Transcription Intermediary Factor 2 Protein-Protein Interactions

Yun Hua, Daniel P. Camarco, Christopher J. Strock, and Paul A. Johnston

Abstract

Transcriptional Intermediary Factor 2 (TIF2) is a key Androgen receptor (AR) coactivator that has been implicated in the development and progression of castration resistant prostate cancer (CRPC). This chapter describes the implementation of an AR-TIF2 protein-protein interaction (PPI) biosensor assay to screen for small molecules that can induce AR-TIF2 PPIs, inhibit the DHT-induced formation of AR-TIF2 PPIs, or disrupt pre-existing AR-TIF2 PPIs. The biosensor assay employs high content imaging and analysis to quantify AR-TIF2 PPIs and integrates physiologically relevant cell-based assays with the specificity of binding assays by incorporating structural information from AR and TIF2 functional domains along with intracellular targeting sequences using fluorescent protein reporters. Expression of the AR-Red Fluorescent Protein (RFP) "prey" and TIF2-Green Fluorescent Protein (GFP) "bait" components of the biosensor is directed by recombinant adenovirus (rAV) expression constructs that facilitated a simple co-infection protocol to produce homogeneous expression of both biosensors that is scalable for screening. In untreated cells, AR-RFP expression is localized predominantly to the cytoplasm and TIF2-GFP expression is localized only in the nucleoli of the nucleus. Exposure to DHT induces the co-localization of AR-RFP within the TIF2-GFP positive nucleoli of the nucleus. The AR-TIF2 biosensor assay therefore recapitulates the ligand-induced translocation of latent AR from the cytoplasm to the nucleus, and the PPIs between AR and TIF2 result in the colocalization of AR-RFP within TIF2-GFP expressing nucleoli. The AR-TIF2 PPI biosensor approach offers significant promise for identifying molecules with potential to modulate AR transcriptional activity in a cell-specific manner that may overcome the development of resistance and progression to CRPC.

Key words Protein-protein interaction biosensors, High content screening, Imaging, Image analysis

---

1 Introduction

Protein-protein interactions (PPIs) are integral to all cellular functions and therefore represent a large number of potential therapeutic targets for drug discovery [1–8]. Despite the critical importance of PPIs and the existence of many assay formats compatible with

Paul A. Johnston and Oscar J. Trask (eds.), *High Content Screening: A Powerful Approach to Systems Cell Biology and Phenotypic Drug Discovery*, Methods in Molecular Biology, vol. 1683, DOI 10.1007/978-1-4939-7357-6_13,
© Springer Science+Business Media LLC 2018

HTS/HCS, the dearth of approved PPI inhibitor/disruptor drugs suggests that the discovery of such molecules is not trivial [1–8]. Although PPI targets are often characterized as "undruggable" [9], the structural analysis of protein-protein complexes suggests that protein-binding interfaces contain discrete "hot spots" and that relatively small numbers of amino acids at the PPI interface contribute the majority of the binding energy [1–4, 6–8]. Furthermore, PPI contact surfaces exhibit some degree of flexibility with cavities, pockets, and grooves available for small molecule binding [1–4, 6–8]. Existing small molecule PPI inhibitors appear to bind to hotspots with much higher efficiencies and deeper within the target protein than do the contact atoms of the native protein partner [1–4, 6–8]. This chapter describes the implementation of a high content positional biosensor (PPIB) assay that we have developed to measure the PPIs between the Androgen Receptor (AR) and a key coactivator Transcriptional Intermediary Factor 2 (TIF2) (Fig. 1) [4]. The biosensor assay employs high content imaging and analysis to quantify AR-TIF2 PPIs and integrates physiologically relevant cell-based assays with the specificity of binding assays by incorporating structural information from AR and TIF2 functional domains along with intracellular targeting sequences and fluorescent reporters (Figs. 1, 2 and 3) [4].

The subcellular localization of macromolecules in specific cellular compartments is a tightly regulated process [10–12]. For example, although molecules <40 kDa passively diffuse through the nuclear pore complexes (NPCs) in the nuclear envelope, cargos ≥40 kDa require an active transport process facilitated by specific receptor proteins to enter the nucleus from the cytoplasm [10–12]. Protein cargos ≥40 kDa bearing a suitable nuclear localization sequence (NLS) bind to an importin-α adaptor receptor that recognizes the NLS and forms a complex with an importin-β transport receptor that facilitates docking interactions with the nucleoporins that line the NPC [10–12]. Protein export from the nucleus is also mediated by the assembly of a complex between exportin-1 (CRM-1), Ran-GTP, and protein cargos bearing a leucine-rich nuclear export sequence (NES) [10–12]. The steady-state localization of a protein that moves between the cytoplasm and the nucleus is a function of the balance between the operational strengths and/or accessibility of its NLS and NES sequences [10–12]. Since the nucleus and the cytoplasm can be readily identified, separated, and quantified independently by image analysis methods, the regulated nucleo-cytoplasmic localization of proteins is frequently used in HCS assays as a surrogate for signaling pathway activation [13–18]. Additionally, the specific targeting sequences that direct proteins to specific subcellular sites or compartments have also been exploited to design positional biosensors to quantify PPIs [2–4, 11, 12, 19, 20]. Positional PPI biosensors typically consist of two parts, a "bait" biosensor that is targeted and anchored to a specific cellular

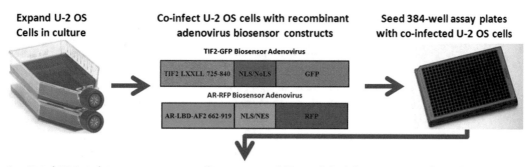

1. **Novel AR Agonist screen** – expose cells to compound library & look for AR-TIF2 PPI formation
2. **AR-TIF2 PPI Blocker screen** – pre-expose cells to compounds then treat with DHT
3. **AR-TIF2 PPI Disruptor screen** – pre-treat with DHT then expose cells to compounds

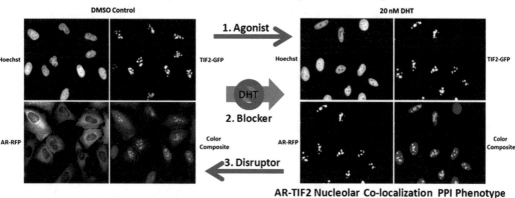

Fig. 1 AR-TIF2 protein-protein interaction biosensor design, grayscale and color composite images of maximum and minimum plate controls, and potential screening formats. Recombinant adenovirus (rAV) AR and TIF2 biosensor constructs were created to co-infect and express the AR and TIF2 protein-protein interaction partners in cells. The AR-RFP "prey" protein interaction partner shuttles between the cytoplasm and the nucleus in a ligand-dependent manner and the rAV construct is composed of AR residues 662–919 encoding the AR-LBD and AF2 surface as a chimeric fusion protein with red fluorescent protein (RFP) and both nuclear localization and nuclear export sequences that are part of the chimera, and not specific to AR. The central region of TIF2 contains three α-helical LXXLL motifs that mediate the binding to ligand-bound AR, and a rAV construct was created to express TIF2 residues 725–840 as a chimeric fusion protein with green fluorescent protein (GFP) and a high affinity nuclear/nucleolar localization (NLS/NoLS) sequence derived from HIV Rev. The TIF2-GFP "bait" protein interaction partner is targeted to and anchored in the nucleoli within the cell nucleus. U-2 OS cells cultured in tissue culture flasks are harvested after exposure to trypsin, counted, co-infected with the AR-RFP and TIF2-GFP adenoviruses, seeded at 2500 cells per well in 384-well collagen-coated assay plates, cultured overnight at 37 °C, 5% CO_2 and 95% humidity, and then treated for 30 min with 0.5% DMSO or 20 nM DHT in 0.5% DMSO prior to formaldehyde fixation and Hoechst staining as described above. Individual gray-scale images of three fluorescent channels (Hoechst Ch1 blue, FITC Ch2 green, and Texas Red Ch3) were sequentially acquired on the IXU automated imaging platform using a 20×/0.45 NA objective, the 405 nm Ch1, 488 nm Ch2, and 561 nm Ch3 laser lines, and a Quad emission filter set as described above. Individual 20× gray-scale and color composite images are presented; Ch1 Hoechst—*blue*, Ch2 TIF2-GFP—*green*, and Ch3 AR-RFP—*red*. In untreated U-2 OS cells expressing both biosensors, AR-RFP expression is localized predominantly to the cytoplasm and TIF2-GFP expression is localized only to nucleoli as indicated by the color composite images of cells with diffuse red cytoplasm and blue nuclei containing bright green TIF2-GFP puncta. After exposure to DHT for 30 min the AR-RFP colocalizes with the TIF2-GFP partner in nucleoli as indicated by the *bright yellow* AR-TIF2 puncta within the blue stained nuclei of color composite

location and a "prey" biosensor designed to shuttle between distinct subcellular compartments [2–4, 11, 12, 19, 20]. Colocalization of both the biosensors to the same site in the cell indicates the formation of productive PPI complexes. Stauber and colleagues pioneered the design of positional PPI biosensors that target "bait" PPI partners to the nucleolus using expression constructs that incorporate a NES deficient HIV-1 Rev. sequence and a fluorescent reporter protein [11, 12, 19, 20]. The matching "prey" PPI biosensor partners are designed to shuttle between the cytoplasm and the nucleus by integrating both the NLS and NES sequences with the fluorescent reporter protein [11, 12, 19, 20].

Prostate cancer (CaP) is the most common solid tumor and second leading cause of cancer death among men in Western countries [21–25]. Existing frontline androgen ablation therapies (AAT) either target androgen production or function. Although the initial responses to these therapies are generally favorable, approximately 20% of patient's progress to castration-resistant metastatic CaP (CRPC) for which there currently is no cure [21–25]. The AR is a nuclear hormone receptor (NR) that is a ligand-dependent and DNA-sequence specific transcriptional regulator involved in prostate growth, terminal differentiation, and function [21–28]. Latent AR in the cytoplasm exists in a complex with heat-shock proteins (Hsp) 90 and 70 that maintain the NR in a stable, partially unfolded state primed for high affinity interactions with androgens [21–24]. Agonist binding induces AR homo-dimerization, trafficking to the nucleus, binding to specific DNA response element sequences in the promoter/enhancer regions of AR target genes, recruitment of coactivators, assembly of the core transcriptional machinery, and activation of transcription [21–24]. Coactivators recruited by ligand activated AR amplify the assembly of the transcription complex and regulate tissue specific spatiotemporal gene expression [27, 28]. Elevated coactivator levels reduce ligand concentration requirements and elicit a more rapid transcriptional response [27, 28]. Overexpression of AR and/or its co-activators, or shifts in the balance between coactivators and core-pressors, are thought to play a role in the emergence of resistance in CRPC [21–24, 29–31]. TIF2 (SRC-2) is a member of the steroid receptor coactivator SRC/p160 family that has been implicated in

Fig. 1 (continued) images. The AR-TIF2 biosensor therefore recapitulates the ligand-induced translocation of AR from the cytoplasm to the nucleus, and the PPIs between AR and TIF2 results in the colocalization of AR-RFP and TIF2-GFP within the nucleolus. The AR-TIF2 PPIB HCS assay can be screened in three distinct formats: format 1, cells are exposed to compounds to identify novel AR agonists capable of inducing the formation of AR-TIF2 PPIs; format 2, cells are exposed to compounds prior to the addition of DHT to screen for compounds that block DHT-induced formation of AR-TIF2 PPIs; and format 3, cells are pretreated with DHT before compound addition to identify small molecules capable of disrupting pre-existing AR-TIF2 complexes

the development and progression of CRPC [21, 22, 25, 27, 28, 31–36]. TIF2 stabilizes the AR and AR-ligand binding, facilitates AR N/C interactions, promotes chromatin remodeling enzyme recruitment, and AR transcriptional activation [21, 22, 27, 28]. There is a significant correlation between tumor TIF2 expression and CaP aggressiveness [25, 32–34]. Transient TIF2 overexpression increased AR responses to adrenal androgens and non-AR ligands, while TIF2 antisense oligo or siRNA knockdowns reduced AR target gene expression and slowed the proliferation of androgen-dependent and independent CaP cells [25, 32]. Prolonged AR localization and TIF2 recruitment to AR target gene promoters has been associated with the development of CRPC, and it was suggested that small molecules that block AR-TIF2 PPIs might have therapeutic value [25, 29–32, 35].

Agonist binding to AR induces a conformational change in the AR ligand-binding domain (AR-LBD) to form the Activation Function 2 (AF2) surface that binds the LXXLL motifs of SRC/p160 coactivators including TIF2 [27–30, 36]. The chimeric AR and TIF2 biosensor components were cloned into separate recombinant adenovirus (rAV) expression constructs (Fig. 1), a high efficiency co-expression system that we have exploited previously for other PPIB HCS assays [2–4]. The AR "prey" biosensor was created to express AR residues 662-919 that encompass the AR-LBD as a chimeric fusion protein with red fluorescent protein (RFP) and included both an NLS and an NES sequence (Fig. 1) [4]. The TIF2 "bait" biosensor was created to express residues 725-840 of TIF2 that contains three α-helical LXXLL motifs as a chimeric fusion protein with GFP and a high affinity nuclear/nucleolar localization (NLS/NoLS) sequence derived from HIV Rev (Fig. 1) [4]. Anchoring the "bait" biosensor in the nucleolus facilitates both the image acquisition process and the subsequent quantification of the colocalization of the "prey" biosensor by image analysis (Figs. 1 and 2) [4]. The use of the recombinant adenovirus expression vectors enables the development of a relatively simple co-infection protocol to produce homogeneous expression of both biosensor components that is scalable for screening (Fig. 1) [4]. The U-2 OS osteosarcoma cell line was acquired from American Type Culture Collection and was maintained in McCoy's 5A medium with 2 mM L-glutamine (Invitrogen, Carlsbad, CA) supplemented with 10% fetal bovine serum (Gemini Bio-Products, West Sacramento, CA), and 100 U/mL penicillin and streptomycin (Invitrogen, Carlsbad, CA) in a humidified incubator at 37 °C, 5% CO_2 and 95% humidity. In untreated U-2 OS cells co-infected with both rAVs, AR-RFP expression is localized predominantly to the cytoplasm and TIF2-GFP expression is localized only in the nucleoli of the nucleus, as indicated by the diffusely red cytoplasm and blue nuclei containing bright green TIF2-GFP puncta of the corresponding composite images (Fig. 1) [4].

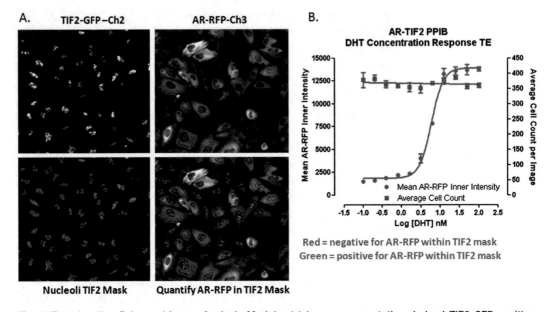

A.

TIF2-GFP–Ch2   AR-RFP-Ch3

Nucleoli TIF2 Mask   Quantify AR-RFP in TIF2 Mask

B.

AR-TIF2 PPIB
DHT Concentration Response TE

Red = negative for AR-RFP within TIF2 mask
Green = positive for AR-RFP within TIF2 mask

Fig. 2 Translocation Enhanced Image Analysis Module. (a) Image segmentation-derived TIF2-GFP positive nucleoli masks in the FITC and Texas Red channels. Enlarged and cropped gray-scale images for presentation purposes of TIF2-GFP (Ch2) and AR-RFP (Ch3) from U-2 OS cells co-infected with both biosensor adenoviruses and then cultured overnight without further treatment. The translocation enhanced (TE) image analysis module utilizes the TIF2-GFP biosensor component in Ch2 to create a mask of the nucleoli. The bright fluorescent puncta in Ch2 with TIF2-GFP fluorescent intensities >750 gray levels over background, an approximate width of 4.0 μm, a minimum area of 5.0 μm^2, and a maximum area < 150 μm^2 are classified by the image segmentation as TIF2-GFP positive nucleoli and used to create translocation masks. AR-RFP images from Ch3 are segmented into nucleoli regions using the masks derived from the detected TIF2-GFP positive nucleoli in Ch2. The *red* or *green* color of the nucleoli masks indicates whether the correlation coefficient for colocalization of the AR-RFP signal within the TIF2-GFP positive nucleoli was below (*red*) or above (*green*) a preset threshold (typically ≥0.25). (b) Quantitative data extracted by the Translocation Enhanced image analysis module: average inner fluorescent intensity of the Ch3 AR-RFP signal within Ch2-derived masks of TIF2-GFP positive nucleoli; average cell count per image determined from the number of Hoechst stained nuclei quantified in Ch1 images. U-2 OS cells were co-infected with the AR-RFP and TIF2-GFP rAV biosensors, 2500 cells were seeded into the wells of 384-well assay plates, cultured overnight at 37 °C, 5% CO_2 and 95% humidity, and then treated with the indicated concentrations of DHT for 30 min. Cells were then fixed and stained with Hoechst, 20× images in three fluorescent channels were acquired on the IXU automated imaging platform, and the extent of DHT-induced AR-TIF2 PPIs was quantified using the TE image analysis module using the average inner intensity of AR-RFP within TIF2-GFP positive nucleoli parameter. To control for differences in cell numbers, the average number of Hoechst stained nuclei per image was also quantified by the TE image analysis module. The mean ± sd ($n = 3$) average inner intensity of AR-RFP within the TIF2-GFP positive nucleoli (●) and cell counts per image (■) at DHT concentrations ranging between 0.001 and 100 nM are presented

Exposure to DHT induces the colocalization of AR-RFP within the TIF2-GFP positive nucleoli of the nucleus, as indicated by the bright yellow of the AR-TIF2 puncta within the blue stained nuclei of the corresponding composite images (Fig. 1) [4]. The AR-TIF2 biosensor assay therefore recapitulates the ligand-induced translocation of latent AR from the cytoplasm to the nucleus, and the PPIs between AR and TIF2 result in the colocalization of AR-RFP within TIF2-GFP expressing nucleoli [4]. Variations in AR coregulator expression levels between normal tissues and CaP cell lines contribute to their altered androgen responsiveness [37, 38]. NRs selectively recruit coactivator complexes, and diverse ligands preferentially recruit different coregulator cohorts [27, 28, 39]. Some NR ligands only activate a subset of target genes, and it is believed that such gene selectivity is cell or tissue specific and reflects the ratio of coactivators to corepressors which determines whether an on or off signal is processed [36, 39]. Coregulators may also exhibit different functions depending upon the specific promoter context [26, 27, 36]. NR coactivator recruitment profiles therefore influence the tissue-specific spatiotemporal gene expression responses to NR ligands [27, 36, 37].

To analyze and quantify AR-TIF2 PPIs in digital images acquired on either the ImageXpress Ultra (IXU) confocal or ImageXpress Micro (IXM) wide-field automated imaging platforms, we utilized the translocation enhanced (TE) image analysis module (Fig. 2) [4]. The bright fluorescent TIF2-GFP puncta apparent in the Ch2 images were used to define a translocation mask of the nucleoli within the Hoechst nuclei (Ch1 images, not shown) of U-2 OS cells (Fig. 2a). The TIF2-GFP puncta objects in Ch2 that had fluorescent intensities above a background threshold with suitable morphologic characteristics (width, length, and area) were classified by the image segmentation as nucleoli and used to generate a TIF2 mask (Fig. 2a, red and green masks). For images acquired on the IXU platform, these settings were generally applicable [4]; TIF2-GFP positive nucleoli were defined as objects with fluorescent intensities >750 gray levels over background, an approximate width of 4.0 μm, a minimum area of 5.0 μm^2, and a maximum area $\leq$ 150 μm^2. AR-RFP images in Ch3 were then segmented into an "Inner" nucleolus region using the masks generated from the detected TIF2-GFP positive nucleoli in Ch2 (Fig. 2a). The color of the nucleoli masks correspond to whether the correlation coefficient for colocalization of the AR-RFP signal within the TIF2-GFP masks was below (red) or above (green) a preset threshold (typically $\geq$0.25) (Fig. 2a). The TE image analysis module outputs quantitative data including: the selected TIF2-GFP positive nucleoli count in Ch2; the average fluorescent intensities of the TIF2-GFP positive nucleoli in Ch2; and the average and integrated fluorescent intensities of AR-RFP signals in Ch3 within TIF2-GFP positive masks (Fig. 2b). In U-2 OS cells expressing both biosensors, 30 min exposure to DHT at the indicated concentrations induced a concentration-dependent increase in the average inner intensity of the AR-RFP signal colocalized within TIF-2-GFP positive nucleoli (Fig. 2b). In the same cells, DHT exposure did not alter the average cell count per image determined from the number of Hoechst

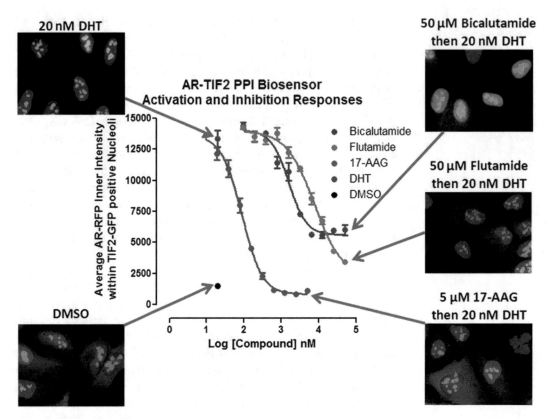

Fig. 3 Inhibition of DHT-induced AR-TIF2 PPI responses and corresponding AR-TIF2 distribution phenotypes. U-2 OS cells were co-infected with the AR-RFP and TIF2-GFP rAV biosensors, 2500 cells were seeded into the wells of 384-well assay plates, cultured overnight at 37 °C, 5% CO_2 and 95% humidity, and then exposed to compounds at the indicated concentrations for 1 h. Cells were then treated with 20 nM DHT for 30 min, fixed and stained with Hoechst, 20× images in three fluorescent channels were acquired on the IXU automated imaging platform, and the AR-TIF2 PPIs were quantified using the TE image analysis module as described above. The mean ± sd ($n = 3$) average inner intensity of AR-RFP within the TIF2-GFP positive nucleoli in cells exposed to the indicated concentrations of Bicalutamide (●), Flutamide (●), or 17-AAG (●) for 1 h and then treated with 20 nM DHT (●) or 0.5% DMSO (●) are presented. Experimental data from one of five independent experiments are presented. Representative 40× color composite images of the AR-TIF2 biosensor phenotypes of co-infected U-2 OS cells pre-exposed to 0.5% DMSO, 50 µM bicalutamide in 0.5% DMSO, 50 µM flutamide in 0.5% DMSO, or 5 µM 17-AAG in 0.5% DMSO for 1 h prior to 30 min treatment ± 20 nM DHT are shown

stained nuclei quantified in Ch1 images (Fig. 2b). In four independent concentration response experiments, DHT exhibited an EC_{50} of 5.33 ± 1.0 nM for the induction of AR-TIF2 PPIs, values that correlate closely with previously published DHT EC_{50} values determined in other AR cell-based assay formats [40].

Pre-exposure of biosensor expressing U-2 OS cells to the indicated concentrations of anti-androgens (flutamide and bicalutamide) or a Hsp 90 inhibitor (17-N-allylamino-17-demethoxygeldanamycin, 17-AAG) for 1 h prior to treatment with 20 nM DHT inhibited the colocalization of AR-RFP within TIF2-GFP positive nucleoli in a concentration-dependent manner (Fig. 3). 17-AAG completely inhibited the DHT-induced formation of AR-TIF2 PPIs, while flutamide and bicalutamide were only partial inhibitors. 17-AAG exhibited an IC_{50} of 88.5 ± 12.5 nM and flutamide exhibited an IC_{50} of 7.6 ± 2.4 μM for DHT-induced AR-TIF2 PPIs. In high-resolution 40× images acquired on the IXM, both 17-AAG and flutamide produced a cytoplasmic AR-RFP distribution phenotype similar to DMSO controls indicating that both of these compounds blocked DHT-induced AR-RFP nuclear translocation (Fig. 3). Bicalutamide exhibited an IC_{50} of 1.6 ± 0.4 μM for DHT-induced AR-TIF2 PPIs, but produced a diffuse AR-RFP nuclear distribution phenotype indicating that it prevented AR-RFP recruitment into TIF2-GFP positive nucleoli, presumably by inhibiting the PPIs between AR and TIF2. The AR-TIF2 PPIB assay was able to identify and quantify the concentration-dependent inhibitory effects of two FDA-approved anti-androgen CaP drugs, and an Hsp 90 inhibitor that prevents AR from assuming a folded state primed for high affinity interactions with androgenic ligands. High-resolution images of the AR-RFP distribution phenotype allowed us to distinguish between compounds that block AR translocation and those that block AR-TIF2 PPIs. The EC_{50} values for DHT and the IC_{50} values for 17-AAG, flutamide, and bicalutamide in the AR-TIF2 PPIB assay correlate closely with published values from other assay formats [40–42], indicating that the rAV expression system does not significantly alter the concentration responses of known AR modulators, and that the AR and TIF2 subdomains of the biosensors faithfully recapitulate the responses of the full length proteins.

The AR-TIF2 PPIB HCS assay can be configured to conduct screening campaigns in three distinct formats [4]: format 1, cells are exposed to compounds to identify novel AR agonists capable of inducing the formation of AR-TIF2 PPIs; format 2, cells are exposed to compounds prior to the addition of DHT to screen for compounds that block DHT-induced formation of AR-TIF2 PPIs; and format 3, cells are pretreated with DHT before compound addition to identify small molecules capable of disrupting pre-existing AR-TIF2 complexes (Fig. 1). In our experience, compound exposures of 3 h have typically been sufficient for us to measure compound effects.

2 Materials

1. AR agonist dihydrotestosterone (DHT).

2. Anti-androgen inhibitor test compounds, e.g., flutamide and bicalutamide.

3. Hsp 90 inhibitor, 17-N-allylamino-17-demethoxygeldanamycin (17-AAG).

4. Formaldehyde solution, 37%.

5. Dimethyl sulfoxide (DMSO) (99.9% high performance liquid chromatography-grade, under argon).

6. Hoechst 33342.

7. U-2 OS osteosarcoma cell line.

8. Culture medium: McCoy's 5A medium, 2 mM L-glutamine, 10% fetal bovine serum, and 100 U/mL penicillin and streptomycin.

9. Humidified incubator at 37 °C, 5% CO_2, and 95% humidity.

10. Dulbecco's Mg^{2+} and Ca^{2+} free phosphate-buffered saline (PBS).

11. Trypsin 0.25%, 1 g/L EDTA solution.

12. Recombinant adenovirus TIF2-GFP and AR-RFP Biosensors: Recombinant adenovirus expression constructs bearing the individual TIF2-GFP (TagGFP, Evrogen, Inc.) and AR-RFP (Tag RFP, Evrogen, Inc.) protein-protein interaction partners were obtained from Cyprotex.

13. 384-well collagen-I-coated barcoded assay microplates.

3   Methods

1. Aspirate spent tissue culture medium from U-2 OS cells in tissue culture flasks that are <70% confluent (*see* **Note 1**), wash cell monolayers 1× with PBS, and expose cells to trypsin-EDTA until they detached from the surface of the tissue culture flasks. Add serum containing tissue culture medium to neutralize the trypsin. Transfer the cell suspension to a 50 mL capped sterile centrifuge tube and centrifuge at 500 × g for 5 min to pellet the cells. Resuspend cells in serum containing tissue culture medium and count the number of trypan blue excluding viable cells using a hemocytometer.

2. Co-infect 1×10^7 U-2 OS cells with TIF2-GFP and AR-RFP adenovirus by incubating cells with the manufacturer's recommended volume of virus (*see* **Note 2**), typically 5 μL/10^6 cells, in 1.0 mL of culture medium for 1 h at 37 °C, 5% CO_2 in a humidified incubator with periodic inversion (every 10 min) to maintain cells in suspension.

3. Dilute co-infected cells to 6.25×10^4 cells/mL in culture media and 40 μL (2500 cells) were seeded in each well of a 384-well collagen-I-coated barcoded microplate using an automated bulk liquid handler (*see* **Note 3**).

4. Incubate assay plates overnight at 37 °C, 5% CO_2 in a humidified incubator (*see* **Note 4**).

5. As described above, the AR-TIF2 PPIB HCS assay can be configured to conduct screening campaigns in three distinct formats. Format 1: To identify novel AR agonists capable of inducing the formation of AR-TIF2 PPIs, use an automated liquid-handling device to transfer 5 μL of diluted compounds or plate controls (DHT 20 nM final, or DMSO 0.2% final) to appropriate wells for a final screening concentration of 20 μM and incubate plates at 37 °C, 5% CO_2 in a humidified incubator for 30 min (see Notes 5–7). Format 2: To identify compounds that block DHT-induced formation of AR-TIF2 PPIs, use an automated liquid-handling device to transfer 5 μL of diluted compounds to appropriate wells for a final screening concentration of 20 μM, or 5 μL of diluted DMSO (0.2% final) to plate controls, and incubate plates at 37 °C, 5% CO_2 in a humidified incubator for 1–3 h. After the appropriate compound exposure period use an automated liquid-handling device to transfer 5 μL of DHT (20 nM final) to compound treated wells and maximum plate controls, minimum plate controls receive media alone, and incubate assay plates at 37 °C, 5% CO_2 in a humidified incubator for 30 min (see Notes 5–7). Format 3: To identify small molecules capable of disrupting pre-existing AR-TIF2 complexes, use an automated liquid-handling device to transfer 5 μL of DHT (20 nM final) to compound treated wells and maximum plate controls, minimum plate controls receive media alone, and incubate assay plates at 37 °C, 5% CO_2 in a humidified incubator for 30 min. After 30 min use an automated liquid-handling device to transfer 5 μL of diluted compounds to appropriate wells for a final screening concentration of 20 μM or 5 μL of diluted DMSO (0.2% final) to plate controls and incubate plates at 37 °C, 5% CO_2 in a humidified incubator for 1–3 h (see Notes 5–7).

6. Fix and stain the nuclei of cells by the adding 50 μL of pre-warmed (37 °C) 7.4% formaldehyde (final is 3.7%) and 2 μg/mL Hoechst 33342 in PBS using a liquid handler and incubate at room temperature for 30 min (see Note 8).

7. Aspirate the liquid and wash the plates twice with 75 μL of PBS using a liquid handler (see Note 9). Seal with adhesive aluminum plate seals with the last 75 μL wash of PBS in place.

8. Acquire fluorescent images in three independent channels on an automated imaging platform (see Note 10) (e.g., ImageXpress Ultra or Micro automated imaging platforms) (Figs. 1, 2 and 3). The ImageXpress Ultra (IXU) platform (Molecular Devices LLC, Sunnyvale, CA) is a fully integrated point-scanning confocal automated imaging platform configured with four independent solid-state lasers providing four excitation wavelengths of 405, 488, 561, and 635 nm. The IXU was equipped with a Quad filter cube providing emission ranges of

417–477 nm, 496–580 nm, 553–613 nm, and 645–725 nm and four independent photomultiplier tubes (PMTs) each dedicated to a single detection wavelength. The IXU utilizes a dedicated high-speed infra-red laser auto-focus system, has a 4-position automated objective changer with air objectives ($10\times$, $20\times$, $40\times$, and $60\times$), and the detection pinhole diameter of the confocal optics was configurable in the software. For the AR-TIF2 HCS assay the IXU was set up to sequentially acquire two images per well using a $20\times/0.45$ NA ELWD objective in each of three fluorescent channels. In the Hoechst channel (Ch1) the 405 laser was set at 10% power and the PMT gain was 550. In the TIF2-GFP channel (Ch2) the 488 laser was set at 10% power, and the PMT gain was 625. In the AR-RFP channel (Ch3) the 561 laser was set at 10% power, and the PMT gain was 625. On average, the IXU scanned a single 384-well plate, two images per channel, in 90 min using these settings. The ImageXpress Micro (IXM) is an automated field-based high content imaging platform integrated with the MetaXpress software. The IXM optical drive includes a 300 W Xenon lamp broad spectrum white light source and $2/3''$ chip Cooled CCD Camera and optical train for standard field of view imaging and an IXM transmitted light option with phase contrast. The IXM is equipped with a $4\times$ Plan Apo 0.20 NA objective, a $10\times$ Plan Fluor 0.3 NA objective, a $20\times$ Ph1 Plan Fluor ELWD DM objective, a $20\times$ S Plan Fluor ELWD 0.45 NA objective, a $40\times$, S Plan Fluor ELWD 0.60 NA objective, and a single slide holder adaptor. The IXM is equipped with the following ZPS filter sets; DAPI, FITC/ALEXA 488, CY3, CY5, and Texas Red.

9. Analyze the images of the three fluorescent channels (Hoechst Ch1, TIF2-GFP Ch2, and AR-RFP Ch3) of the AR-TIF2 PPIB using the translocation enhanced image analysis module of the MetaXpress software (Molecular Devices LLC, Sunnyvale, CA), as described above (Fig. 2) (*see* **Note 11**). The Ch3 average inner fluorescent intensity of AR-RFP within the TIF2-GFP positive nucleoli masks generated from the Ch2 images is utilized to quantify the formation or disruption of AR-TIF2 PPIs (Fig. 2b).

4   Notes

1. Typically better responses are obtained when the AR and TIF2 adenovirus biosensors are used to co-infect U-2 OS cells harvested from tissue culture flasks that are ≤70% confluent.

2. The optimal volume of each lot of recombinant adenovirus biosensor per 10^6 U-2 OS cells is determined empirically in virus titration experiments [4]. Increasing amounts of virus are

incubated with the same number of cells and then the levels of biosensor expression and % of cells that are co-infected are determined on the HCS platform after 24 h in culture. Performing infections in cells suspended in a low volume of media combined with periodic inversion (every 10 min) to maintain cells in suspension enhances both the rate of infection and expression levels. In addition to the U-2 OS osteosarcoma cell line, the adenovirus biosensors have been used to co-infect several prostate cancer cell lines; DU-145, PC-3, LNCaP, C4-2, and 22Rv1. Compared to U-2 OS cells, the infection of CaP cell lines required significantly more (>10-fold) adenovirus and the rates of co-infection with both viruses for DU-145, LNCaP, and 22RV1 cell lines were typically <20%, compared to >90% in U-2 OS cells. Since the AR-TIF2 PPIB assay requires co-expression of both biosensors, the much lower co-infection rates in these CaP cell lines are a serious limitation. In addition, LNCaP cells are poorly adherent and were not retained after cell-washing procedures. We successfully titrated the adenoviruses high enough to achieve >50% co-infection in the PC-3 cell line and demonstrated that the assay can be adapted to this CaP cell background.

3. Determining the optimal cell seeding density is a critical assay development parameter for all cell-based assays, including HCS assays. The goal is to minimize the cell culture burden while ensuring that sufficient cells are captured per image to give statistical significance to the image analysis parameters of interest. Typically, variability is inversely related to the number of cells captured and analyzed; in HCS assays variability generally increases as the number of cells analyzed decreases.

4. The optimal length of time in culture post viral infection should be determined empirically for each adenovirus biosensor. In general, we have found 24 h post infection to be optimal for most recombinant adenovirus biosensor constructs.

5. Agonist concentration response and time-course experiments should be conducted to determine the optimal DHT concentration and length of exposure required to induce AR-TIF2 PPIs, and to determine how stable AR-TIF2 PPIs are over time. These data are used to select a DHT concentration and treatment time for the maximum plate controls to provide a robust and reproducible assay signal window, relative to DMSO minimum plate controls, with acceptable signal-to-background ratios, typically $\geq$3-fold, and Z'-factor coefficients $\geq$0.5.

6. Determining the DMSO tolerance is a critical assay development step for all cell-based assays including HCS assays. DMSO has two notable effects on HCS assays [2–4, 13–18].

At DMSO concentrations $\geq 5\%$ there is a significant cell loss due to cytotoxicity and/or reduced cell adherence. At DMSO concentrations $>1\%$ but $<5\%$, cell morphologies can be significantly altered from a well spread and attached morphology to a more rounded loosely attached morphology that interferes with the ability of the image analysis algorithm to segment images into distinct cytoplasm and nuclear regions. The DMSO tolerance of the HCS assay and the compound library stock concentration are the major factors that influence the selection of the compound concentrations for primary screening, confirmation, and follow-up studies.

7. The optimal compound exposure period is a critical assay development parameter for all cell-based assays that should be empirically determined for each new assay. Longer compound exposure periods can result in elevated cytotoxicity levels that may significantly hinder the ability to make reliable measurements, or may obscure target-based activity.

8. Determining appropriate cell fixation and nuclear staining conditions is a critical assay development parameter for all end point HCS assays [2–4, 13–18]. Combining cell fixation with Hoechst nuclear staining in a single procedure saves time and reduces the number of steps in the protocol. Cell fixation is important because the scanning and acquisition of a full 96-well or 384-well assay plate on an automated imaging platform depends upon the number of fluorescent channels and images captured per well and the focusing options selected. Scanning times may range anywhere from 10–15 min to 2–3 h per plate depending upon the complexity of the image acquisition procedure and a time-dependent drift in the assay signal may occur in unfixed samples.

9. To control and reduce environmental exposure levels from formaldehyde and for long-term storage of fixed assay plates, we recommend using an automated dispensing and plate washing platform for dispensing formaldehyde and for the aspiration and washing steps.

10. Although we have described the image acquisition process on the ImageXpress Ultra and Micro platforms, most automated imaging platforms designed for HCS with similar light sources, objective lens magnification, and detectors should be capable of capturing images.

11. Although we have described the translocation enhanced image analysis module of the MetaXpress software provided with the ImageXpress Ultra and Micro platforms (Fig. 2), most automated imaging platforms designed for HCS provide similar image analysis algorithms that should be capable of analyzing these images. Alternatively, images could be analyzed using

third-party image analysis software such as the open-source CellProfiler image analysis software.

Acknowledgments

The studies reported herein were funded by grant support from the National Institutes of Health (NIH); R21NS073889 Johnston (PI) from the NINDS, and R01CA160423 and RO1CA183882 Johnston (PI) from the NCI.

References

1. Colas P (2008) High-throughput screening assays to discover small-molecule inhibitors of protein interactions. Curr Drug Discov Technol 5:190–199

2. Dudgeon D, Shinde SN, Shun TY, Lazo JS, Strock CJ, Giuliano KA, Taylor DL, Johnston PA, Johnston PA (2010) Characterization and optimization of a novel protein-protein interaction biosensor HCS assay to identify disruptors of the interactions between p53 and hDM2. Assay Drug Dev Technol 8:437–458

3. Dudgeon D, Shinde SN, Hua Y, Shun TY, Lazo JS, Strock CJ, Giuliano KA, Taylor DL, Johnston PA, Johnston PA (2010) Implementation of a 220,000 compound HCS campaign to identify disruptors of the interaction between p53 and hDM2, and characterization of the confirmed hits. J Biomol Screen 15:152–174

4. Hua Y, Shun TY, Strock CJ, Johnston PA (2014) High-content positional biosensor screening assay for compounds to prevent or disrupt androgen receptor and transcriptional intermediary factor 2 protein-protein interactions. Assay Drug Dev Technol 12:395–418

5. Lalonde S, Ehrhardt DW, Loqué D, Chen J, Rhee SY, Frommer WB (2008) Molecular and cellular approaches for the detection of protein-protein interactions: latest techniques and current limitations. Plant J 53:610–635

6. Pagliaro L, Felding J, Audouze K, Nielsen SJ, Terry RB, Krog-Jensen C, Butcher S (2004) Emerging classes of protein-protein interaction inhibitors and new tools for their development. Curr Opin Chem Biol 8:442–449

7. Reilly M, Cunningham KA, Natarajan A (2009) Protein-protein interactions as therapeutic targets in neuropsychopharmacology. Neuropsychopharmacology 34:247–248

8. Wells J, McClendon CL (2007) Reaching for high-hanging fruit in drug discovery at protein-protein interfaces. Nature 450:1001–1009

9. Koehler A (2010) A complex task? Direct modulation of transcription factors with small molecules. Curr Opin Chem Biol 14:331–340

10. Kumar S, Saradhi M, Chaturvedi NK, Tyagi RK (2006) Intracellular localization and nucleocytoplasmic trafficking of steroid receptors: an overview. Mol Cell Endocrinol 246:147–156

11. Stauber R, Afonina E, Gulnik S, Erickson J, Pavlakis GN (1998) Analysis of intracellular trafficking and interactions of cytoplasmic HIV-1 Rev mutants in living cells. Virology 251:38–48

12. Stauber R (2002) Analysis of nucleocytoplasmic transport using green fluorescent protein. Methods Mol Biol 183:181–198

13. Johnston P, Sen M, Hua Y, Camarco D, Shun TY, Lazo JS, Grandis JR (2014) High-content pSTAT3/1 imaging assays to screen for selective inhibitors of STAT3 pathway activation in head and neck cancer cell lines. Assay Drug Dev Technol 12:55–79

14. Johnston PA, Shinde SN, Hua Y, Shun TY, Lazo JS, Day BW (2012) Development and validation of a high-content screening assay to identify inhibitors of cytoplasmic Dynein-mediated transport of glucocorticoid receptor to the nucleus. Assay Drug Dev Technol 10:432–456

15. Nickischer D, Laethem C, Trask OJ, Williams RG, Kandasamy R, Johnston PA, Johnston PA (2006) Development and implementation of three mitogen-activated protein kinase (MAPK) signaling pathway imaging assays to provide MAPK module selectivity profiling for kinase inhibitors: MK2-EGFP translocation, c-Jun, and ERK activation. Methods Enzymol 414:389–418

16. Trask O, Baker A, Williams RG, Nickischer D, Kandasamy R, Laethem C, Johnston PA, Johnston PA (2006) Assay development and case

history of a 32K-biased library high-content MK2-EGFP translocation screen to identify p38 mitogen-activated protein kinase inhibitors on the ArrayScan 3.1 imaging platform. Methods Enzymol 414:419–439

17. Trask O, Nickischer D, Burton A, Williams RG, Kandasamy RA, Johnston PA, Johnston PA (2009) High-throughput automated confocal microscopy imaging screen of a kinase-focused library to identify p38 mitogen-activated protein kinase inhibitors using the GE InCell 3000 analyzer. Methods Mol Biol 565:159–186

18. Williams R, Kandasamy R, Nickischer D, Trask OJ, Laethem C, Johnston PA, Johnston PA (2006) Generation and characterization of a stable MK2-EGFP cell line and subsequent development of a high-content imaging assay on the Cellomics ArrayScan platform to screen for p38 mitogen-activated protein kinase inhibitors. Methods Enzymol 414:364–389

19. Knauer S, Moodt S, Berg T, Liebel U, Pepperkok R, Stauber RH (2005) Translocation biosensors to study signal-specific nucleocytoplasmic transport, protease activity and protein-protein interactions. Traffic 6:594–606

20. Knauer S, Stauber RH (2005) Development of an autofluorescent translocation biosensor system to investigate protein-protein interactions in living cells. Anal Chem 77:4815–4820

21. Burd C, Morey LM, Knudsen KE (2006) Androgen receptor corepressors and prostate cancer. Endocr Relat Cancer 13:979–994

22. Chmelar R, Buchanan G, Need EM, Tilley W, Greenberg NM (2006) Androgen receptor coregulators and their involvement in the development and progression of prostate cancer. Int J Cancer 120:719–733

23. Culig Z, Hobisch A, Bartsch G, Klocker H (2000) Androgen receptor–an update of mechanisms of action in prostate cancer. Urol Res 28:211–219

24. Culig Z, Klocker H, Bartsch G, Hobisch A (2002) Androgen receptors in prostate cancer. Endocr Relat Cancer 9:155–170

25. Gregory C, Johnson RT Jr, Mohler JL, French FS, Wilson EM (2001) Androgen receptor stabilization in recurrent prostate cancer is associated with hypersensitivity to low androgen. Cancer Res 61:2892–2898

26. Evans RM (1988) The steroid and thyroid hormone receptor family. Science 240:889–895

27. McKenna N, O'Malley BW (2002) Minireview: nuclear receptor coactivators–an update. Endocrinology 143:2461–2465

28. McKenna N, O'Malley BW (2002) Combinatorial control of gene expression by nuclear receptors and coregulators. Cell 108:465–474

29. Culig Z, Santer FR (2012) Androgen receptor co-activators in the regulation of cellular events in prostate cancer. World J Urol 30:297–302

30. Culig Z, Santer FR (2013) Molecular aspects of androgenic signaling and possible targets for therapeutic intervention in prostate cancer. Steroids 78(9):851–859

31. Fujimoto N, Miyamoto H, Mizokami A, Harada S, Nomura M, Ueta Y, Sasaguri T, Matsumoto T (2007) Prostate cancer cells increase androgen sensitivity by increase in nuclear androgen receptor and androgen receptor coactivators; a possible mechanism of hormone-resistance of prostate cancer cells. Cancer Investig 25:32–37

32. Agoulnik I, Vaid A, Nakka M, Alvarado M, Bingman WE 3rd, Erdem H, Frolov A, Smith CL, Ayala GE, Ittmann MM, Weigel NL (2006) Androgens modulate expression of transcription intermediary factor 2, an androgen receptor coactivator whose expression level correlates with early biochemical recurrence in prostate cancer. Cancer Res 66:10594–10602

33. Feng S, Tang Q, Sun M, Chun JY, Evans CP, Gao AC (2009) Interleukin-6 increases prostate cancer cells resistance to bicalutamide via TIF2. Mol Cancer Ther 8:665–671

34. Nakka M, Agoulnik IU, Weigel NL (2013) Targeted disruption of the p160 coactivator interface of androgen receptor (AR) selectively inhibits AR activity in both androgen-dependent and castration-resistant AR-expressing prostate cancer cells. Int J Biochem Cell Biol 45:763–772

35. Shi X, Xue L, Zou JX, Gandour-Edwards R, Chen H, deVere White RW (2008) Prolonged androgen receptor loading onto chromatin and the efficient recruitment of p160 coactivators contribute to androgen-independent growth of prostate cancer cells. Prostate 68:1816–1826

36. Xu J, Li Q (2003) Review of the in vivo functions of the p160 steroid receptor coactivator family. Mol Endocrinol 17:1681–1692

37. Bebermeier J, Brooks JD, DePrimo SE, Werner R, Deppe U, Demeter J, Hiort O, Holterhus PM (2006) Cell-line and tissue-specific signatures of androgen receptor-coregulator transcription. J Mol Med 84:919–931

38. Heemers H, Schmidt LJ, Kidd E, Raclaw KA, Regan KM, Tindall DJ (2010) Differential regulation of steroid nuclear receptor coregulator expression between normal and neoplastic prostate epithelial cells. Prostate 70:959–970

39. Huang P, Chandra V, Rastinejad F (2010) Structural overview of the nuclear receptor superfamily: insights into physiology and therapeutics. Annu Rev Physiol 72:247–272

40. Sonneveld E, Jansen HJ, Riteco JA, Brouwer A, van der Burg B (2005) Development of androgen- and estrogen-responsive bioassays, members of a panel of human cell line-based highly selective steroid-responsive bioassays. Toxicol Sci 83:136–148

41. Festuccia C, Gravina GL, Angelucci A, Millimaggi D, Muzi P, Vicentini C, Bologna M (2005) Additive antitumor effects of the epidermal growth factor receptor tyrosine kinase inhibitor, gefitinib (Iressa), and the nonsteroidal antiandrogen, bicalutamide (Casodex), in prostate cancer cells in vitro. Int J Cancer 115:630–640

42. Luo S, Martel C, LeBlanc G, Candas B, Singh SM, Labrie C, Simard J, Belanger A, Labrie F (1996) Relative potencies of flutamide and casodex: preclinical studies. Endocr Relat Cancer 3:229–241

Chapter 14

High Content Imaging Assays for IL-6-Induced STAT3 Pathway Activation in Head and Neck Cancer Cell Lines

Paul A. Johnston, Malabika Sen, Yun Hua, Daniel P. Camarco, Tong Ying Shun, John S. Lazo, and Jennifer R. Grandis

Abstract

In the canonical STAT3 signaling pathway, IL-6 receptor engagement leads to the recruitment of latent STAT3 to the activated IL-6 complex and the associated Janus kinase (JAK) phosphorylates STAT3 at Y705. pSTAT3-Y705 dimers traffic into the nucleus and bind to specific DNA response elements in the promoters of target genes to regulate their transcription. However, IL-6 receptor activation induces the phosphorylation of both the Y705 and S727 residues of STAT3, and S727 phosphorylation is required to achieve maximal STAT3 transcriptional activity. STAT3 continuously shuttles between the nucleus and cytoplasm and maintains a prominent nuclear presence that is independent of Y705 phosphorylation. The constitutive nuclear entry of un-phosphorylated STAT3 (U-STAT3) drives expression of a second round of genes by a mechanism distinct from that used by pSTAT3-Y705 dimers. The abnormally elevated levels of U-STAT3 produced by the constitutive activation of pSTAT3-Y705 observed in many tumors drive the expression of an additional set of pSTAT3-independent genes that contribute to tumorigenesis. In this chapter, we describe the HCS assay methods to measure IL-6-induced STAT3 signaling pathway activation in head and neck tumor cell lines as revealed by the expression and subcellular distribution of pSTAT3-Y705, pSTAT3-S727, and U-STAT3. Only the larger dynamic range provided by the pSTAT3-Y705 antibody would be robust and reproducible enough for screening.

Key words STAT3 pathway activation, Head and neck cancer, High content screening, Imaging, Image analysis

1 Introduction

The Signal Transducer and Activator of Transcription 3 (STAT3) signaling pathway is hyper-activated in many cancers [1–5]. Activation of the canonical STAT3 pathway involves the phosphorylation of a conserved tyrosine residue (Y705) of the C-terminal transactivation domain (TAD) downstream of growth factor, cytokine, or G-protein-coupled (GPCRs) receptor ligand binding [5–13] (Fig. 1). Src family kinases (SFKs) can also phosphorylate STAT3-Y705 either directly or downstream of receptor tyrosine kinase or

Paul A. Johnston and Oscar J. Trask (eds.), *High Content Screening: A Powerful Approach to Systems Cell Biology and Phenotypic Drug Discovery*, Methods in Molecular Biology, vol. 1683, DOI 10.1007/978-1-4939-7357-6_14,
© Springer Science+Business Media LLC 2018

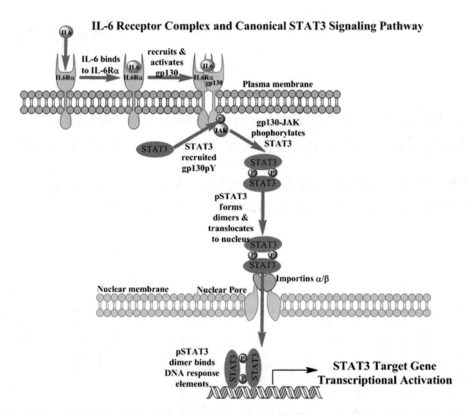

Fig. 1 Interleukin-6 receptor complex activation and canonical STAT3 signaling pathway. Interleukin-6 (IL-6) first engages and binds to IL-6-receptor-α subunit (IL-6Rα, gp80) in the plasma membrane of cells and the IL-6-IL-6Rα complex then recruits the gp130 signaling subunit. The gp130 subunit and its constitutively associated Janus kinases (JAKs) are rapidly activated as indicated by the auto-phosphorylation and activation of the JAK together with phosphorylation of a number of tyrosine residues on gp130 (pYs 767, 814, 905 and 915). Latent STAT3 in the cytoplasm is recruited to specific pY docking sites on the gp130 subunits of IL-6-activated receptor complexes through interactions with the STAT3 SH domain. The associated JAK of the activated IL-6 receptor complex phosphorylates STAT3 at pY705. Reciprocal interactions between SH-2 domains of pSTAT3 monomers and the pY705 residue of another pSTAT3 partner lead to dimerization and translocation to the nucleus. Large protein complexes like activated pSTAT3-Y705 dimers may only enter the nucleus through the nuclear pore complex in a process facilitated by importin α and β proteins. pSTAT3-Y705 dimers interact with importins α5 and α7 and in conjunction with importin β these proteins mediate the entry of pSTAT3-Y705 through the nuclear pore. Inside the nucleus pSTAT3 dimers bind specific DNA response elements in the promoters of target genes to regulate their transcription

GPCR engagement [5, 7, 8, 10–12, 14]. After Y705 phosphorylation, pSTAT3-Y705 dimers traffic into the nucleus where they bind to specific DNA response elements in the promoters of target genes to regulate transcription [2, 4, 5, 8–10] (Fig. 1). Elevated pSTAT3-Y705 levels are often detected in tumor cell lines and are frequently associated with a poor clinical prognosis when observed in human tumor biopsy samples [1, 2, 4, 6, 10, 15]. Several STAT3 signaling pathway components that are commonly altered in many cancers result in elevated pSTAT3-Y705 expression levels; overexpression/

amplification of EGFRs, receptor tyrosine kinase (RTK) mutations that result in constitutive activation, overexpression/amplification of SFKs, and hyper-activating mutations of Janus kinases (JAKs) [4–6, 8–11, 13]. Increased levels of growth factors and cytokines, such as transforming growth factor alpha (TGFα) or interleukin-6 (IL-6), may also lead to inappropriate or sustained activation of the STAT3 pathway that in turn promotes tumor development and/or survival [1, 5, 10, 15, 16].

In addition to the canonical STAT3 signaling pathway depicted in Fig. 1, cytokine or growth factor stimulation also induces phosphorylation of a serine 727 (S727) residue in a conserved (LPMSP) sequence of the STAT3 C-terminal TAD [17–20]. The differential splice variant STAT3β lacks the C-terminal region that contains the LPMSP motif and exhibits markedly reduced transcriptional activity compared to wild-type STAT3α [18–20]. Overexpression of a STAT3 mutant in which S727 was mutated to A727 did not alter ligand-induced DNA binding, but dramatically reduced its transcriptional activity relative to wild-type STAT3 [18–20]. In contrast, overexpression of a STAT3 mutant in which S727 was mutated to an D727 did not affect either IL-6-induced STAT3 DNA binding or transcriptional activation, but transcriptional activation by this mutant was independent of Rac-1-mediated signal transduction and the MEKK-1 and MKK-4 kinases [18]. IL-6 receptor activation induces the phosphorylation of both the Y705 and S727 residues of STAT3 and produces a strong association with the p300 transcriptional coactivator [18]. Overexpression of p300 enhances the transcriptional activity of wild-type STAT3α but not the STAT3β isoform or the STAT3 S727A mutant, and the STAT3 S727D mutant forms a strong constitutive association with p300 [18]. Collectively, these data suggest that the p300 coactivator may be involved in the recruitment and assembly of the basal transcriptional machinery by pSTAT3-S727. While Y705 phosphorylation is obligatory for STAT3 dimerization, nuclear translocation, and DNA binding, S727 phosphorylation is required to achieve maximal STAT3 transcriptional activity [17–20].

The transcriptional activity of STAT3 also depends upon its ability to gain entry into the nucleus and to bind to specific DNA sequences [21–23]. The passage of large molecules between the cytoplasm and the nucleus is restricted and only proteins bearing appropriate nuclear localization signals (NLS) and/or nuclear export signals (NES) can be escorted through the nuclear pore complexes by the importin and exportin family of cargo transporters [21–23]. Unlike other members of the STAT family, STAT3 continuously shuttles between the nucleus and the cytoplasm and maintains a prominent nuclear presence that is independent of Y705 phosphorylation [22, 23]. While the constitutive nuclear entry of un-phosphorylated STAT3 (U-STAT3) is mediated by importin-α3 [22, 23], the nuclear-cytoplasmic shuttling of v-Src-activated pSTAT3-Y705 is mediated by importins α5 and α7 [21]. The STAT3 gene itself is one of the target genes activated by the IL-6-induced formation of pSTAT3-Y705 and the resulting increase in U-STAT3 drives expression of a second round of genes (*RANTES, IL-6, IL-8, MET,* and MRAS) that do not respond to pSTAT3-Y705 [24–26]. U-STAT3 drives gene expression by a mechanism distinct from that used by pSTAT3-Y705 dimers [24–26]. U-STAT3 competes with IκB for the un-phosphorylated NFκB (p65/p50, U-NFκB)) and the STAT3 NLS of the resulting U-STAT3::U-NFκB complexes promotes their accumulation in the nucleus where they active a subset of κB-responsive target genes [24–26]. The abnormally elevated levels of U-STAT3 produced by the constitutive activation of pSTAT3-Y705 observed in many tumors drive the expression of an additional set of pSTAT3-independent genes that contribute to tumorigenesis [24–26].

Head and neck cancer (HNC) is the eighth leading cause of cancer worldwide with a projected incidence of 540,000 new cases and 271,000 deaths per annum [27–29]. The front line therapies for HNC are surgical resection and chemo-radiotherapy. Surgical therapy can be disfiguring and there can be significant deleterious effects on swallowing, speech, and appearance. Radiation and chemotherapy treatment has produced limited improvement in prognosis and the 5-year survival rate for HNC has remained at 50% for over 30 years. There is, therefore, a need for new effective therapies. Elevated levels of activated STAT3 are frequently detected in HNC tumor samples and in head and neck squamous cell carcinoma cell lines used in mouse xenograft models that respond to STAT3 inhibition [4, 5, 7, 10, 15, 30, 31]. STAT3 constitutively activated in an EGFR-independent manner by the autocrine/paracrine activation of the IL-6 receptor complex in HNSCC cells provides growth and survival benefits and may contribute to their resistance to EGFR-targeted therapies [16]. We recently described the development, optimization, and validation of high content imaging (HCS) assays to measure IL-6-induced pSTAT3-Y705 and interferon-gamma (IFNγ)-induced pSTAT1-Y701 levels in HNC cell lines [32], and used these phenotypic assays to screen for compounds that selectively inhibited STAT3 but not STAT1 pathway activation in HNSCC cells [33]. The IL-6-induced pSTAT3-Y705 and IFNγ-induced pSTAT1-Y701 HCS assays have subsequently been used to support the chemical lead optimization of hits that selectively inhibited STAT3 but not STAT1 pathway activation in HNSCC cells [33–35]. In this chapter, we describe the HCS assay method to measure IL-6-induced pSTAT3-Y705 activation, and because STAT3 S727 phosphorylation is required to achieve maximal STAT3 transcriptional activity and elevated levels of U-STAT3 drive the expression of genes that may contribute to tumorigenesis, we also describe methods to measure the expression and subcellular distribution of pSTAT3-S727 and U-STAT3.

Cal33 HNSCC cells that had been seeded into 384-well plates in serum containing medium were cultured overnight and then serum starved for 24 h prior to a 15 min treatment ±50 ng/mL IL-6. The cells were then fixed in 3.7% paraformaldehyde containing Hoechst, permeabilized with methanol, and then stained with the indicated primary anti-STAT3 antibodies and a FITC-conjugated anti-species secondary antibody. We then utilized the ImageXpress Ultra (IXU) confocal automated imaging platform to sequentially acquire images for the Hoechst (Ch 1) and FITC (Ch 2) fluorescent channels using a 20× 0.45 NA ELWD objective (Fig. 2). Fluorescent images of Hoechst stained nuclei from Cal33 cells acquired on the IXU were unaffected by treatment with IL-6 (Fig. 2a). In untreated Cal33 cells immuno-stained with a rabbit polyclonal antibody that recognizes total-STAT3 independently of its phosphorylation status, uniform STAT3 staining was observed throughout the cytoplasm and nuclear compartments in both gray-scale and color composite images (Fig. 2a, b). In Cal33 cells that had been exposed to IL-6 however, there was an apparent increase in STAT3 staining in the nuclear compartment even though STAT3 was also still apparent in the cytoplasm (Fig. 2a, b). In Cal33 cells immuno-stained with mouse monoclonal antibodies that recognize pSTAT3-Y705 or pSTAT3-S727, the indirect immuno-fluorescent staining was clearly higher in images acquired of Cal33 cells treated with IL-6 compared to those of un-stimulated cells (Fig. 2b). The color composite overlays of the Ch 1 and Ch 2 fluorescent images indicate that the increased immuno-fluorescent pSTAT3-Y705 and pSTAT3-S727 staining are predominantly localized within the nuclei of IL-6-treated Cal33 cells (Fig. 2b).

To extract and analyze quantitative data from these digital images, we used the translocation enhanced (TE) image analysis module of the MetaXpress software on the IXU platform (Fig. 3). Hoechst DNA-stained objects in Ch 1 that exhibited fluorescent intensities above a background threshold and had suitable morphology (width, length, and area) characteristics were classified by the TE image segmentation as nuclei and were used to create nuclear masks for each cell (Fig. 3a). The Ch 1 nuclear mask was then eroded 1 μm in from the edge of the detected nucleus and the reduced Ch 2 "inner" mask was established to quantify the target STAT3 Ch 2 fluorescence within the nucleus (Fig. 3a, light green and red). A Ch 2 "outer" cytoplasm mask was then established 1 μm out from the edge of the Ch 1 detected nucleus and the width was set at 3 μm, to cover a region of the cytoplasm within the cell boundary (Fig. 3a, dark green and red). The Ch 2 "outer" mask was used to quantify the amount of target STAT3 fluorescence within this region of the cytoplasm (Fig. 3a). The TE image analysis module outputs quantitative individual cell or well-averaged data including the selected object or cell counts per image in Ch 1 (Fig. 3b), and the average fluorescent intensities of the STAT3 Ch

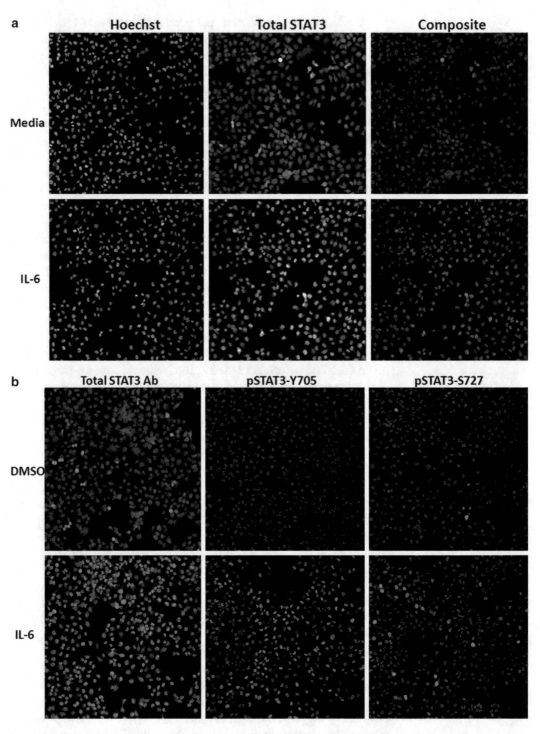

Fig. 2 Images of IL-6-induced STAT3 signaling pathway activation. (**a**) Grayscale and color composite images of total-STAT3 in Cal33 Cells ± IL-6 treatment. 20× grayscale images of Hoechst 33342 stained nuclei (Ch 1) and total-STAT3 staining (Ch 2) and the corresponding color composite images of non-stimulated and IL-6-treated (50 ng/mL,15 min) Cal33 HNSCC cells. (**b**) Color composite images of total STAT3, pSTAT3-Y705, and

2 signal in the nucleus (Fig. 3c). While exposure to 50 ng/mL of IL-6 for 15 min had no significant effect on the cell counts per image (Fig. 3b), the average fluorescent intensities of nuclear total-STAT3, pSTAT3-Y705, and pSTAT3-S727 were higher in IL-6-treated Cal33 cells compared to untreated cells (Figs. 2b and 3c). Although IL-6 treatment induced increases in the average fluorescent intensities of total-STAT3 and pSTAT3-S727 in the nuclei of Cal33 cells by 50% and 45% respectively, the corresponding increase in nuclear pSTAT3-Y705 was much greater, typically >5-fold higher than in untreated cells (Figs. 2b and 3c). Based on the IL-6-induced assay signal windows of the three STAT3 antibodies (Fig. 3c), only the larger dynamic range provided by the pSTAT3-Y705 antibody would be robust and reproducible enough to use as a screening assay for inhibitors of STAT3 pathway activation.

2   Materials

1. Methanol.

2. Formaldehyde.

3. Triton X-100.

4. Tween 20.

5. Hoechst 33342.

6. Dimethyl sulfoxide (DMSO) (99.9% high performance liquid chromatography-grade, under argon).

7. Recombinant human interleukin-6 (IL-6).

8. Mouse monoclonal anti-pSTAT3-Y705 primary antibody.

9. Mouse monoclonal ant-pSTAT3-S727 primary antibody.

Fig. 2 (continued) pSTAT3-S727 in Cal33 cells ± IL-6 treatment. 20× color composite images of Hoechst 33342 stained nuclei (Ch 1) and total STAT3, pSTAT3-Y705, and pSTAT3-S727 staining (Ch 2) of non-stimulated and IL-6-treated (50 ng/mL,15 min) Cal33 HNSCC cells. The ImageXpress Ultra (IXU) platform (Molecular Devices LLC, Sunnyvale, CA) is a fully integrated point-scanning confocal automated imaging platform configured with four independent solid-state lasers providing four excitation wavelengths of 405, 488, 561, and 635 nm. The IXU was equipped with a Quad filter cube providing emission ranges of 417–477, 496–580, 553–613, and 645–725 nm and four independent photomultiplier tubes (PMTs) each dedicated to a single detection wavelength. The IXU utilizes a dedicated high-speed infra-red laser auto-focus system, has a 4-position automated objective changer with air objectives (10×, 20×, 40×, and 60×), and the detection pinhole diameter of the confocal optics was configurable in the software. For the STAT3 HCS assays the IXU was set up to acquire two images using a 20× 0.45 NA ELWD objective in each of two fluorescent channels that were acquired sequentially. The Hoechst channel laser autofocus Z-offset was −6.98 μm, the 405 laser was set at 10% power, and the PMT gain was 550. The pSTAT3-Y705 FITC channel Z-offset from W1 (the Hoechst channel) was 12.96 μm, the 488 laser was set at 10% power, and the PMT gain was 625. On average, the IXU scanned a single 384-well plate, two images per channel, in 90 min using these settings. Similar settings were used to acquire images of total-STAT3 and pSTAT3-S727 immuno-staining

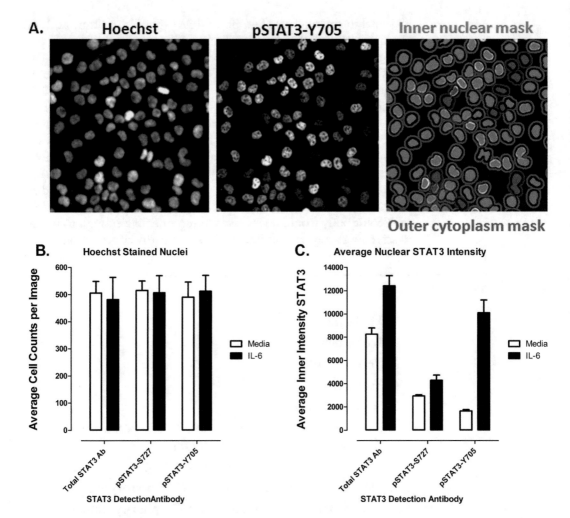

Fig. 3 Translocation enhanced image analysis module. (**a**) Translocation enhanced image segmentation. Hoechst 33342 was used to stain and identify the nuclei of Cal33 cells, and this fluorescent signal in the images of channel 1 (Ch 1) was used by the translocation enhanced (TE) image analysis module to define an "inner" nuclear mask and an "outer" cytoplasm mask in channel 2 (Ch 2). Hoechst stained objects in Ch 1 that exhibited fluorescent intensities above an established background threshold intensity value and had the suitable size (width, length, and area) characteristics were identified and classified by the image segmentation as nuclei. For Hoechst stained Cal33 cells the following settings typically proved effective; objects defined as nuclei had to exhibit fluorescent intensities >1500 gray levels over background, and they had to have an approximate width of 10 μm with a minimum area of 20 μm^2, and did not exceed a maximum area of 1000 μm^2. Objects with these criteria were used to create "inner" nuclear masks for each cell. The nuclear mask was eroded by 1 μm in from the edge of the detected nucleus to reduce cytoplasmic contamination within the nuclear area, and the reduced "inner" mask (*light green* or *red*) was used to quantify the amount of target channel (Ch 2) fluorescence (Total-STAT3, pSTAT3-S727, or pSTAT3-Y705) within the nucleus. The "outer" cytoplasm mask (*dark green* or *red*) was then established 1 μm out from the edge of the detected nucleus and width of the outer mask was set at 3 μm, to cover as much of the cytoplasm region as possible without going outside the cell boundary. The "outer" mask was used to quantify the amount of target Ch 2 fluorescence (Total-STAT3, pSTAT3-S727, or pSTAT3-Y705) within the cytoplasm. (**b**) Cell counts. The TE image analysis module outputs quantitative data including the average fluorescent intensities of the Hoechst

10. Rabbit polyclonal anti-STAT3 antibody primary antibody.

11. Goat anti mouse-IgG conjugated with Alexa Fluor 488 secondary antibody.

12. Goat anti-rabbit-IgG conjugated with Alexa Fluor 488 secondary antibody.

13. Washing and antibody dilution buffer - Dulbecco's Mg^{2+} and Ca^{2+} free phosphate buffered saline (PBS).

14. Blocking buffer—0.1% (v/v) Tween 20 in Mg^{2+} and Ca^{2+} free PBS.

15. Cal33 head and neck squamous cell carcinoma (HNSCC) cell line [36, 37].

16. Dulbecco's Modified Eagle Medium (DMEM) with 2 mM L-glutamine supplemented with 10% fetal bovine serum, 100 μM nonessential amino acids, 100 μM sodium pyruvate, and 100 U/mL penicillin and streptomycin.

17. Trypsin 0.25%, 1 g/L EDTA solution (trypsin-EDTA).

18. Uncoated black walled clear-bottomed 384-well tissue culture-treated microtiter plates.

3 Methods

1. Aspirate spent tissue culture medium from Cal33 cells in tissue culture flasks that are <70% confluent (*see* **Note 1**), wash cell monolayers 1× with PBS, and expose cells to trypsin-EDTA until they detached from the surface of the tissue culture flasks. Add serum containing tissue culture medium to neutralize the trypsin. Transfer the cell suspension to a 50 mL capped sterile centrifuge tube and centrifuge at $500 \times g$ for 5 min to pellet the cells. Resuspend cells in serum containing tissue culture medium and count the number of trypan blue excluding viable cells using a hemocytometer.

2. Use an automated bulk reagent dispenser to seed Cal33 HNSCC cells at 2000 cells per well in 45 μL of DMEM tissue

---

Fig. 3 (continued) stained objects in Ch 1, and the selected object or cell count in Ch 1. The average cell counts per image ± sd from 32 wells each for Cal33 cells ± IL-6 treatment are presented. (c) Nuclear STAT3 Intensity. The average fluorescent intensities of the Ch2 signals in the nuclear (inner) regions of Cal33 cells ± IL-6 stained with three different primary STAT3 antibodies are presented; Total-STAT3, pSTAT3-S727, and pSTAT3-Y705. The average inner STAT3 intensity ± sd from 32 wells each for Cal33 cells ± IL-6 treatment are presented. The IL-6-induced translocation of STAT3 from the cytoplasm to the nucleus is indicated by an increase in the average inner intensity parameter output by the translocation enhanced image analysis module

culture medium into 384-well assay plates and culture cells overnight at 37 °C, 5% CO_2, and 95% humidity (*see* **Note 2**).

3. Use an automated liquid handling platform to aspirate serum containing medium from Cal33 cell monolayers and to exchange with 45 µL of serum free medium (SFM) twice and return the assay plates to an incubator at 37 °C, 5% CO_2, and 95% humidity for 24 h (*see* **Note 3**).

4. Use an automated liquid handling platform to transfer 5 µL of compounds or controls pre-diluted in SFM to the wells of the assay plates (0.2% DMSO final), centrifuge the plates at 50 × *g* for 1-2 min, and return them to the incubator for 1–3 h at 37 °C, 5% CO_2, and 95% humidity (*see* **Note 4**).

5. Use an automated liquid-handling platform to transfer 10 µL of IL-6 (50 ng/mL final) diluted in SFM to compound wells and maximum (Max) control wells, minimum (Min) control wells receive SFM, centrifuge the assay plate at 50 × *g* for 1-2 min, and return to the incubator at 37 °C, 5% CO_2, and 95% humidity for 15 min (*see* **Note 5**).

6. Use an automated bulk reagent dispenser to add 50 µL of pre-warmed (37 °C) 7.4% paraformaldehyde containing 2 µg/mL Hoechst 33342 to fix cells and stain nuclei (*see* **Note 6**).

7. After 30 min at room temperature use an automated plate washer-dispenser to aspirate the fixative and then wash the fixed cell monolayers 2× with PBS (*see* **Note 7**).

8. Use an automated plate washer-dispenser to aspirate the PBS and permeabilize the fixed cells by adding 50 µL of 95% ice-cold methanol and place the plates on ice for 30 min (*see* **Note 8**).

9. Use an automated plate washer-dispenser to aspirate the 95% methanol from the permeabilized cell monolayers, wash 2× with 50 µL of Tween 20 blocking buffer, and incubate the permeabilized cells in 50 µL of Tween 20 blocking buffer at room temperature for 15 min (*see* **Note 9**).

10. Aspirate the blocking buffer and using an automated bulk reagent dispenser add 25 µL of the primary antibody (mouse anti-pSTAT3-Y705, mouse anti-pSTAT3-S727, or rabbit anti-STAT3) diluted 1:100 in dilution in PBS to the wells of the 384-well assay plates and incubate the plate for 1 h in the dark at room temperature (*see* **Note 10**).

11. Use an automated plate washer-dispenser to aspirate the primary antibody solution and wash cell monolayers 2× with Tween 20 blocking buffer, and incubate the cells in 50 µL of Tween 20 blocking buffer at room temperature for 15 min (*see* **Note 11**).

12. Aspirate the blocking buffer and using an automated bulk reagent dispenser add 25 µL of secondary antibody (Goat

anti-mouse-IgG conjugated with Alexa-488 or Goat anti-rabbit-IgG conjugated with Alexa-488) diluted 1:500 in PBS to the wells of the 384-well assay plates and incubate the plate for 45 min in the dark at room temperature (*see* **Note 12**).

13. Use an automated plate washer-dispenser to aspirate the secondary antibody and wash cell monolayers 2× with PBS, and seal assay plates with aluminum foil seals (*see* **Note 13**).

14. Acquire fluorescent images in two channels on an automated imaging platform (*see* **Note 14**). On the ImageXpress Ultra (IXU) confocal automated HCS platform, the images of the Hoechst (Ch1) and STAT3 (Ch2) channels are sequentially acquired using a 20× 0.45 NA objective, the 405 and 488 nm excitation laser lines, a Quad filter cube set, and individual PMTs for each channel (Fig. 2).

15. Analyze the acquired digital images using an appropriate image analysis algorithm to extract quantitative data (*see* **Note 15**). In our case, the images of pSTAT3-Y705, pSTAT3-S727 and total STAT3 in Cal33 cells ± IL-6 treatment are analyzed using the Translocation Enhanced image analysis module of the MetaXpress™ software to analyze (Fig. 3).

4   Notes

1. Typically better responses are obtained when Cal33 cells are harvested from tissue culture flasks that are <70% confluent.

2. Optimizing the cell seeding density is a critical component of assay development for any cell-based assay, and perhaps most especially HCS assays [32, 38–43]. The objective is to find a balance between the efforts expended in cell culture support while making sure that sufficient numbers of cells are acquired per image to give statistical significance to the image analysis parameters of interest. Typically variability increases as the number of cells captured and analyzed decreases.

3. HNSCC cell lines have previously been shown to synthesize and secrete TGFα and IL-6 into a culture medium that can constitutively activate the STAT3 pathway in an autocrine or paracrine manner [7, 10, 15, 16, 32]. To control for potential autocrine/paracrine activation of the STAT3 signaling pathway, we introduced a medium exchange and serum starvation protocol [32]. The IL-6-induced assay signal window collapsed from >5-fold using this protocol to around twofold in Cal33 cells cultured "undisturbed" in a serum-containing medium, because un-stimulated pSTAT3-Y705 levels were higher and the response to IL-6 was smaller in a serum-containing medium [32].

4. Determining the DMSO tolerance of an assay is critical because together with the concentration of the screening library it limits the maximum compound concentration that can be tested. DMSO has two major effects on HCS assays [32, 38–43]; at concentrations >5% there is significant cell loss due to cytotoxicity and/or reduced cell adherence, and at concentrations >1% but <5%, cells change from a well-spread and well-attached morphology to a more rounded loosely attached morphology that interferes with the image analysis segmentation. An appropriate compound exposure period should be determined empirically for each new assay. Although compounds with poor cell permeability may benefit from extended compound exposure times, prolonged compound exposures can also produce significant levels of cytotoxicity that may reduce the reliability of measurements.

5. During the development HCS assays it is critical to define the concentration and time-dependent responses for stimuli that are used to activate signaling pathways or to induce cellular phenotypes [32, 38–43]. The time course of deactivation is equally important since signal stability has a direct impact on assay throughput and capacity. In situations where the stimulus is expensive or of limited availability, the dead volume of a bulk reagent dispenser may be prohibitive, and either a hand-held multichannel pipettor or liquid handler outfitted with a 384-well transfer head may be preferable for transferring the reagent to the assay plate.

6. The selection of cell fixation and nuclear staining protocols are important sample preparation/assay development parameters for end point HCS assays. Cell fixation methods need to preserve the biological features of interest and are required because scanning an assay plate may range anywhere from 10 to 15 min to 2–3 h per plate, depending upon the complexity of the image acquisition procedure. The excitation and emission spectra of the nuclear stain selected will impact a number of image acquisition parameters; light source, filter sets, detector settings, exposure times and which other fluorescent probes can be imaged. Combining cell fixation with nuclear staining in a single procedure saves time and reduces the number of protocol steps.

7. To reduce/control environmental exposure levels to formaldehyde, we recommend dispensing, aspirating, and washing assay plates on an automated plate dispenser-washer platform.

8. The selection of the cell permeabilization method can have a significant impact on the performance of HCS assays that use antibodies for indirect immuno-fluorescence. In our hands, cell permeabilization with 95% ice-cold methanol typically works

best for phospho-specific antibodies, but we recommend that different cell permeabilization methods should be evaluated for each new antigen and/or antibody pair.

9. Blocking buffers are typically used to reduce or control non-specific antibody binding in indirect immuno-fluorescence assays and thereby lower the background staining. A variety of different components are typically incorporated into blocking buffers including bovine serum albumin, nonfat dry milk, fetal bovine serum, serum from the same species as the secondary antibody, or nonionic detergents like Tween 20. In our hands, 0.1% (v/v) Tween 20 in Mg^{2+} and Ca^{2+} free PBS works well as blocking buffer, and can also be used as the diluent for primary and secondary antibodies.

10. The selection of suitable primary antibodies, working dilutions, and incubation times needs to be empirically determined for each antigen and cell type. Typically, this process involves the pairwise cross titration testing of several different primary and secondary antibody pairs at a variety of different antibody dilutions, ranging from 1:50 for lower affinity antibodies all the way up to 1:2000 for higher affinity antibodies. The length of incubation for primary antibodies can be ≤ 1 h with higher affinity antibodies, or as long as overnight with lower affinity antibodies. In situations where the primary antibody is expensive or needs to be applied at lower dilutions, the dead volume of a bulk reagent dispenser may be prohibitive, and either a hand-held multichannel pipettor or liquid handler outfitted with a 384-well transfer head may be preferable for transferring the antibody to the assay plate.

11. Given the number and importance of the reagent addition, aspiration and washing steps involved in an indirect immuno-fluorescence HCS assay protocol, the selection of an automated plate washer-dispenser platform is probably the most critical automation decision that impacts the performance of the assay.

12. There are many sources of anti-species-IgG fluorophore-conjugated antibodies and typically these higher affinity antibodies can be used at larger dilutions ($\geq 1:500$). Since nonspecific binding fluorophore-conjugated secondary antibodies can be a major component of background staining, some investigators use serum from the same species as the secondary antibody in their blocking buffers.

13. For the long-term storage of fixed and stained assay plates, we recommend aspirating the secondary antibody, washing the cell monolayers with PBS, adding PBS to the wells, and sealing the plates with adhesive foil seals for storage at 4 °C in the dark.

14. Although we have described the image acquisition process on the ImageXpress Ultra confocal platform, most automated imaging platforms designed for HCS should be capable of capturing these images.

15. Although we have described the translocation enhanced image analysis module of the MetaXpress™ software that is integrated with the IXU platform, most automated imaging systems designed for HCS provide similar image analysis algorithms that should be capable of analyzing these images.

Acknowledgments

This project has been funded in part with Federal Funds from the National Cancer Institute, National Institutes of Health, under Contract No. HSN261200800001E. The content of this publication does not necessarily reflect the views or policies of the Department of Health and Human Services, nor does mention of trade names, commercial products, or organizations imply endorsement by the U.S. Government." NExT-CBC Project ID #1015, S08-221 Task Order 6 "STAT3 Pathway Inhibitor HCS" (Grandis, PI), NCI Chemical Biology Consortium, Pittsburgh Specialized Application Center (PSAC) (Lazo & Johnston, co-PIs). The project was also supported in part by funds from the American Cancer Society (Grandis) and a Head and Neck Spore P50 award (Grandis, CA097190).

References

1. Frank DA (2007) STAT3 as a central mediator of neoplastic cellular transformation. Cancer Lett 251:199–210

2. Germain D, Frank DA (2007) Targeting the cytoplasmic and nuclear functions of signal transducers and activators of transcription 3 for cancer therapy. Clin Cancer Res 13:5665–5669

3. Jing N, Tweardy DJ (2005) Targeting Stat3 in cancer therapy. Anti-Cancer Drugs 16:601–607

4. Johnston PA, Grandis JR (2011) STAT3 signaling: anticancer strategies and challenges. Mol Interv 11:18–26

5. Quesnelle KM, Boehm AL, Grandis JR (2007) STAT-mediated EGFR signaling in cancer. J Cell Biochem 102:311–319

6. Aggarwal BB, Kunnumakkara AB, Harikumar KB, Gupta SR, Tharakan ST, Koca C, Dey S, Sung B (2009) Signal transducer and activator of transcription-3, inflammation, and cancer: how intimate is the relationship? Ann N Y Acad Sci 1171:59–79

7. Egloff AM, Grandis JR (2009) Improving response rates to EGFR-targeted therapies for head and neck squamous cell carcinoma: candidate predictive biomarkers and combination treatment with Src inhibitors. J Oncol 2009:896407, 12 pages

8. Heinrich PC, Behrmann I, Müller-Newen G, Schaper F, Graeve L (1998) Interleukin-6-type cytokine signalling through the gp130/Jak/STAT pathway. Biochem J 334:297–314

9. Heinrich PC, Behrmann I, Haan S, Hermanns HM, Müller-Newen G, Schaper F (2003) Principles of interleukin (IL)-6-type cytokine signalling and its regulation. Biochem J 374:1–20

10. Leeman RJ, Lui VW, Grandis JR (2006) STAT3 as a therapeutic target in head and neck cancer. Expert Opin Biol Ther 6:231–241

11. Ram PT, Iyengar R (2001) G protein coupled receptor signaling through the Src and Stat3

pathway: role in proliferation and transformation. Oncogene 20:1601–1606

12. Silva CM (2004) Role of STATs as downstream signal transducers in Src family kinase-mediated tumorigenesis. Oncogene 23:8017–8023

13. Wilks AF (2008) The JAK kinases; not just another kinase drug discovery target. Semin Cell Dev Biol 19:319–328

14. Murray P (2007) The JAK-STAT signaling pathway: input and output integration. J Immunol 178:2632–2629

15. Seethala RR, Gooding WE, Handler PN, Collins B, Zhang Q, Siegfried JM, Grandis JR (2008) Immunohistochemical analysis of phosphotyrosine signal transducer and activator of transcription 3 and epidermal growth factor receptor autocrine signaling pathways in head and neck cancers and metastatic lymph nodes. Clin Cancer Res 14:1303–1309

16. Sriuranpong V, Park JI, Amornphimoltham P, Patel V, Nelkin BD, Gutkind JS (2003) Epidermal growth factor receptor-independent constitutive activation of STAT3 in head and neck squamous cell carcinoma is mediated by the autocrine/paracrine stimulation of the interleukin 6/gp130 cytokine system. Cancer Res 63:2948–2956

17. Lufei C, Koh TH, Uchida T, Cao X (2007) Pin1 is required for the Ser727 phosphorylation-dependent Stat3 activity. Oncogene 26:7656–7664

18. Schuringa J, Schepers H, Vellenga E, Kruijer W (2001) Ser727-dependent transcriptional activation by association of p300 with STAT3 upon IL-6 stimulation. FEBS Lett 495:71–76

19. Wen Z, Zhong Z, Darnell JE Jr (1995) Maximal activation of transcription by Stat1 and Stat3 requires both tyrosine and serine phosphorylation. Cell 82:241–250

20. Wen Z, Darnell JE Jr (1997) Mapping of Stat3 serine phosphorylation to a single residue (727) and evidence that serine phosphorylation has no influence on DNA binding of Stat1 and Stat3. Nucleic Acids Res 25:2062–2067

21. Herrmann A, Vogt M, Mönnigmann M, Clahsen T, Sommer U, Haan S, Poli V, Heinrich PC, Müller-Newen G (2007) Nucleocytoplasmic shuttling of persistently activated STAT3. J Cell Sci 120:3249–3261

22. Liu L, McBride KM, Reich NC (2005) STAT3 nuclear import is independent of tyrosine phosphorylation and mediated by importin-alpha3. Proc Natl Acad Sci U S A 102:8150–8155

23. Reich N, Liu L (2006) Tracking STAT nuclear traffic. Nat Rev Immunol 6:602–612

24. Yang J, Chatterjee-Kishore M, Staugaitis SM, Nguyen H, Schlessinger K, Levy DE, Stark GR (2005) Novel roles of unphosphorylated STAT3 in oncogenesis and transcriptional regulation. Cancer Res 65:939–947

25. Yang J, Stark GR (2008) Roles of unphosphorylated STATs in signaling. Cell Res 18:443–451

26. Ynag J, Liao X, Agarwal MK, Barnes L, Auron PE, Stark GR (2010) Unphosphorylated STAT3 accumulates in response to IL-6 and activates transcription by binding to NFkB. Genes Dev 21:1396–1408

27. Brockstein B (2011) Management of recurrent head and neck cancer: recent progress and future directions. Drugs 71:1551–1559

28. Goerner M, Seiwert TY, Sudhoff H (2010) Molecular targeted therapies in head and neck cancer—an update of recent developments. Head Neck Oncol 2:8–12

29. Stransky N, Egloff AM, Tward AD, Kostic AD, Cibulskis K, Sivachenko A, Kryukov GV, Lawrence MS, Sougnez C, McKenna A, Shefler E, Ramos AH, Stojanov P, Carter SL, Voet D, Cortés ML, Auclair D, Berger MF, Saksena G, Guiducci C, Onofrio RC, Parkin M, Romkes M, Weissfeld JL, Seethala RR, Wang L, Rangel-Escareño C, Fernandez-Lopez JC, Hidalgo-Miranda A, Melendez-Zajgla J, Winckler W, Ardlie K, Gabriel SB, Meyerson M, Lander ES, Getz G, Golub TR, Garraway LA, Grandis JR (2012) The mutational landscape of head and neck squamous cell carcinoma. Science 333:1157–1160

30. Boehm AL, Sen M, Seethala R, Gooding WE, Freilino M, Wong SM, Wang S, Johnson DE, Grandis JR (2008) Combined targeting of epidermal growth factor receptor, signal transducer and activator of transcription-3, and Bcl-X(L) enhances antitumor effects in squamous cell carcinoma of the head and neck. Mol Pharmacol 73:1632–1642

31. Leeman-Neill RJ, Wheeler SE, Singh SV, Thomas SM, Seethala RR, Neill DB, Panahandeh MC, Hahm ER, Joyce SC, Sen M, Cai Q, Freilino ML, Li C, Johnson DE, Grandis JR (2009) Guggulsterone enhances head and neck cancer therapies via inhibition of signal transducer and activator of transcription-3. Carcinogenesis 30:1848–1856

32. Johnston P, Sen M, Hua Y, Camarco D, Shun TY, Lazo JS, Grandis JR (2014) High-content pSTAT3/1 imaging assays to screen for selective inhibitors of STAT3 pathway activation in head and neck cancer cell lines. Assay Drug Dev Technol 12:55–79

33. Johnston P, Sen M, Hua Y, Camarco DP, Shun TY, Lazo JS, Wilson GM, Resnick LO, LaPorte

MG, Wipf P, Huryn DM, Grandis JR (2015) HCS campaign to identify selective inhibitors of IL-6-induced STAT3 pathway activation in head and neck cancer cell lines. Assay Drug Dev Technol 13:356–376

34. LaPorte M, da Paz Lima DJ, Zhang F, Sen M, Grandis JR, Camarco D, Hua Y, Johnston PA, Lazo JS, Resnick LO, Wipf P, Huryn DM (2014) 2-Guanidinoquinazolines as new inhibitors of the STAT3 pathway. Bioorg Med Chem Lett 24:5081–5085

35. LaPorte M, Wang Z, Colombo R, Garzan A, Peshkov VA, Liang M, Johnston PA, Schurdak ME, Sen M, Camarco DP, Hua Y, Pollock NI, Lazo JS, Grandis JR, Wipf P, Huryn DM (2016) Optimization of pyrazole-containing 1,2,4-triazolo-[3,4-b]thiadiazines, a new class of STAT3 pathway inhibitors. Bioorg Med Chem Lett 26:3581–3585

36. Bauer V, Hieber L, Schaeffner Q, Weber J, Braselmann H, Huber R, Walch A, Zitzelsberger H (2010) Establishment and molecular cytogenetic characterization of a cell culture model of head and neck squamous cell carcinoma (HNSCC). Genes 1:338–412

37. Gioanni J, Fischel JL, Lambert JC, Demard F, Mazeau C, Zanghellini E, Ettore F, Formento P, Chauvel P, Lalanne CM et al (1988) Two new human tumor cell lines derived from squamous cell carcinomas of the tongue: establishment, characterization and response to cytotoxic treatment. Eur J Cancer Clin Oncol 24:1445–1455

38. Dudgeon D, Shinde SN, Shun TY, Lazo JS, Strock CJ, Giuliano KA, Taylor DL, Johnston PA, Johnston PA (2010) Characterization and optimization of a novel protein-protein interaction biosensor HCS assay to identify disruptors of the interactions between p53 and hDM2. Assay Drug Dev Technol 8:437–458

39. Johnston PA, Shinde SN, Hua Y, Shun TY, Lazo JS, Day BW (2012) Development and validation of a high-content screening assay to identify inhibitors of cytoplasmic dynein-mediated transport of glucocorticoid receptor to the nucleus. Assay Drug Dev Technol 10:432–456

40. Nickischer D, Laethem C, Trask OJ Jr et al (2006) Development and implementation of three mitogen-activated protein kinase (MAPK) signaling pathway imaging assays to provide MAPK module selectivity profiling for kinase inhibitors: MK2-EGFP translocation, c-Jun, and ERK activation. Methods Enzymol 414:389–418

41. Trask O, Nickischer D, Burton A, Williams RG, Kandasamy RA, Johnston PA, Johnston PA (2009) High-throughput automated confocal microscopy imaging screen of a kinase-focused library to identify p38 mitogen-activated protein kinase inhibitors using the GE InCell 3000 analyzer. Methods Mol Biol 565:159–186

42. Trask OJ Jr, Baker A, Williams RG et al (2006) Assay development and case history of a 32K-biased library high-content MK2-EGFP translocation screen to identify p38 mitogen-activated protein kinase inhibitors on the ArrayScan 3.1 imaging platform. Methods Enzymol 414:419–439

43. Williams RG, Kandasamy R, Nickischer D et al (2006) Generation and characterization of a stable MK2-EGFP cell line and subsequent development of a high-content imaging assay on the Cellomics ArrayScan platform to screen for p38 mitogen-activated protein kinase inhibitors. Methods Enzymol 414:364–389

Chapter 15

Single Cell and Population Level Analysis of HCA Data

David Novo, Kaya Ghosh, and Sean Burke

Abstract

High Content Analysis instrumentation has undergone tremendous hardware advances in recent years. It is now possible to obtain images of hundreds of thousands to millions of individual objects, across multiple wells, channels, and plates, in a reasonable amount of time. In addition, it is possible to extract dozens, or hundreds, of features per object using commonly available software tools. Analyzing this data provides new challenges to the scientists. The magnitude of these numbers is reminiscent of flow cytometer, where practitioners have long been taking what effectively amounted to very low resolution, multi-parametric measurements from individual cells for many decades. Flow cytometrists have developed a wide range of tools to effectively analyze and interpret these types of data. This chapter will review the techniques used in flow cytometry and show how they can easily and effectively be applied to High Content Analysis.

Key words Single cell analysis, Flow cytometry, Segmentation, Gating, Data analysis, Histogram, Scatter plot, Region of interest, Population statistics, Cell cycle analysis

---

1 Introduction

High Content Analysis (HCA) instrumentation has undergone tremendous hardware advances in recent years. Three major technological advances have been responsible for the rapid increase in the number of cellular images that can be acquired in a given unit of time. The first is the ability of HCA instruments to capture images from multiple wavelengths by using improved LED or laser light with either multiple bandpass filters with a single camera or in parallel by splitting the light via various filter configurations onto multiple cameras at the same time. The latter allows multiple fluorescent probes to be simultaneously imaged without sacrificing acquisition speed. The second advance is that the size of the optical detection sensors and therefore the size of the images captured have increased dramatically, by a factor of approximately four, since the initial commercial HCA instruments. The largest chips are now greater than 2500×2000 pixels in full resolution with a chip size of 13.5 mm^2. Lastly, the specimen plate/slide stages and robotics

Paul A. Johnston and Oscar J. Trask (eds.), *High Content Screening: A Powerful Approach to Systems Cell Biology and Phenotypic Drug Discovery*, Methods in Molecular Biology, vol. 1683, DOI 10.1007/978-1-4939-7357-6_15,
© Springer Science+Business Media LLC 2018

have improved in both speed and accuracy to allow more rapid XY movement between different fields of view (FOV) and different wells. This improved accuracy has led to large montage images that are equivalent to the tissue-based scanners used in digital pathology.

The net result is that it is now possible to obtain images of hundreds of thousands to millions of individual objects, across multiple wells, channels and plates, in a reasonable amount of time (minutes to hours). Coupled with improvements in software that easily allow for the extraction of a multiplicity of feature data on a per pixel object basis, HCA is undergoing a rapid explosion of generating "big data" that impacts almost every chapter of this book.

The magnitude of these numbers is reminiscent of flow cytometry (FC), where practitioners have long been taking what effectively amounted to very low resolution, multi-parametric measurements from individual cells for many decades [1]. The FC community has developed a specific "standard" terminology for a wide variety of display methodologies, workflows, and software to facilitate the analysis and interpretation of large numbers of multi-parametric measurements to assess single cell and population responses. In spite of the similarities of technologies in both the instrumentation and resulting data, the HCA community at large (with some notable exceptions) is still in its infancy with regards to single cell analysis (SCA) when compared to the FC community. The power of SCA in HCA is under-utilized, even though HCA capture measurements at the single cell level, it has traditionally been used to report summarized field averages or more commonly well averages that are similar to a plate reader. The goal of this chapter is to highlight the importance and techniques of analyzing single cell data, showcasing the additional power than can be obtained by combining the imaging capabilities of HCA instrumentation with classical techniques commonly used in the FC community.

2 Reasons for Slow Uptake of SCA in HCA

2.1 Terminology

In spite of the underlying similarity in the principles of FC and HCA, the two fields have evolved somewhat in isolation. Furthermore to complicate the emerging technologies of HCA, there is no agreement upon image format, or image analysis terminology or nomenclature standards that have been adopted across the commercial platforms [2]. Together, this has resulted in a divergence in the terminology and/or nomenclatures used to describe similar, or even identical, concepts that produce unnecessary confusion when practitioners from one field attempt to read papers or apply methodologies to the other. This has likely been a barrier to HCA practitioners being able to effectively apply some analysis strategies common in FC. Table 1 describes some of these terms.

Table 1
Terms commonly used in FC and HCA

HCA	FC	Meaning
Features	Parameters	An individual measurement made on a single object. In HCA these are derived from segmentation and feature extraction and can refer to intensity, morphological or other measurements. In FC, these are generally direct readouts from a photodetector
Channels	Colors/detectors	Describes the detectors in the instrumentation. In HCA these are digital cameras. In FC these are generally PMTs or Photodiodes, i.e., in FC a "six color system" would mean a FC that can measure six fluorescence values on a single object
Intensity values	Channels	Refers to a specific value of a measurement
Thresholding	Thresholding	In HCA it generally refers to a minimum or maximum value required for the segmentation algorithm to classify something as an object. In FC it is the minimum value in a particular detector to record that event as part of the resulting dataset
Object	Event	In HCA this is collection of pixels classified by the segmentation algorithm. In FC, it is an object that was above the threshold setting. In both, these are often individual cells, but not always
Dynamic range	Resolution	The number of possible values that the detector/electronics can use to represent an intensity of photons In FC resolution is also used to describe the number of possible bins underlying a histogram or dot plot
Scatter plot	Dot plot/2D Plot/ scattergram	2D plot of different features showing the position of each object in this feature space
Filter/threshold	Gate	A concept of selecting only certain objects within the data set for analysis
Thresholding	Marker gating	Filtering the object to only consider objects that have values between a certain upper and lower bound in a particular parameter
Region of interest	Shape gating	In HCA this described a 2D shape drawn on an image. In FC it is 2D shape drawn on a scatter plot. Generally used to perform analysis only on objects within this region
*Gating	Boolean gating	Combing various gates using Boolean logic (and/or/not) to define a subset of objects within the dataset *Capabilities in HCA is limited compared to FC
Cross-talk/spectral bleed-through	Spillover	The fact that fluorescence from a single label may be detected in multiple detectors. This is generally not desirable

(continued)

Table 1
(continued)

HCA	FC	Meaning
Unmixing	Compensation	The mathematical treatment of the resulting fluorescence data in an attempt to remove artifacts resulting from cross-talk. In HCA implementation is limited
Cell size	Scatter	In HCA cell size is determined by image analysis of the segmented image and is generally described in pixels or microns. Many "size" measurements are possible in HCA. In FC this refers to light scattered at various angles to the incident laser beam. Forward scatter has a loose correlation to cell size and is dependent on differences in refractive index between the object and surrounding fluid

The remainder of this chapter will utilize HCA terminology whenever possible.

2.2 Focus on Screening Large Numbers of Samples

Another impediment to utilizing single cell approaches with HCA data is that traditional HCA assays have been focused toward providing single readouts on a preset minimum number of cells typically determined during assay development, optimization, and validation. This reflects that the initial, and still important, market of HCA devices is the drug screening community. The focus of this community is primarily on increasing the number of compounds that can be screened in the shortest possible time. The complexity of screening such a large number of compounds often necessitates simplifying other aspects of the assay, with data analysis being one of them. Often, the databases that contain the analyzed HCA results from these screens are challenged to handle multiple data-point features per well. Since the primary purpose of the screen is to identify hits that might be developed into leads and eventually drug candidates, the interpretation of differential drug effects on multiple populations within a well may complicate or obscure matters and can lead to either undiscovered findings or ambiguous results, which does not directly help drive the drug discovery process forward.

2.3 Software Tools: Chicken or the Egg

A significant practical impediment is the paucity of software tools designed to perform SCA on HCA data and the difficulty in exporting data in a way that these tools can use. There has been significant effort to standardize the description of the instrument configurations and experimental configuration [3]. The OME group has achieved remarkable progress in providing tools to operate on the wide variety of image formats via the Bioformats library, which has helped alleviate a problem in reading the wide variety of image data file formats that have been produced [4]. The OME-TIFF standard is also widely used and has standardized a description of the instrument and experimental pipeline [4]. However, a precise description of the experimental/instrument metadata is not a requirement for SCA. In contrast, the Flow Cytometry Standard (FCS) has been used in the FC field for decades [5] and is currently on version 3.1 [6]. In spite of the fact that only limited metadata is stored with the raw FC data, the standard has enjoyed widespread acceptance and is currently utilized by every instrument manufacturer and FC data analysis software. The International Society for the Advancement of Cytometry (ISAC) has recognized that a similar standard, focusing on data only, should be useful for the microscopy community wishing to perform SCA. As such, the society developed the Image Cytometry Experiment (ICE) format [7]. The ICE format provides a simple, standardized approach to store the original microscopy images and their relationships to each other (which well, field of view, channel, etc. each image came from), segmentation results (within each image, where each object is located) and the per-cell feature data.

Currently, many instruments provide a way to export numerical data as some sort of text-based flat file. However, analyzing the raw numerical values without being able to visualize the images of the objects that underlie the data underutilizes the true power of imaging, as will be shown later. In addition, without dedicated SCA software, users often have to resort to importing such data into third-party software packages such as Spotfire, Matlab, R, JMP, or other tools that are either costly or require considerable expertise to operate. Experience within the FC community has shown that in order for software to obtain wide acceptance for SCA it must not only be capable of the required statistical analytics, it should also: be intuitive and user friendly; contain a variety of reporting options; be accessible to all levels of users; and have specific workflows that support applying analysis protocols to a large number of samples with results easily exportable to common formats (such as spreadsheets, database, etc.). The SCA tools provided currently to the HCA community lag behind those commonly available to the FC community. However, new tools are emerging from both HC instrument manufacturers and third-party software vendors.

2.4 Hard to Know How Cells Are Supposed to Behave

SCA in HCA is also complicated by the large numbers of features that are produced by the image analysis algorithms and the difficulty knowing, a priori, which morphological and/or fluorescence measurements will delineate different subpopulations within a sample. FC generally relies on staining with fluorescent molecules, which can vary by 4 or more orders of magnitude in expression among different subpopulations within a sample. The dynamic ranges of the current HCA imaging CCD and sCMSO 16-bit cameras (65,535 channels) differ from FC PMT detectors [1]. An additional challenge to distinguishing subsets of cells based solely on morphology features is that often these parameters only change by small increments. Thus, although creative approaches are being attempted, it is still a challenge for the HCA community to determine reproducible, reliable criteria to accurately characterize subpopulations based on morphology measurement alone within a sample.

3 Why Is Single Cell Analysis Gaining in Popularity?

In spite of the historical and technological difficulties described above, SCA is becoming more prevalent in HCA and where applicable will displace the traditional well-based approaches and analysis. As SCA approaches are utilized, even in assays that were designed to be "homogenous" assays the heterogeneity of cellular responses is being increasingly appreciated. In addition, the proper application of SCA is allowing researchers to obtain more meaningful results by eliminating confounding cell populations that are present in the wells. But perhaps most importantly, the sophistication of the instruments allows researchers to obtain biological results that go far beyond a "glorified plate reader" and investigate biological phenomena that are intractable to other methodologies.

4 Obtaining Single Cell Data

4.1 Importance of Proper Segmentation

Like any other data analysis, the adage of "garbage in- garbage out" applies. In order to perform useful SCA, accurate single cell data is required. Of course, this requires proper experimental procedure and methodologies, and it starts with the best possible image. From a data analysis point of view, the key component is proper segmentation. By definition, the image analysis segmentation algorithm defines the objects and parameters that will be extracted and analyzed at the single cell level, and its importance cannot be overstated. Most instrument vendors provide segmentation and feature extraction software of various sophistications and there are several open-source packages available [8–10] . Most segmentation algorithms have a variety of settings that determine how the algorithms will define objects. It is important for the end user to experiment with the various settings to determine the optimum settings for their particular assay. HCA, along with SCA, provides a means to do so graphically, numerically, and objectively. The difficulties associated with determining proper segmentation can be seen in Fig. 1. The left-hand panel shows the same images segmented with three different segmentation algorithms. It is clearly apparent that Algorithm 1 performs poorly, both by visual inspection of the images and the apparent broad distribution of nuclear staining in the histograms in Panel A. Algorithms 2 and 3 seem very similar

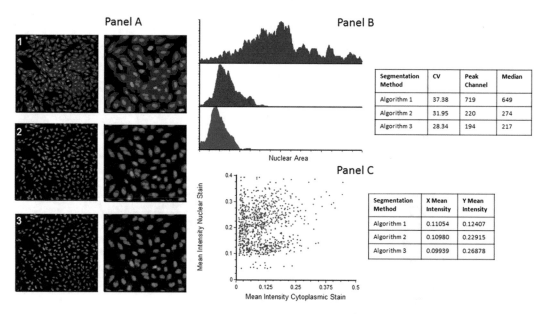

Fig. 1 Effect of segmentation algorithm on numerical results. Images were obtained from the Broad Bioimage Benchmark Collection (http:/www.broadinstitute.org/bbbc/BBBC013/). Human osteosarcoma cells, U2OS, are assayed for cytoplasm-nucleus translocation of the Forkhead (FKHR-EGFP) fusion protein in the stably transfected cell line. FKHR is localized within the cytoplasm during proliferation and accumulated in the nucleus upon inhibition. Three examples of basic segmentation algorithms as implemented by CellProfiler are shown on the same image: Algorithm 1 (Background), Algorithm 2 (Otsu with 2 class thresholding), and Algorithm 3 (Otsu with 3 class thresholding). Second column of Panel (**a**) is a magnified section of corresponding image in the first column. The SCA data in Panels (**b**) and (**c**) clearly show differences between Algorithms 1 (*blue*), 2 (*red*), and 3 (*green*)

when inspected visually. However, SCA clearly shows the quantitative differences between the algorithms (Fig. 1, Panels b, c) and highlights that the lowest CV of nuclear staining is found using Algorithm 3.

Single cell analysis techniques are generally designed to display all the objects in a dataset and to visually observe distinct subpopulations in some manner. This is opposed to the more common HCA technique of requesting objects that fulfill a certain set of criteria (i.e., find all the cells in a certain size range with a given fluorescence range). This latter approach requires some a priori knowledge of the parameters of interest and their putative values, whereas the former displays all the data and allows the user to select subpopulations based on the entire dataset.

4.2 Histograms

Histograms display the data from a single feature. The data from the specified parameter is binned into a user-specified number of bins. Distinct peaks in the histogram generally reflect different subpopulations within the sample and histograms provide a clear and simple way of visualizing and comparing the magnitude and distribution of the parameter of interest as well as the number of

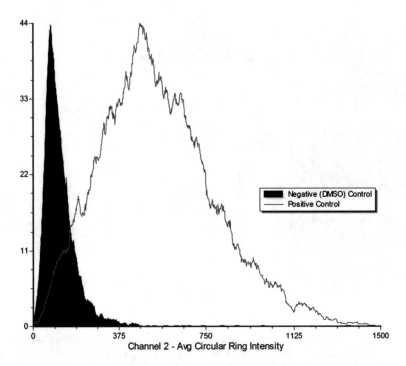

Fig. 2 Histogram representation of average circular ring Intensity on positive and negative control cells. Differences in relative cell numbers, distribution spread, and intensity are readily apparent

events within the different subpopulations (Fig. 2). Often several "overlays" are displayed on top of each other to highlight characteristics of different subpopulations.

Cultured cells were incubated in the presence of DMSO (negative control) or Progesterone (positive control) to elicit the highest and lowest cellular responses measurable via the channel 2 intensity parameter. The apparent response and separation between the two peaks indicates the controls may be used to assay further compounds inducing responses between the apparent ranges of the histogram overlays. Cells were imaged and data exported from the HCS Studio software (ThermoFisher Scientific).

Histograms also provide the basis for specialized analysis whereby mathematical models are applied to the histogram data in order to extract underlying parameters regarding the biological state of the sample. The canonical example is cell cycle analysis (Fig. 3), in which cells are stained with fluorescent DNA-binding dyes and models extract the fraction of the sample that is in the G0/ G1, S, and G2/M phases of the cell cycle (among other information) [11–13]. There are several examples of cell cycle measurements using High Content instrumentation [14–17] although the number that utilize modeling approaches are more limited [18]. Although the cell cycle models used in FC are similarly applicable to HCA data, the experimental challenges associated with HCA data,

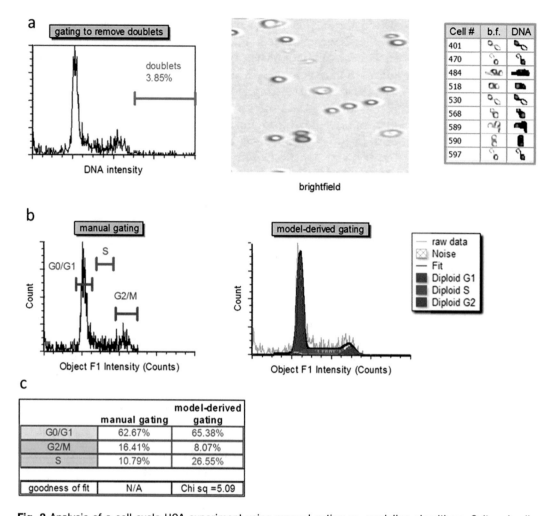

Fig. 3 Analysis of a cell cycle HCA experiment using manual gating vs. modeling algorithms. Cultured cells were stained with Propidium iodide under normal, untreated conditions. Imaging was performed by the CBA Vision instrument. First, high-intensity DNA events were identified on a 1D–plot (**a**, *left*) and backgated onto the original image (**a**, *middle, right*) to reveal a population consisting mostly of doublets. The image gallery on the *right* is gated to only show events in the doublets marker, which confirms that most cells in this marker are indeed doublets. Next, cell cycle phases were identified either manually (**b**, *left*) or automatically by applying a modeling algorithm to the data (**b**, *right*). Images and segmentation from CBA Vision (Nexcelom). Cell cycle modeling was performed by the Multicycle AV (Phoenix Flow Systems) plugin to FCS Express 4 Image Cytometry (De Novo Software)

such as fixation methodologies and the lower numbers of cells typically acquired in HCA, make it more challenging of a model to accurately fit the data. Improvements in this regard should result in more accurate and biologically relevant results.

Figure 3 shows an example of an experiment in which traditional manual gating strategies were combined with cell cycle modeling approaches. First, doublets were identified on the basis of a high-DNA intensity marker (Fig. 2a, *left*). Examining the cells within the marker in the original image confirmed that most of these cells were doublets (Fig. 2a, *middle*, *right*). These events were excluded from the downstream analysis. Figure 2b shows the identification of the G0/G1, S, and G2/M phases of the population via subjective manual gating (*left*) and by cell cycle modeling (*right*) [12]. Advantages of the modeling approach include the ability to discriminate signal from noise, and to examine the goodness of fit both by eye (Fig. 2b, *right*, black "fit" overlay vs. gray "raw data" overlay) and statistically (Fig. 2c, Chi square statistic). In contrast, drawing regions by eye is entirely subjective, and the raw data is not "fitted" to any particular reference frame for comparison. Determining the S-phase boundaries is particularly problematic. The general tendency is to define mutually exclusive G0/G1, S, and G2/M regions, which may significantly underestimate the S phase (11% by manual gating vs. 26% determined by model, Fig. 2c, third row in table).

4.3 Dot Plots

The most common representation of FC data is on 2D plots (*see* Table 1 for synonyms) and these are extremely valuable for the display of HCA data as well. These plots represent each object as a dot whose location in the plot is based on the object's value in two of the features. 2D plots provide all of the information of the single parameter histograms described above, and in addition can show correlations between the selected features that single parameter histograms cannot.

Jurkat cells were supplemented with camptothecin (CPT) to induce a response from Caspase-FAM marker. Caspases are a family of enddoproteases that are involved in apoptotic pathways and the detection of activated caspases within the cell indicates that the cell is undergoing an apoptotic process. Cells were imaged and data exported from the NucleoCounter NC-3000 instrument (Chemometec).

Several different kinds of 2D plots are commonly used. Standard dot plots (Fig. 3) simply show the positions of the objects on the plot in a single color. Density plots (Fig. 4a) are similar to dot plots, except that the color of a particular dot represents the number of objects at that coordinate. Contour plots (Fig. 4b) are similar to common topological maps, except that instead of displaying elevation lines, the lines represent the boundary between different levels on the density plots.

The type of plot to choose is primarily a function of personal preferences and what information one is trying to display. Density and Contour plots do have an advantage over simpler dot plots in that they also convey the number of cells associated with different regions of the 2-parameter space. Although there are some subtleties in terms of understanding how the plot resolution, smoothing, and number/coloring of the density levels can affect the interpretation of the data, for the most part 2D plots present the data

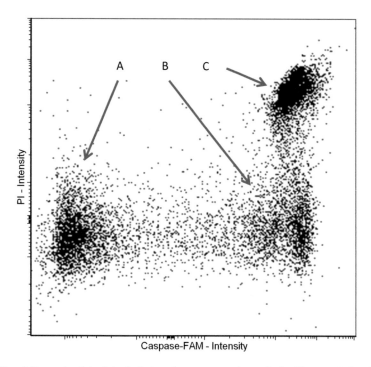

Fig. 4 Example dot plot. Jurkat cells were supplemented with camptothecin (CPT) to induce a response from Caspase-FAM marker. Cells were imaged and data exported from the NucleoCounter NC-3000 instrument (Chemometec). Graph of per cell Caspase mean intensity vs per cell Caspase maximum intensity. Feature values were calculated using CellProfiler software from the images of cells stained with anti-caspase antibody. Several different subpopulations of cells are readily apparent: (**a**) viable cells with low PI and low caspase staining, (**b**) Apoptotic cells with low PI and high caspase staining, (**c**) necrotic cells with both high PI and caspase staining

equivalently, in a highly intuitive manner, and as such are commonly used.

4.4 Alternative Visualization Strategies

In addition to the basic visualization of SCA data, alternative plotting techniques have arisen in an attempt to allow users to directly visualize higher dimensional data. 3D plots, which draw a point representing each object based on its location in three-parameter space, do offer the possibility of visualizing subpopulations that would be impossible on 2D plots (Fig. 5). However, routine use of these plots is rare, since they generally require rotating each data file differently even on the same plot to properly visualize the populations.

Principle Component Analysis (PCA) has also been used to help improve the separation between different populations in both FC [19, 20] and HCA [21–23] . PCA transforms the single object data into values that represent its location in a transformed dimensional space. The transformation is calculated such that each

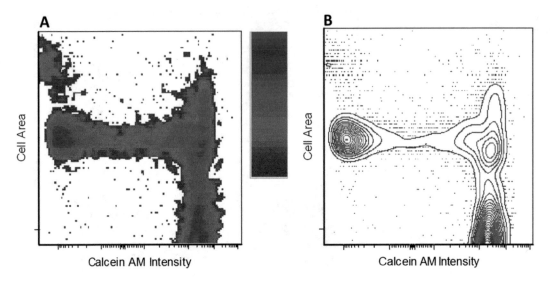

Fig. 5 Dot and contour plots. These graphs display the per-object value of calcein AM intensity vs cell area. *Colors* are based on density from low (*red*) to high (*purple*). Several populations of cells are evident. It is clear that there is a large population of cells near the *X* axis, representing cells with a low area, perhaps indicated that the segmentation algorithm classified debris as objects

axis of this transformed space is orthogonal to each other and maximizes the variance of the entire population along that axis. An advantage of this technique is that first few components can capture most of the variation in the dataset and be plotted in traditional manners described above. Although PCA is used in HCA to analyze well level data, its use in SCA both in FC and HCA is rare, since the principal components of different datasets can be different from each other and lead to difficulties in comparing different datasets. In addition, the assumption of Gaussian variance implicit in PCA is rarely realized in experimental data and appropriate variance stabilization techniques must be applied [24]. However, intriguing results have been obtained using these methodologies [25, 26].

4.5   Gates

The cornerstone of single cell analysis is gating, i.e., selecting objects of interest from the larger dataset based on certain criteria. Sometimes, gates are referred to in the imaging/database terminology as a filter. Any criteria can be chosen, including well location on a plate, physical location of the object within a well, time of acquisition, etc.; however, typically, objects are selected based on their location within 1D or 2D plots. The advantage of gating over standard database queries is that, in general, one visualizes the entire population and selects the subset visually against the backdrop of all the data as opposed to having preconceived guesses of what is the correct value to filter on and continually refining those criteria. An example of typical gates is shown in Fig. 5.

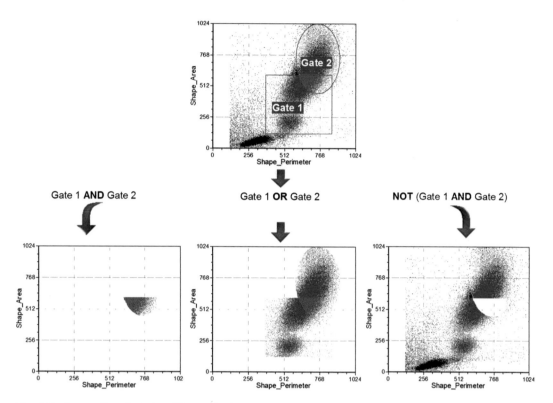

Fig. 6 Boolean gating. Applying different Boolean combinations individual gates allows defining different cell populations

Individual gates are rarely sufficient to completely identify a population of interest. For example, identification of CD4+ T-cells generally requires a size gate to eliminate debris, a CD3+ gate to identify T-cells, and a CD4+ gate to obtain the CD4+ T-cells. Individual gates are used to define populations by combining them together with the standard Boolean logic operators: AND, OR, NOT (Fig. 5).

Due to the ubiquity of the AND Boolean operator in terms of defining populations, the concept of a gating hierarchy has become popular in recent years (Fig. 6). This allows for easy visualization of the relationship between the gates and allows for the construction of complex Boolean conditions more intuitively and efficiently, with minimal typing and errors.

Once gates (or combinations thereof) are defined, subsequent plots (and statistics derived from them) are restricted to only consider objects that fall within particular gates, allowing for easy visual and statistical comparisons of the properties of various cell populations.

4.6 Markers

Populations are visually defined on histograms by drawing markers (Fig. 3), which specify the lower and upper boundaries of the values that object may achieve in order to be included within the marker.

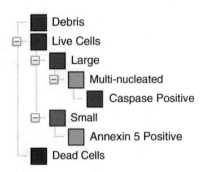

Fig. 7 Gate hierarchy. When cells lower in the hierarchy are applied to data, they are always combined with their parents with an implicit AND gate

Markers are often combined with gates in a Boolean manner, as described above, in order to further refine the definition of a population of interest. For populations that lie close together, markers drawn on histograms often allow a more precise delineation of the population than 2D plots in cases where it is possible to discriminate the population in a single dimension (Fig. 7).

4.7 Backgating

A particularly powerful plotting technique is "backgating" whereby the entire dataset is shown on a 2D plot in one color, and populations of interest are shown in another color. By displaying both the ungated and gated populations in the same plot, subtle relationships between the populations are more easily observed (*see* Figs. 8 and 9).

4.8 Basic Statistics

Basic descriptive statistics (mean, median, standard deviation, etc.) are by far the most common statistics used in SCA. Nonparametric statistics (such as median) are generally preferred over parametric ones (such as arithmetic mean) since biological populations are rarely Gaussian distributed. The geometric mean is often used for data displayed on a logarithmic scale, although this statistic does have the limitation that it is undefined if the dataset contains a zero or negative value. The interpretation of these statistics is highly dependent on the particular experiment being conducted. For example, the change in the mean of a particular feature of population with drug treatment could indicate a change in this parameter across all the cells in the population, or a larger shift in a subset of the population. Careful inspection and validation of the results is critical.

The fraction of a certain population compared to another population (or the entire dataset) is also a frequently used measure. While percentages are easily obtained and interpreted when populations are well separated, more sophisticated techniques are required if the populations overlap to any significant degree [27–30].

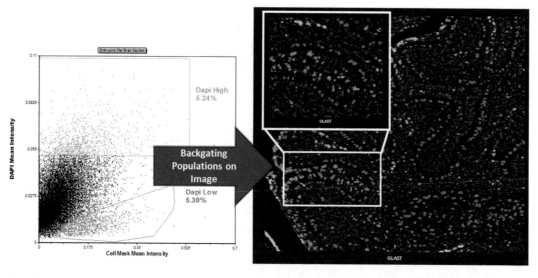

Fig. 8 A developing rat embryo brain section stained for DAPI, Cell Mask, and GLAST. The dot plot on the *left* highlights DAPI high intensity staining (*light blue*) and DAPI low intensity staining (*green*). Events that fall within either gate are highlighted on the GLAST-stained image on the *right* (and *inset zoomed area*) in *light blue* or *green*. Backgating reveals spatial information such that the DAPI dim population falls toward the interior of invaginations in the section while the DAPI bright population falls toward the periphery. Images courtesy of Dragan Maric, Ph.D.

4.9   Automated Cluster Analysis

With the proliferation of measured features, both in FC and HCA, automated methods for finding subpopulations within a complex multi-parametric dataset have been investigated [31–33]. Significant research effort has been invested into developing techniques capable of replacing time-consuming and potentially subjective gating based on manual manipulation with computer-driven data processing.

In some ways, the research efforts in both fields have diverged somewhat due to the nature of the technology. In FC, there is no way to assist the cluster-finding algorithms by providing a priori knowledge about an individual object, so much focus has been placed on so-called unsupervised clustering approaches. These rely on the underlying structure of the data point-clouds formed in multi-dimensional space defined by extracted features to find unimodal populations of measured instances, i.e., clusters. Simple heuristic clustering approaches, such as k-means, were introduced into cytometry as early as mid-1980; however, owing to lack of robustness they did not gain popularity despite a number of reports describing the technique were subsequently published in the 1990s [34–36]. Recently, model-based approaches have been reported by cytometry data analysts. Among these, the finite mixture models seem to be especially promising. This technique assumes that the multidimensional point-cloud of FC data can be modeled as a mixture of well-characterized distributions. Gaussian distributions,

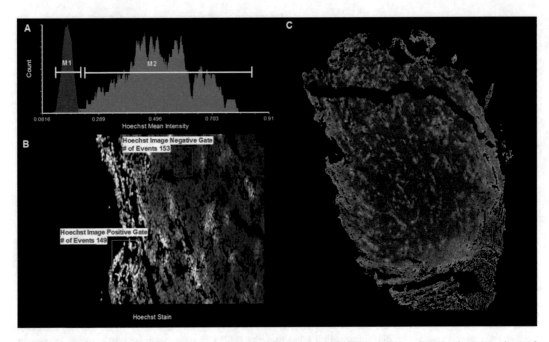

Fig. 9 Image gates were defined on Hoechst positive (*green* gate) and Hoechst negative (*red* gate) regions of the image (Panel **b**). Gates in the image were adjusted to contain similar numbers of events. Events from each gate were overlaid in a histogram (Panel **a**) and Markers M1 and M2 were created based on the distribution of Hoechst staining in each gate. The markers were then converted to gates and backgated onto the entire tumor section (Panel **c**) to better highlight the distribution through image backgating. Gates in Panel (**c**) are the same as in Panel (**b**). Images acquired on Axiovert 200 M (Zeiss), Segmentation performed by CellProfiler, SCA performed using FCS Express 4 Image Cytometry, courtesy of Sean Carlin Ph.D.

Student's t-distribution, or skewed t-distributions are typically employed [37, 38].

Although the references above describe many successful applications of these techniques, in practice they have yet to come into common use among the vast majority of FC practitioners due to the lack of robustness, owing mostly to heteroskedasticity of the collected data [39]. Also, many of the algorithms are extremely sensitive to changes in staining, instrumentation, etc., and require additional transformations to condition them appropriately for cluster analysis. Finding the "best" transformation is a nontrivial problem, the solution of which can crucially impact the clustering results [40, 41]. In spite of their shortcomings, automated methods are critical for continued advancements in SCA and should find wide applicability in SCA-based HCA.

While all of the unsupervised techniques above are applicable toward the numeric feature values obtained using HCA and image analysis, HCA has the additional advantage that the experimenter can manually inspect images of cells and classify them based on their experience and knowledge of the assay system. Thus automated clustering in HCA has relied primarily on supervised approaches, which rely on "training" the algorithms based on the data and have the algorithm determine which of the extracted features most closely correlate with those selected by the investigator. The supervised classification systems can employ a variety of techniques, but neural networks [42–44] and support vector machine [45–48] approaches are the most popular in image and flow cytometry applications. Supervised methods are quite commonly used in HCA since many sophisticated algorithms are implemented and readily available in popular image-analysis and feature-extraction packages such as CellProfiler [9, 49]. Once critical parameters are identified using these approaches subsequent experiments can focus on only extracting and storing the relevant parameters, which decreases the image analysis time and space requirements to store the extracted features.

5   Examples of Utility of Single Cell Analysis

Below are examples of how harnessing the single cells data from an HCA experiment can provide insights beyond those provided by traditional well-based analyses.

5.1   Tissue Sections

The benefits of SCA with HCA image cytometry-derived data are particularly apparent when working in biological models that contain discrete structures such as tissue sections, which are stained with multiple fluorescent antibodies against different antigens. In a traditional approach, "representative" images would be selected to highlight the biology of the sample. Of course, these images would represent a very small fraction of the total image data acquired in the experiment. In classic SCA, data from all the cells in the image are compared to each other on 2D plots to define phenotypic subsets. In the experiment shown in Fig. 8, slices of a developing rat embryo brain were stained with fluorescent markers for DAPI (nuclear counterstain), CellMask Orange (Plasma membrane counterstain), and anti-Glial Glutamate Aspartate Transporter (GLAST) antibody. Figure 8a shows a 2D plot of CellMask vs DAPI allowing gating of cells with high and low DAPI intensities. This may be indicative of dividing and quiescent cells due to more DAPI binding to cells with 4 N DNA content. Backgating these populations on the GLAST stained image (Fig. 8b) reveals that cells with a low intensity of DAPI (green) appear primarily within invaginations and high intensity DAPI populations (Blue) cluster toward the periphery. Such techniques and results may be useful to determine the relationship of dividing and quiescent populations relative to a third marker, in this case, GLAST.

Instead of gating on a 2D plot, another approach is to gate on morphological or phenotypic characteristics directly in a representative section of an image and backgate those populations on the entire image to better visualize distributions. A xenografted tumor was fluorescently stained with Hoechst (blood perfusion marker) (Fig. 9) and two gates were drawn in different locations on the image, based on Hoechst staining intensity. These gates were subsequently applied to the histogram to identify the Hoechst distribution in each region as controls and markers set and converted to gates. Backing all events that fall under M1 or M2 on the entire image clearly shows the distribution of Hoechst in the entire image. The distribution difference is most likely due to greater blood perfusion at the extremity of the tumor than in the interior of the tumor although small patchy regions of Hoechst staining can be found throughout the interior.

5.2 Heterogenous Populations in a Single Sample

Another example of the utility of backgating image gates onto 1D or 2D gates is in the analysis of data derived from in vitro or ex vivo tissue culture experiments. Such artificial environments present their own challenges in identifying populations of interest. The presence of "feeder cells," serving as a growth substrate, may complicate identification of the actual cells of interest. Under these circumstances, it is useful to cross-reference image gates with parametric 1D or 2D gates. Figure 10 shows data from an experiment in which rat hepatic progenitor cells (HPCs) were cocultured with hepatic mesenchymal precursor "feeder" cells (MPCs) on a hyaluronan growth substrate. Over time, HPCs that are positive for albumin and cadherin form colonies that are surrounded by a monolayer of MPC feeder cells. Although clusters of HPCs are readily recognizable (Fig. 10a, *left*, white brackets) interspersed among a sparse bed of feeder cells (Fig. 10a, *left*, red gate), the morphological differences between the cells themselves are not sufficient to distinguish feeder cells that lie within the clumps. Thus, the feeder cells represent a contaminant during the data analysis phase of the experiment that must be removed, or "gated out," before the cells of interest can be identified and further analyzed. This required a two-step process. First, a presumptive feeder cell gate is drawn on an area of the image absent of clumps (Fig. 10a, *left*, red gate). Next, this gate is backgated onto a 2D dot plot of albumin vs. DNA intensity, revealing that this population is albumin- and DNA-low (Fig. 10a *right*, red dots). A 2D gate is subsequently drawn on the basis of this backgated population Fig. 10b, *left*), which serves to eliminate the feeder cells experiment-wide; note a few feeder cells have been identified within the clumps of HPCs (Fig. 10a, b, *right*, arrows). The remaining cells, pseudocolored green, comprise the bulk of the clumps (Fig. 10b, *right*), and may be reasonably considered as such for downstream analysis of additional markers (e.g., cadherin, Fig. 10c). This gating strategy clearly exposes the difference in cadherin intensity between the two identified populations. Reciprocal backgating of image and parametric gates thus affords a

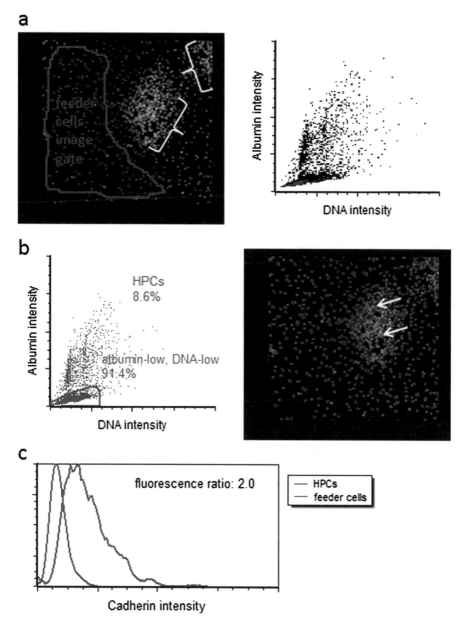

Fig. 10 Reciprocal backgating for the identification of populations in coculture. Hepatocytes (**a**, *left*, *white brackets*) cocultured with "feeder cells" (**a**, *left*, *red* image gate) cannot be definitively identified on the basis of morphology or position alone. A feeder cell gate was defined on the image and backgated onto a 2D plot (**a**). A parametric gate was drawn (**b**, *left*) to "gate out" these backgated feeder cells, and the remaining cells (*pseudocolored green*) are seen to comprise the bulk of the clumps when backgated onto the image (**b**, *right*, *green* cells). These cells are herein considered hepatocytes in the downstream analysis of additional markers (e.g., cadherin, **c**). Note a few feeder cells located within the clump (**b**, *right*, *arrows*), identified successfully with this gating strategy. (Data acquired BD Pathway 354; Segmentation and image analysis using Cellprofiler; SCA using FCS Express 4 Image Cytometry, courtesy of Joshua Harrill, Ph.D., The Hamner Institutes for Health Sciences)

simple yet powerful method for methodically discerning populations before advancing to next level of analysis. This is of critical importance in any kind of experiment where heterogeneous populations are known or suspected.

6 Summary

The data generated from a typical HCA experiment is an ideal candidate for SCA techniques. HCA has the advantage compared to FC by being able to visually confirm that the cells within gates regions actually correspond to the target cells of interest. Increased use of SCA data analysis methodologies to HCA data should unlock the potential of the extremely powerful instrumentation and allow an explosion in the complexity and utility of this cutting edge tool.

Acknowledgments

The authors would like to acknowledge all of the people listed in the chapter who contributed excellent example data to illustrate the power of HCA. We would also like to acknowledge Dr. Bartek Rajwa for editing sections of the manuscript and excellent discussions regarding automated cluster analysis. We would especially like to thank the editors for inviting us to prepare this chapter and for their patience with the authors for their difficulties in meeting the submission deadline.

References

1. Shapiro HM (2003) Practical flow cytometry, 4th edn. Wiley-Liss, New York, NY

2. Trask OJ, Johnston PA (2015) Standardization of high content imaging and informatics. Assay Drug Dev Technol 13:341–346. doi:10.1089/adt.2015.29022.ojt

3. Allan C, Burel J-M, Moore J, Blackburn C, Linkert M, Loynton S, Macdonald D et al (2012) OMERO: flexible, model-driven data management for experimental biology. Nat Methods 9:245–253. doi:10.1038/nmeth.1896

4. Linkert M, Rueden CT, Allan C, Burel J-M, Moore W, Patterson A, Loranger B et al (2010) Metadata matters: access to image data in the real world. J Cell Biol 189:777–782. doi:10.1083/jcb.201004104

5. Murphy RF, Chused TM (1984) A proposal for a flow cytometric data file standard. Cytometry 5:553–555. doi:10.1002/cyto.990050521

6. Spidlen J, Moore W, Parks D, Goldberg M, Bray C, Bierre P, Gorombey P et al (2010) data file standard for flow cytometry, version FCS 3.1. Cytometry A 77:97–100. doi:10.1002/cyto.a.20825

7. Spidlen J, Novo D (2012) ICEFormat-the image cytometry experiment format. Cytometry A 81:1015–1018. doi:10.1002/cyto.a.22212

8. Schindelin J, Arganda-Carreras I, Frise E, Kaynig V, Longair M, Pietzsch T, Preibisch S et al (2012) Fiji: an open-source platform for biological-image analysis. Nat Methods 9:676–682. doi:10.1038/nmeth.2019

9. Carpenter AE, Jones TR, Lamprecht MR, Clarke C, Kang IH, Friman O, Guertin DA et al (2006) CellProfiler: image analysis software for identifying and quantifying cell phenotypes. Genome Biol 7:R100. doi:10.1186/gb-2006-7-10-r100

10. Schneider CA, Rasband WS, Eliceiri KW (2012) NIH Image to ImageJ: 25 years of image analysis. Nat Methods 9:671–675. doi:10.1038/nmeth.2089

11. Dean PN, Jett JH (1974) Mathematical analysis of DNA distributions derived from flow microfluorometry. J Cell Biol 60:523–527

12. Rabinovitch PS (1994) DNA content histogram and cell-cycle analysis. Methods Cell Biol 41:263–296

13. Bagwell B (1993) Theoretical Aspects of Data Analysis. In: Bauer KD, Duque RE, Vincent Shankey T (eds) Clinical flow cytometry: principles and application. Williams & Wilkins, Baltimore, MA, pp 41–61

14. Chan GKY, Kleinheinz TL, Peterson D, Moffat JG (2013) A simple high-content cell cycle assay reveals frequent discrepancies between cell number and ATP and MTS proliferation assays. PLoS One 8:e63583. doi:10.1371/journal.pone.0063583

15. Sutherland JJ, Low J, Blosser W, Dowless M, Engler TA, Stancato LF (2011) A robust high-content imaging approach for probing the mechanism of action and phenotypic outcomes of cell-cycle modulators. Mol Cancer Ther 10:242–254. doi:10.1158/1535-7163.MCT-10-0720

16. Gasparri F (2006) Multiparametric cell cycle analysis by automated microscopy. J Biomol Screen 11:586–598. doi:10.1177/1087057106289406

17. Barabasz A, Foley B, Otto JC, Scott A, Rice J (2006) The use of high-content screening for the discovery and characterization of compounds that modulate mitotic index and cell cycle progression by differing mechanisms of action. Assay Drug Dev Technol 4:153–163. doi:10.1089/adt.2006.4.153

18. Zhang L, He M, Zhang Y, Nilubol N, Shen M, Kebebew E (2011) Quantitative high-throughput drug screening identifies novel classes of drugs with anticancer activity in thyroid cancer cells: opportunities for repurposing. J Clin Endocrinol Metab 97:E319–E328. doi:10.1210/jc.2011-2671

19. Kosugi Y, Sato R, Genka S, Shitara N, Takakura K (1988) An interactive multivariate analysis of FCM data. Cytometry 9:405–408. doi:10.1002/cyto.990090419

20. Lugli E, Pinti M, Nasi M, Troiano L, Ferraresi R, Mussi C, Salvioli G et al (2007) Subject classification obtained by cluster analysis and principal component analysis applied to flow cytometric data. Cytometry A 71:334–344. doi:10.1002/cyto.a.20387

21. Zhou X, Cao X, Perlman Z, Wong STC (2006) A computerized cellular imaging system for high content analysis in Monastrol suppressor screens. J Biomed Inform 39:115–125. doi:10.1016/j.jbi.2005.05.008

22. Nagano R, Akanuma H, Qin X-Y, Imanishi S, Toyoshiba H, Yoshinaga J, Ohsako S, Sone H (2012) Multi-parametric profiling network based on gene expression and phenotype data: a novel approach to developmental neurotoxicity testing. Int J Mol Sci 13:187–207. doi:10.3390/ijms13010187

23. Perlman ZE, Mitchison TJ, Mayer TU (2005) High-content screening and profiling of drug activity in an automated centrosome-duplication assay. Chembiochem 6:145–151. doi:10.1002/cbic.200400266

24. Sanguinetti G, Milo M, Rattray M, Lawrence ND (2005) Accounting for probe-level noise in principal component analysis of microarray data. Bioinformatics (Oxford) 21:3748–3754. doi:10.1093/bioinformatics/bti617

25. Janes KA, Albeck JG, Gaudet S, Sorger PK, Lauffenburger DA, Yaffe MB (2005) A systems model of signaling identifies a molecular basis set for cytokine-induced apoptosis. Science (New York, NY) 310:1646–1653. doi:10.1126/science.1116598

26. Jensen KJ, Janes KA (2012) Modeling the latent dimensions of multivariate signaling datasets. Phys Biol 9:045004. doi:10.1088/1478-3975/9/4/045004

27. Watson JV (2001) Proof without prejudice revisited: immunofluorescence histogram analysis using cumulative frequency subtraction plus ratio analysis of means. Cytometry 43:55–68

28. Lampariello F (1994) Evaluation of the number of positive cells from flow cytometric immunoassays by mathematical modeling of cellular autofluorescence. Cytometry 15:294–301. doi:10.1002/cyto.990150404

29. Overton WR (1988) Modified histogram subtraction technique for analysis of flow cytometry data. Cytometry 9:619–626. doi:10.1002/cyto.990090617

30. Bagwell B (1996) A journey through flow cytometry immunofluorescence analyses. Clin Immunol Newsl 16:33–37

31. Aghaeepour N, Finak G, FlowCAP Consortium, DREAM Consortium, Hoos H, Mosmann TR, Brinkman R, Gottardo R, Scheuermann RH (2013) Critical assessment of automated flow cytometry data analysis techniques. Nat Methods 10:228–238. doi:10.1038/nmeth.2365

32. Lugli E, Roederer M, Cossarizza A (2010) Data analysis in flow cytometry: the future just started. Cytometry A 77:705–713. doi:10.1002/cyto.a.20901

33. Bashashati A, Brinkman RR (2009) A survey of flow cytometry data analysis methods. Adv Bioinforma 584603. doi:10.1155/2009/584603

34. Salzman GC, Beckman RJ, Parson JD, Nauman AM, Stewart SJ, Stewart CC (1996) Flow cytometric immunophenotyping using cluster analysis and cluster editing. In: Blackwell L (ed) Flow and image cytometry. Springer, New York, NY, pp 191–212

35. Murphy RF (1985) Automated identification of subpopulations in flow cytometric list mode data using cluster analysis. Cytometry 6:302–309. doi:10.1002/cyto.990060405

36. Demers S, Kim J, Legendre P, Legendre L (1992) Analyzing multivariate flow cytometric data in aquatic sciences. Cytometry 13:291–298. doi:10.1002/cyto.990130311

37. Pyne S, Hu X, Wang K, Rossin E, Lin T-I, Maier LM, Baecher-Allan C et al (2009) Automated high-dimensional flow cytometric data analysis. Proc Natl Acad Sci U S A 106:8519–8524. doi:10.1073/pnas.0903028106

38. Lo K, Brinkman RR, Gottardo R (2008) Automated gating of flow cytometry data via robust model-based clustering. Cytometry A 73:321–332. doi:10.1002/cyto.a.20531

39. Novo D, Grégori G, Rajwa B (2013) Generalized unmixing model for multispectral flow cytometry utilizing nonsquare compensation matrices. Cytometry A 83:508–520. doi:10.1002/cyto.a.22272

40. Hahne F, Khodabakhshi AH, Bashashati A, Wong C-J, Gascoyne RD, Weng AP, Seyfert-Margolis V et al (2010) Per-channel basis normalization methods for flow cytometry data. Cytometry A 77:121–131. doi:10.1002/cyto.a.20823

41. Finak G, Perez J-M, Weng A, Gottardo R (2010) Optimizing transformations for automated, high throughput analysis of flow cytometry data. BMC Bioinformatics 11:546. doi:10.1186/1471-2105-11-546

42. Quinn J, Fisher PW, Capocasale RJ, Achuthanandam R, Kam M, Bugelski PJ, Hrebien L (2007) A statistical pattern recognition approach for determining cellular viability and lineage phenotype in cultured cells and murine bone marrow. Cytometry A 71:612–624. doi:10.1002/cyto.a.20416

43. Kothari R, Cualing H, Balachander T (1996) Neural network analysis of flow cytometry immunophenotype data. IEEE Trans Biomed Eng 43:803–810. doi:10.1109/10.508551

44. Agatonovic-Kustrin S, Beresford R (2000) Basic concepts of artificial neural network (ANN) modeling and its application in pharmaceutical research. J Pharm Biomed Anal 22:717–727

45. Rajwa B, Venkatapathi M, Ragheb K, Banada PP, Daniel Hirleman E, Lary T, Paul Robinson J (2008) Automated classification of bacterial particles in flow by multiangle scatter measurement and support vector machine classifier. Cytometry A 73:369–379. doi:10.1002/cyto.a.20515

46. Toedling J, Rhein P, Ratei R, Karawajew L, Spang R (2006) Automated in-silico detection of cell populations in flow cytometry readouts and its application to leukemia disease monitoring. BMC Bioinformatics 7:282. doi:10.1186/1471-2105-7-282

47. Loo L-H, Lani FW, Altschuler SJ (2007) Image-based multivariate profiling of drug responses from single cells. Nat Methods 4:445–453. doi:10.1038/nmeth1032

48. Bashashati A, Lo K, Gottardo R, Gascoyne RD, Weng A, Brinkman R (2009) A pipeline for automated analysis of flow cytometry data: preliminary results on lymphoma sub-type diagnosis. Conf Proc IEEE Eng Med Biol Soc 2009:4945–4948. doi:10.1109/IEMBS.2009.5332710

49. Logan DJ, Carpenter AE (2010) Screening cellular feature measurements for image-based assay development. J Biomol Screen 15:840–846. doi:10.1177/1087057110370895

Chapter 16

Utilization of Multidimensional Data in the Analysis of Ultra-High-Throughput High Content Phenotypic Screens

Judith Wardwell-Swanson and Yanhua Hu

Abstract

High Content Screening (HCS) platforms can generate large amounts of multidimensional data. To take full advantage of all the rich contextual information provided by these screens, a combination of traditional as well as nontraditional hit identification and prioritization strategies is required. Here, we describe the workflow and analytics of multidimensional high content data to differentiate, group, and prioritize hits.

Key words Phenotypic drug discovery, PDD, High content screening, HCS, High-throughput screen, HTS, Image analysis, Multiparametric data, Hierarchical clustering, Data analysis, Data correction

1 Introduction

Phenotypic Drug Discovery is enjoying a resurgence, due in part to the recent recognition that more New Molecular Entities (NME) were identified by phenotypic screens over the past decade than by target-based approaches [1, 2]. Phenotypic drug discovery, also called forward pharmacology, is the search for substances that cause a desirable phenotype in cellular or animal models. Compared to target-based drug discovery (TDD) also known as reverse pharmacology, phenotypic drug discovery (PDD) is agnostic to both drug target and molecular mechanism of action (MOA) and can theoretically self-select the optimal drug target or combination of targets, as well as identify the pharmacokinetic properties needed to achieve the most desirable cellular response. In other words, phenotypic screens are tailor-made for the identification of small molecules with novel and potentially superior mechanism(s) of action. Realizing the full mechanistic potential of a given phenotypic screen depends not only on the size, quality, and structural diversity of the input chemical library but also on the screener's ability to preserve diversity throughout the screen triage process. In our experience,

Paul A. Johnston and Oscar J. Trask (eds.), *High Content Screening: A Powerful Approach to Systems Cell Biology and Phenotypic Drug Discovery*, Methods in Molecular Biology, vol. 1683, DOI 10.1007/978-1-4939-7357-6_16, © Springer Science+Business Media LLC 2018

maintaining mechanistic diversity during hit selection is paramount to the success of a phenotypic screen and is rarely achieved by selecting and prioritizing the initial hit list exclusively on potency. Therefore, we developed analysis methods aimed at discriminating and selecting hits based on characteristics other than, or in addition to, potency. For example, the multidimensional output of high content analysis can be used to further characterize hits. Such datasets are amenable to hierarchical clustering and line similarity analysis which can serve to group compounds with shared target activity and, in a similar manner, differentiate those with divergent mechanisms of action [3].

One advantage of using an HCS platform for phenotypic screening is the large selection of compatible detection reagents; another is the large number of multiplexed measurements high content platforms can accommodate. Compatible detection reagents include fluorescently labeled antibodies, fluorescent probes, and GFP-containing fusion proteins, to name a few. The average high content assay multiplexes between two and four such detection reagents in a single assay. Independent images are typically acquired for each detection reagent and the resulting image files are then subjected to image analysis and quantitative feature extraction. As a result of endpoint multiplexing and the vast array of image analysis features extracted at both the cell and population levels, a large number of independent and interdependent measures can be extracted from a single sample. The exact number of measurements can range from one to thousands depending on the sophistication of both the assay and the image analysis algorithm. Hits are generated using these measurements, either by applying a threshold to a single high content feature, or by utilizing a combination of high content features or a feature profile. There are a number of compelling reasons to utilize multidimensional feature profiles in phenotypic screen analysis, which we address in the following paragraphs.

Utilizing multidimensional data enables the identification of subtle phenotype variations, including the identification of unexpected phenotypes [4, 5]. Cellular processes are complex and involve multiple pathways. Compounds that act through different mechanisms may have a similar effect on the primary measure (or feature) but may differ in their effect on secondary measures. By utilizing a full collection of features, we strive to capture as many phenotypic variants as possible during the initial hit selection.

Phenotypic profiles generated from a collection of features can also be instrumental in hit list stratification and can help preserve mechanistic diversity during triage [6]. When hit lists are too large to prosecute through the secondary screening tier, it is customary to filter the hit list by potency, usually based on the single most important endpoint. We have found that this practice significantly reduces the mechanistic diversity contained within a phenotypic screen hit list and in some cases, selects compounds representing only the most well-established pharmaceutical targets. The identification and preservation of diverse phenotypes has important implications for PDD where the initial goal is to identify an optimal and novel mechanism of action and not necessarily to identify the most potent and/or efficacious lead. Hit stratification strategies that incorporate phenotypic profiles can enable differentiation of hits by mechanism of action, and therefore preserve a greater level of diversity while filtering the hit list.

Similarly, phenotypic profiling can be used to prioritize chemical leads at the conclusion of the screening tier. Following confirmation and secondary screening, a selection of chemotypes exhibiting the appropriate phenotypic and selectivity characteristics often are subjected to an intense pursuit of their molecular target or targets. Target deconvolution approaches such as SILAC proteomics [4] are relatively low throughput and time intensive; therefore, only a small number of compounds can be prosecuted. Choosing compounds for target deconvolution with distinct phenotypic profiles should improve the chances of progressing compounds with unique and potentially novel MOAs.

Phenotypic profiles can also help gain early insight into MOA. When reference compound profiles (i.e., compounds with known MOA) are included in the analysis, a potential MOA can often be inferred for hits that closely phenocopy a reference compound [7, 8]. The MOA hypotheses generated from this approach are often sufficient to initiate target validation, thereby rendering a resource-intensive proteomics-based deconvolution approach unnecessary.

And finally, by utilizing the mechanistic discrimination enabled by multidimensional phenotypic data, it becomes possible to establish structure-activity relationships (SAR) in the setting of phenotypic assay, without resorting to target-based assays [9, 10]. This has important implications for a downstream medicinal chemistry effort.

These insights led us to adopt multidimensional data analysis during the initial hit selection phase of our high content phenotypic screens and to develop hit stratification strategies based on phenotypic clustering to more effectively bin and prioritize hits based on their MOA.

Preparing and analyzing multidimensional high content datasets for phenotypic profiling is not without unique challenges. For example, an HTS campaign performed on a high content platform can easily generate a terabyte of image files and gigabytes of numerical data. Identifying software packages and desktop computing solutions capable of handling such immense datasets can be a significant challenge for the average laboratory. In addition, high content HTS data are particularly prone to systematic errors due to the large number of automated steps involved in assay processing.

Therefore, data normalization and data correction are commonly required prior to subjecting the high content dataset to hit selection or further analysis. Numerous data normalization and data correction software packages have been developed and commercialized for HTS data, but only a few are able to accommodate the scale and multidimensional nature of HCS data [5].

A number of methods for analyzing multidimensional datasets have already been described. For example, methods for clustering multivariate expression profiling data have been utilized for the past two decades to cluster treatments with similar expression profiles in an attempt to identify compounds with shared or differentiating MOA [6, 7, 11]. Clustering using phenotypic profiles is a relatively new twist on this theme but has recently been shown by several labs to be an effective way to group compounds by their MOA [8, 9]. Unsupervised clustering of high content data can be achieved by a number of methods such as K Means, Hierarchical clustering, Mean Shift Clustering [10, 12]. Supervised classification methods, such as SVM (support vector machine), and semi-supervised methods have also been utilized. Supervised classification methods are generally utilized to compare compounds to a known or reference [8, 13, 14], while unsupervised clustering approaches, such as hierarchical clustering, have the ability to identify unexpected phenotypes. Therefore, if the goal of the phenotypic screen is to identify and explore as many novel mechanisms as possible or if suitable reference treatments are not yet known, an unsupervised clustering method is preferable. Once compounds have been differentiated by their respective phenotypic profiles, a selection of chemotypes from each of the putative "mechanistic bins" can be prioritized and selected for further triage without sustaining substantial loss of mechanistic diversity or novelty.

Here, we describe the acquisition of multidimensional high content data set from a phenotypic screen, the selection of high content parameters for phenotypic profiling, and the generation of the initial hit list. We provide the methods employed for the preprocessing of raw HCS data as well as the methods used for hierarchical clustering and line similarity analysis of the dataset.

To illustrate our methods, we provide an example of an ultra-high-throughput multiplexed high content phenotypic screen. The purpose of the screen was to identify mechanisms involved in the endocytosis-mediated uptake of a fluorescently labeled ligand (Ligand X). For the high content assay two fluorescent detection reagents were utilized: an Alexa647-labeled Ligand and Hoechst, a nuclear DNA probe. Several mechanisms of endocytosis modification had been previously described; these as well as other novel mechanisms were of interest to the project team. Hit identification methods based on multidimensional data were employed to maximize our ability to identify novel mechanisms and to differentiate and group the resulting hit list.

The Bristol-Myers Squibb lead discovery group conducted the high content screen in a miniaturized 1536-well high-throughput format utilizing over 900,000 compounds from our screening deck. All compound treatments were performed at a single 10 μM dose. More detailed information on the high-throughput screening methods can be found in the chapter by Nickischer and Weston entitled "High Throughput, Miniaturized High-Content Screening: Reducing it to practice."

In the provided example, an extracellular ligand, which we name Ligand X, binds to a plasma membrane-associated receptor and orchestrates the rapid internalization and lysosomal degradation of the ligand/receptor complex. In one instance, we sought to identify treatments known as "inhibitors" which significantly reduce the cellular entry of Ligand X. In another instance, we sought to identify treatments that enable a specific type of intracellular "accumulation" of Ligand X, broadly referred to as "accumulators." Within the broader phenotypic groups we were keenly interested in treatments that redirect any internalized Ligand X away from lysosomes and to the endosomal recycling compartment (ERC) where presumably the associated receptor can be recovered and recycled back to the cell surface. Lysosomes are small abundant vesicles uniformly distributed throughout the cytoplasm, while ERC presents in our cell line as medium to large vesicles more heavily localized to the nuclear and perinuclear regions of the cell. This inherent difference in the morphology and subcellular localization of internalized Ligand X can be exploited to distinguish between the two intracellular fates of the ligand.

2   Materials

A commercial automated high content imaging platform with the associated image analysis software package was used for the image acquisition and analysis—in the example provided, the Opera automated high content platform and integrated Acapella software (Perkin Elmer, Norwalk, CT) were utilized.

Genedata Screener software (Genedata, AG. Basel, Switzerland) was used for data normalization, automated data correction, and data masking.

TIBCO Spotfire software (TIBCO, Inc. Palo Alto, CA) was utilized for initial raw data QC, and for the cluster analysis and hit selection.

A 64bit dual core processing desktop computer with 16GB RAM was used for all the data analysis.

3 Methods

3.1 Image Analysis   To fully develop a cellular phenotypic profile from an HCS assay, it is necessary to utilize an image analysis algorithm capable of extracting quantitative data related to the relative abundance, distribution, and morphological characteristics of the target proteins and/or intracellular structures. The resulting multiparametric data can then be used to define a cellular phenotype. A carefully designed multiparametric image analysis algorithm is capable of quantifying phenotypic changes, which would otherwise be indiscernible with a single endpoint assay. Figure 1 illustrates the key stages of developing such an image analysis algorithm. In this example, we developed an image analysis algorithm to address cellular phenotypes associated with a particular type of intracellular protein trafficking. We extracted quantitative information from the images related to the

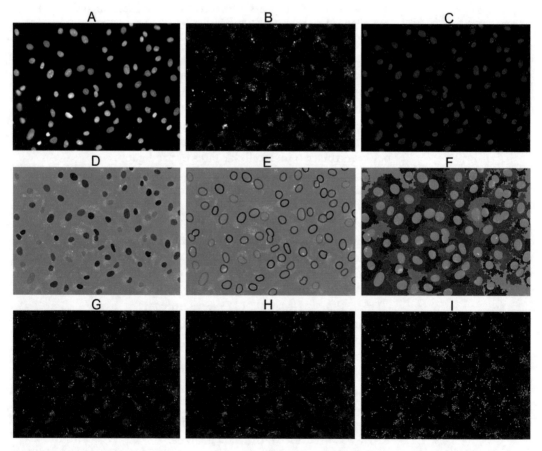

Fig. 1 Illustration of image analysis algorithm. (**a**) Raw nuclear channel image. (**b**) Raw Ligand X channel image. (**c**) Composite image of nuclear channel in *blue* and Ligand X channel in *red*. (**d**) Definition of nuclear region. (**e**) Definition of perinuclear region. (**f**) Definition of cytoplasmic region. (**g**) Spot definition in nuclear region. (**h**) Spot definition in perinuclear region. (**i**) Spot definition in cytoplasmic region

relative abundance of Ligand X in three subcellular compartments that we refer to as nuclear, perinuclear, and cytoplasmic. We also extracted information related to size, shape, number, and distribution of punctate structures that correspond to the accumulation of Ligand X in either lysosomes or the ERC. The general approach to developing such an algorithm is provided in the steps below. The number and type subcellular compartments, labeled targets, and structures will vary according to the biology being studied (*see* **Note 1**).

1. First define the nuclear region. The nuclear region is selected by applying the appropriate intensity and area thresholds to the nuclear image (in most cases this will be the fluorescent channel containing the image of Hoechst or DAPI labeled nuclei) using a standard image analysis software package (Fig. 1a, d).

2. Next define additional regions of the cell. Use the nuclear region defined in the previous step as a seed to define the additional regions. For example, a perinuclear region might be defined as a 5-pixel wide ring surrounding the nuclear region (Fig. 1e). And a cytoplasmic region might start at the outer boundary of the perinuclear region and extend to the cell boundary to cover the entire region between the perinuclear boundary and the extracellular membrane (Fig. 1f).

3. Overlay the subcellular regions of interest obtained in the previous steps onto each fluorescent channel and calculate the relevant intensity and morphological features (such as shape, area, and texture measures) for each subcellular region. In the example shown in Fig. 1d–f the subcellular masks were overlaid onto images that correspond Alex647 labeled Ligand X (Fig. 1b) and the corresponding Mean Average Intensity, Mean Total Intensity measurements were calculated.

4. Next identify any notable intracellular structures such as spots or fibers within the subcellular compartments identified in the steps above. Regions of punctate or high contrast staining commonly characterize such structures. Use intensity, area, and shape thresholds to define the regions of interest corresponding to spot/fiber structures. For each of the subcellular regions (defined in **steps 1** and **2**) calculate all relevant intensity and morphological features for the individual spots and/or fibers. In a protein trafficking assay for example, the relevant features might include the number of spot/fibers, total and average spot/fiber intensity ,and total and average spot/fiber area for each of the individual cell regions. Based on previous observations we postulate that all of these features might be useful for differentiating protein trafficking phenotypes. *See* Fig. 1g–i where intracellular spots were detected and quantified.

5. Include any calculated features that might be relevant to the desired phenotypes. For example, differentiation of protein trafficking phenotypes might be aided by including calculated features such as the ratio of the total spot/fiber intensity values from adjacent cell regions (i.e., mean total spot intensity perinuclear region/mean total spot intensity cytoplasmic region).

6. To capture treatment effects on nuclear health and cell cycle, quantify the number of nuclei per field, mean nuclear intensity, and shape.

7. Table 1 contains a list of 25 image analysis features acquired with an image analysis algorithm designed for the ligand internalization and compartmentalization assay. In this example, the extracted features were preselected for their biological and phenotypic relevance and their combined ability to discriminate the phenotypes of interest.

3.2 Joining the Image Analysis Files with Screen Metadata

For most large screens, the next step after data acquisition involves the joining of endpoint data with the screen annotation. Large multiparametric datasets pose some unique challenges when making these joins such as the manipulation of such large data files and the need to apply different control groups to each of the high content features, to name a couple. Fortunately, both commercial and open-source solutions have been developed to facilitate this process.

1. To properly evaluate screen performance and interpret the screen results the image analysis data needs to be joined with experimental annotation such as Run ID, Plate ID, Well ID, Row ID, Column ID, Channel ID, Treatment Type and Sample ID, also known as the screen metadata.

2. Merge the multi-feature high content data with required screen metadata using commercial software tools such as Genedata, in-house HCS data management solutions such as HCS Road [15] or manual approaches involving cutting and pasting information from multiple sources.

3. Once joined with the metadata, export the completed datasets as text files and open in a downstream data analysis and visualization tool such as TIBCO Spotfire.

3.3 Initial Visualization and QC

TIBCO Spotfire offers a large number of data visualization options making it a good choice for the initial visual inspection of HCS data. Visualize the overall screen performance by creating a heat map visualization of the entire screen.

1. Import the joined data file into TIBCO Spotfire. Organize the data table such that the rows contain treatments and the columns contain the screen metadata and high content features.

Table 1
List of features extracted from the 2-channel assay

Fluorescent probe	Imaging channel	Feature name
Hoechst	Channel 1	Number of cells per field
Hoechst	Channel 1	Mean Nuclear Total Intensity
Hoechst	Channel 1	Mean Nuclear Area
Hoechst	Channel 1	Mean Nuclear Roundness
Alexa647-Ligand X	Channel 2	Nuclear Mean Spot Count
Alexa647-Ligand X	Channel 2	Nuclear Mean Spot Average Intensity
Alexa647-Ligand X	Channel 2	Nuclear Mean Spot Average Area
Alexa647-Ligand X	Channel 2	Nuclear Mean Spot Total Intensity
Alexa647-Ligand X	Channel 2	Nuclear Mean Spot Total Area
Alexa647-Ligand X	Channel 2	Nuclear Mean Average Intensity in region
Alexa647-Ligand X	Channel 2	Nuclear Mean Total Intensity in region
Alexa647-Ligand X	Channel 2	Perinuclear Mean Spot Count
Alexa647-Ligand X	Channel 2	Perinuclear Mean Spot Average Intensity
Alexa647-Ligand X	Channel 2	Perinuclear Mean Spot Average Area
Alexa647-Ligand X	Channel 2	Perinuclear Mean Spot Total Intensity
Alexa647-Ligand X	Channel 2	Perinuclear Mean Spot Total Area
Alexa647-Ligand X	Channel 2	Perinuclear Mean Average Intensity in region
Alexa647-Ligand X	Channel 2	Perinuclear Mean Total Intensity in region
Alexa647-Ligand X	Channel 2	Cytoplasmic Mean Spot Count
Alexa647-Ligand X	Channel 2	Cytoplasmic Mean Spot Average Intensity
Alexa647-Ligand X	Channel 2	Cytoplasmic Mean Spot Average Area
Alexa647-Ligand X	Channel 2	Cytoplasmic Mean Spot Total Intensity
Alexa647-Ligand X	Channel 2	Cytoplasmic Mean Spot Total Area
Alexa647-Ligand X	Channel 2	Cytoplasmic Mean Average Intensity in region
Alexa647-Ligand X	Channel 2	Cytoplasmic Mean total Intensity in region

Four features were extracted from the nuclear channel, including cell count and three nuclear morphology features. Seven features were extracted from each region in the Ligand X channel, including five spot features and two overall intensity features. All features are well-level features: our image analysis program extracted features from each cell and averaged them across all cells in the same well, except for the cell count feature

2. Use Spotfire's visualization tools to create a scatterplot using Column ID for the *x*-axis and Row ID for the *y*-axis.

3. Use the Trellis feature to create a unique scatterplot for each plate. In the Properties menu select Trellis, and then select Panels from the submenu and Split by PlateID.

4. Create a heat map-like view of the data by coloring the markers according to the value range for the selected high content feature (e.g., Mean Cyto Spot Total Intensity).

5. Visualize the results for each high content feature by selecting and viewing the individual features by using the "color by" dropdown menu. Use this simple plate visualization to easily and rapidly review a large list of high content features for response patterns related to either screen performance and/or systematic errors. Figure 2 is an example of a visualization that provides a valuable high-level look at overall screen performance. In this small selection of plates from the screen, the performance of the positive control (located in the last two columns of each plate) can be observed across all the plates. In addition, process-related artifacts can also be observed including plates with row effects, intensity gradients, and significant

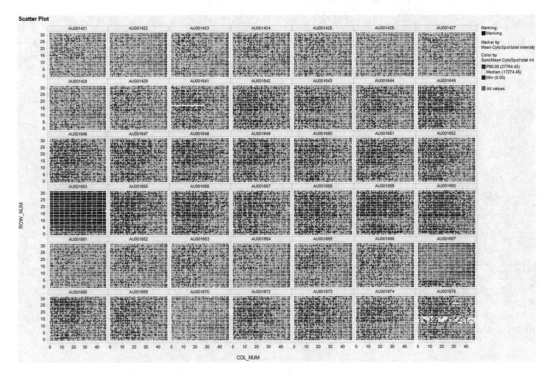

Fig. 2 Heat map visualization in TIBCO Spotfire of the overall screen performance. *Red* represents CytoSpot Total Intensity greater than the median and blue represents CytoSpot Total Intensity less than the median. The pattern of color distribution reveals recurring row effects, a failed plate, and a plate gradient artifact that persists across many consecutive plates

variations in mean intensity. Alternatively, visualization and QC can be performed in a commercial application designed specifically for screening data (*see* **Note 2**).

3.4 Data Preprocessing

High content phenotypic assays are typically complex cell-based assays utilizing one or more treatments/well, an incubation period, sample fixation, and a sequence of staining steps. These steps involve automated aspiration, plate washing, dispensing, and mixing. With complex cellular assays, there are abundant opportunities to introduce systematic errors (such as those illustrated in Fig. 2). The following steps will prepare a high content dataset containing typical systematic errors for high confidence hit calling and for more advanced data operations such as hierarchical clustering or line similarity analysis (*see* **Note 3**).

3.4.1 Data Normalization

Feature normalization is an essential step to eliminate plate-based and batch-based variability, and to scale the different features. Following normalization, features can be compared across plates and batches, and different features will be equally weighted in downstream distance-based analyses, for example hierarchical clustering.

1. Select a data normalization and data correction software or solution suitable for multidimensional datasets. Genedata's Screener product [8] and HCS Analyzer open-source software [16] are examples of data normalization and correction software designed with multidimensional HCS datasets in mind.

2. Import the merged dataset into Genedata Screener (or similar software tool) and subject the data to plate-by-plate normalization.

3. To perform data normalization using the Z score method, compare the feature value for individual treatment wells to the mean and standard deviation of the feature value for the negative control wells. Always use treatments and negative controls from the same plate.

   Z score is calculated per Eq. 1.

$$Z \text{ score} = \frac{x - m}{s}, \tag{1}$$

where x is the feature value for any well, m and s are the mean and standard deviation of the same feature across negative control wells in the same plate respectively. Z score was selected as the normalization method because of the statistical implications, i.e., if the feature in the negative controls has a normal distribution, then a Z score with an absolute value greater than 3 is significant. Figure 3b illustrates how Z score normalization eliminates batch effects from the raw data in Fig. 3a.

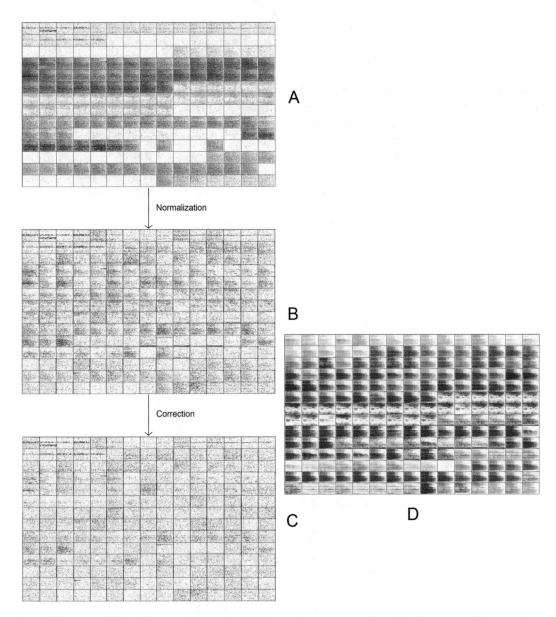

Fig. 3 Heat map of data before and after normalization and correction in Genedata Screener. (**a**) The raw data exhibits feature variability across batches, striping artifacts and patterned artifacts across multiple consecutive plates. (**b**) After normalization, batch effects are removed but intra-plate artifacts remain. (**c**) After data correction, the majority of intra-plate artifacts are removed, although some remain. (**d**) The correction coefficients that were applied during the well-by-well data correction were recorded and can be exported

4. Carry out the normalization for each high content feature within the dataset. With the Genedata software, normalization can be performed across all the features simultaneously. Normalization methods other than Z score may be more suited to certain types of datasets (*see* **Note 4**).

3.4.2 Data Correction

Plate to plate variation can be largely corrected by normalization; however, certain systematic errors within plates, usually characterized by recurring suspect patterns, may still exist. One such example is the row or column effects caused by pipetting or aspiration error. Often such artifacts can be easily identified by looking at the heat map of each single plate. A second and subtler type of artifact is a nonrandom pattern located in a specific area of the plate or screen (e.g., a recurring intensity gradient caused by an uneven plate washer head). Both types of systematic errors are present in the example shown in Fig. 3b. To correct for systematic errors, apply the following steps.

1. Subject the *Z*-score normalized data to a data correction algorithm. The additive data correction tool available in the Screener software from Genedata is adept at recognizing and correcting systematic patterns in batches of plates and generates data files and visualizations with both the corrected data (Fig. 3c) and the correction coefficients for each well (Fig. 3d). The Additive Data Correction tool in Screener uses a proprietary algorithm based on Factor Analysis to determine dominant patterns on a set of reference plates. The strength of each pattern is determined on the target plate and the respective contribution from this pattern is subtracted to obtain a corrected signal (Genedata, personal communication). In addition to the commercially available solution offered by Genedata, several data correction methods for HTS have been recently published (*see* **Note 5**).

2. Review the heat map of the correction coefficients to confirm that the automated pattern recognition agrees with the original visual inspection. *See* the example shown in Fig. 3d.

   Prior to data correction a disproportionate number of wells corresponding to the "artifact region" identified by the Genedata software had very high feature values, and therefore would have met the hit selection criteria. This would have resulted in a significant number of false positives and would have skewed downstream cluster analysis. After data correction, such systematic errors were largely eliminated (*see* **Note 6**).

3.4.3 Data Masking

Many systematic artifacts can be corrected (as above) rendering useful data, but some cannot. In the present example, we observed (Fig. 3c) that a small number of row pattern artifacts remain following the data correction step. In this case, it is necessary to remove or "mask" these wells from further downstream analysis. Genedata Screener software provides two methods of masking these wells: automatic and manual. For large datasets, automated masking is clearly a preferable first step over manual knockouts.

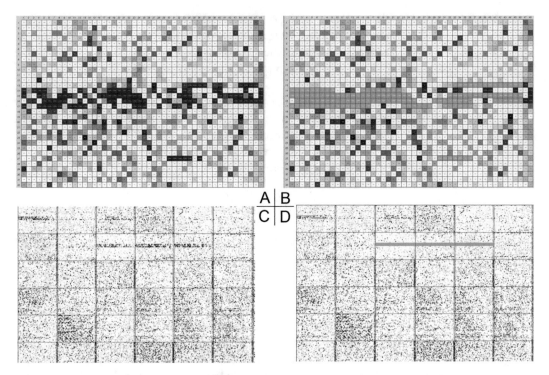

Fig. 4 Masks were applied to artifacts that could not be corrected. The automatic artifact detection in Genedata Screener locates contiguous wells with abnormal feature values (**a**), and masks/removes them from further analysis. The masked areas are illustrated as gray wells in (**b**). The manual masking of "bad" wells is lower throughput but potentially more accurate, for example the downstream impact of the remaining artifacts in (**b**) can be minimized by removing the whole rows in the middle of the plate. For a more efficient manual masking of artifacts that appear in the same positions across multiple plates, we used Genedata's Group Lens feature (**c**), and applied the same masks to every plate in the group (**d**)

1. In Genedata Screener, initiate the automatic masking using the default settings.

2. Select the threshold settings for the automatic masking empirically by subjecting the dataset to incremental threshold adjustments followed by the examination of the results. Use an aggressive threshold to remove all wells with suspicious patterns or a more lenient threshold to remove only the strongest artifacts, as required (*see* **Note 7**). Figure 4a, b illustrate before and after automatic masking for the same plate. The automatic masking algorithm was effective at identifying the majority of the suspect wells in the middle of the plate; however, some outlier wells that were non-contiguous with the stripe were missed. Missed outlier wells can be addressed with the next step.

3. If necessary, remove remaining wells with clear artifacts by performing a manual masking step. For positional artifacts that occur on multiple plates use the Group Lens feature in

Genedata Screener to enable a more efficient and less subjective masking option. In the example shown in Fig. 4c, three plates in a row have similar striping artifacts in the middle of the plate, the three plates were grouped via the Group Lens feature, and the suspect wells on all three plates were masked simultaneously (Fig. 4d).

3.5 Generating the Initial Hit List

3.5.1 Data Reduction

A data reduction step has several positive downstream repercussions. Smaller file sizes enable more facile handling of the data during memory intensive operations such as hierarchical clustering. Further, removing samples with significant liabilities such as cytotoxicity or cell cycle aberrations reduces time and resources spent on compounds/treatments that are ultimately of little or no interest.

1. Import the preprocessed data into TIBCO Spotfire and reduce the dataset to the control treatments and non-cytotoxic hits by using Spotfire's data filters to remove any treatments that meet any one of the following criteria:

2. "Empty wells"—any wells that contain no treatment information.

3. "Cytotoxic wells"—any wells where the cell count has a Z score <-3 or a nuclear morphology feature(s) with a Z score >3.

4. "Inactive wells"—any treatment wells where all the features have a nonsignificant Z score (e.g., Z score between -3 and 3).

The resulting file will include the initial hits, sample wells associated with at least one feature with a significant Z score value and without overt cytotoxicity, and the control treatments.

3.5.2 Feature Selection/ Reduction

1. Visualize the remaining data as a heat map (samples vs. high content features) in TIBCO Spotfire. Remove features from further analysis if they meet the following criteria (see **Note 8**).

2. Non-informative features, features that do not vary significantly (e.g., Z scores between -3 and 3) across any treatments can be removed from the dataset.

3. Redundant features, features that are highly correlated with other features can be replaced with a single feature representative of the correlated group.

4. Non-biologically relevant features, features that are unlikely to be related to the desired phenotype or where the relationship to the phenotype is not well understood may also be removed at the researcher's discretion.

3.6 Hierarchical Clustering

A key objective for phenotypic screens is to generate an initial hit list that represents as many diverse mechanisms as possible. Since the resources for further hit characterization are often limited, a prior-itized and meaningfully stratified hit list can add significant value. For many phenotypic screening projects, more than one MOA can yield a favorable outcome; therefore, the goal is to capture as many mechanisms as possible. In light of this goal, one core objective is to stratify the treatments by their phenotypic profiles, thereby enabling the selection and prioritization of hits from each of the biologically relevant phenotypic clusters. A second objective involves clustering treatments by phenotypic profile for the purpose of understanding their mechanistic relationship. Both objectives rely on the assumption that compounds with similar MOAs will share similar phenotypic response profiles. With this in mind, an unsupervised learning method can be utilized to cluster com-pounds by their similarities in the feature space. Verification of the deduced MOA is required for every screen, using known or refer-ence compounds or by ex post facto validation.

1. Upload the reduced dataset of initial hits and selected features into a new TIBCO Spotfire file.

2. Perform phenotypic clustering by selecting the Hierarchical Clustering method from the Spotfire Tools menu. Use the default parameters—UPGMA clustering method and Euclid-ean distance measure. Cluster both the samples (rows) and features (columns). *See* Fig. 5 for an example heat map of hierarchical clustering results. A tree structure is shown on the top of the heat map depicting the similarity level of the 18 features, and the tree structure on the left shows the simi-larity level of the 10^5 treatments.

3. Move the dendrogram threshold bar across the tree structures to change the cutoff threshold for cluster definition. Move it closer to the root for fewer clusters and broader definition of phenotypic profiles, and move it closer to the leaves to further segregate big clusters into smaller ones with more subtle differ-ences. Clusters are labeled green and blue alternatively along the side of the heat map, to enable easy visualization of the cluster boundaries. In the example shown in Fig. 5 the initial threshold was set such that two broad clusters were identified. Each cluster can be further subdivided and investigated if needed.

3.7 Line Similarity

Scientists familiar with the underlying biology of a phenotypic screen often have a desired phenotype or feature profile in mind that they want to use to prioritize the hit list. One approach to identifying specific phenotypes is to gradually stratify hierarchical cluster data and search each cluster for the expected phenotype. However, significant effort is required to systematically examine a tree of 10^5 nodes. A more direct and effective method is to use a line similarity approach.

Heat Map

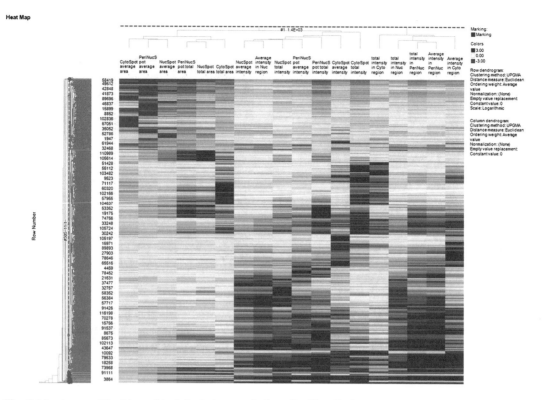

Fig. 5 A heat map of the hierarchical clustering results from Spotfire. Each row represents one treatment, and each column represents one feature. Similar treatments or features are present close to each other in the hierarchical tree structure. Z scored features are colored *red* if greater than 3, *blue* if smaller than −3, and colored in gradient if between −3 and 3. Since the samples have been pre-filtered to have at least one feature significant (> 3 or <−3), we see at least one *solid red* or *blue* in each *row*. The heat map contains two large clusters, one "inhibition" group and one "accumulation" group, along with multiple smaller clusters

1. Select 5–10 features that best describe and most effectively differentiate the desired phenotype from the other phenotypes in the dataset.

2. Create line plots (profiles) for all the treatments, by selecting the features of interest on the Y axis and using "(Column Names)" as the X axis.

3. Select a treatment representative of the "desired profile" to serve as the line of reference.

4. Run Line Similarity algorithm by selecting it from the Spotfire Tools menu and rank the entire collection of hits by their similarity to the line of reference.

5. Repeat this process for all the phenotypic profiles of interest.

In the present example, we were interested in identifying treatments with an "ERC" phenotype. (The "ERC" phenotype is a subtype of the more general "accumulation" and "inhibition" phenotypes, *see* Subheading 1). Treatments that produce this phenotype had been previously described ([17], and unpublished data);

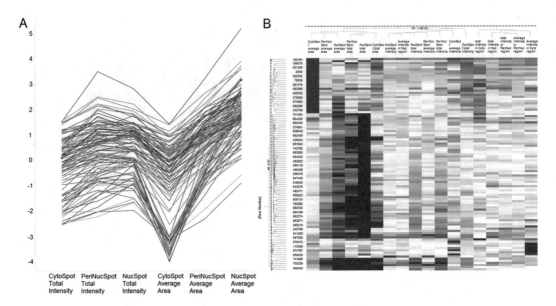

Fig. 6 Results from a line similarity search in Spotfire. (**a**) Top 100 treatments that are most similar to a reference treatment profile in terms of six features. (**b**) Same 100 treatments shown in the heat map within the tree structure built in Fig. 5. All 100 treatments have highly correlated six features, but still diverse overall feature profiles

however no specific reference treatments bearing this signature were included in the screen. The "ERC" profile results from the trafficking of Ligand X to the endocytic recycling compartment rather than the typical lysosomal trafficking, and would be expected to present as a more significant "accumulation" of Ligand X in medium to large vesicles, at or near the nucleus of the cell. Therefore, we would expect treatments with this phenotype to exhibit a distinct profile based on the following six features: higher total spot intensity in the nuclear and/or perinuclear regions, lower total spot intensity in the cytoplasmic region accompanied by higher total spot area in the nuclear and/or perinuclear regions, and lower total area in the cytoplasmic region. From the list of hits, we preselected a treatment with a strong "ERC" profile (above) to serve as the reference profile. Next, we created a line plot for all 10^5 initial hits, using the same six features and ranked the entire collection of hits by their similarity to the reference profile. The top 100 hits are shown in the line graph (Fig. 6a). Mapping the top 100 hits to our hierarchical clustering heat map confirms an earlier observation that the ERC phenotype is a subtype of several clusters. Treatments exhibiting a trend of increased spot area and intensity in the perinuclear region can be members of both the "inhibition" and "accumulation" phenotypic clusters (as shown in Fig. 6b) (*See* **Note 9**). By retrieving the tree structure for the ERC candidates, we have the potential to triage them further by cluster sub-type.

3.8 Quality Control of HCS Hit Calling and Cluster Analysis

The prioritized hit list selected using both hierarchical clustering and line similarity may have the desired feature distribution; however, further quality control (QC) of these hits is required to rule out artifacts introduced during the sample generation, imaging, and image analysis as well as to ascertain the validity of our data preprocessing and hit selection. Therefore, the following QC steps are recommended.

1. First, examine the distribution of the "hits" in the plate map view to identify any potential positional artifacts. Figure 7a illustrates a suspicious distribution of hits from a dataset that had not been subjected to data correction and well masking. The distribution of hits is heavily biased toward the left half of the plates and the pattern is repeated across many plates, an

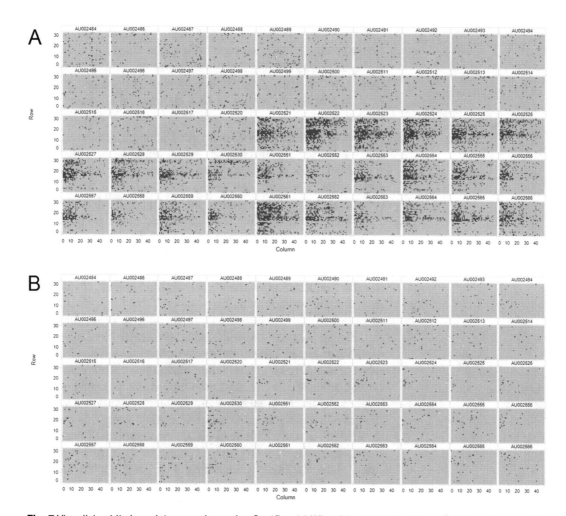

Fig. 7 Visualizing hits in a plate map view using Spotfire. (**a**) When hits appear in either specific patterns and in the same positions in multiple plates, we suspect they are artifacts. (**b**) Feature correction and masking successfully exclude these false positive hits

indication that artifacts are being erroneously scored as hits. Figure 7b shows the hit distribution from our hit list following data correction and masking. The random distribution of hits observed is consistent with the known random distribution of the screening library and therefore no longer appears suspicious.

2. Next, validate the cluster analysis by inspecting the cluster membership of the positive control treatments. The expectation from a robust screen is that majority of positive control treatments will fall into the same cluster (or nearest neighbor clusters) and that negative control treatments will never (or only very rarely) cocluster with the positive controls or hits. Inspecting the cluster membership of the positive and negative controls is also a good way to empirically evaluate alternative clustering methods and settings.

3. Examine the clusters for enrichment of structurally related compounds. The enrichment of a given chemotype (relative to input library) in a phenotypic cluster serves as further validation of the clustering method particularly when accompanied by marked depletion of that chemotype in more distant clusters.

4. Next examine representative image files for each of the major or high interest phenotypic clusters. This is a particularly resource-intensive QC step, but there really is no substitute for the direct examination of the raw image files. The process can be significantly improved by using applications such as Spotfire that enable direct links from the metadata-associated feature files to the corresponding image files. Distant clusters on the tree should contain visually distinct phenotypes; likewise, images from the same cluster should appear similar. During the examination of the representative images, any clusters that contain images with obvious artifacts (e.g., fluorescent compounds) should be noted and possibly removed from the file. Figure 8a is an example of an abnormal phenotype that passed our hit and feature selection criteria and therefore was selected as a hit. The treatment matches the "inhibition" phenotype as shown in the heat map underneath; however, visual inspection of the Hoechst channel image reveals nuclei with abnormal shapes, an indication that this sample should be disqualified as a hit. This example also suggests that our nuclear shape/size criteria were not sufficient to capture all nuclear shape abnormalities. By including additional nuclear features or more stringent thresholds, it is possible to improve the automated detection of this phenotype in the future. Figure 8b is an example of images that match well with the

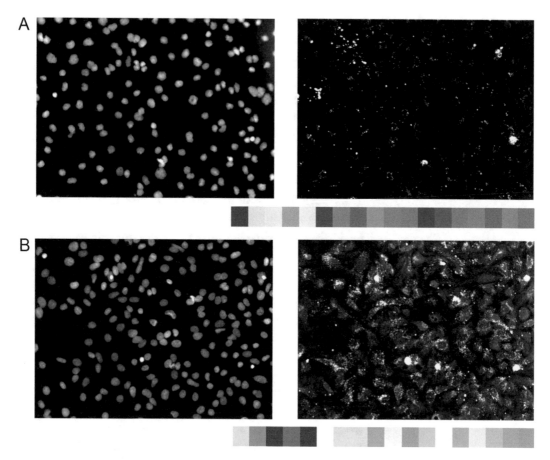

Fig. 8 Comparing image phenotypes with feature profiles. (a) A compound with across the board feature reduction ("inhibition") phenotype is inspected. Nuclear channel on the *left* shows jagged boundary. The compound passed the nuclear feature QC, but apparently the cut off value for nuclear roundness was too lenient. (b) A compound with "redistribution" and "accumulation." The nuclei appear healthy and Ligand X has expected "accumulation" of spots in the perinuclear region

ERC phenotype. Nuclei appear normal; the spots in the Ligand X channel are clearly aggregated around the nuclei, forming a ring-like structure, consistent with the phenotype previously highlighted as high interest.

5. Intensive downstream characterization of hits from each cluster is required to fully validate the initial hit calling and the clustering methods. For example, representative hits from unique clusters can be subjected to expression profiling or to pathway-specific assays to validate the mechanistic similarity of the compounds. Molecular targets deduced from the coclustering of hits with known reference treatments can be verified using genetic approaches such as gene editing or RNAi.

4 Notes

1. Image analysis is a mechanism for extracting a large amount of biologically relevant information in the form of features, from a fluorescently labeled sample. Multiple extracted features can be used to generate a complex cellular response profile for each treatment. The resulting profiles, which can vary according to MOA, enable the discrimination of treatments that might otherwise appear identical in a single endpoint assay. When designing the image analysis algorithm, one should keep in mind the possible cellular responses and make sure that all biological relevant features are captured.

2. The initial data visualization, QC, and the hit calling were conducted in TIBCO Spotfire because the authors had previously established a custom Spotfire template for high content screening data. Potentially all these operations and the data preprocessing tasks could have been conducted in Genedata's Screener software eliminating the need to move the data between two software packages.

3. The requirement for data polishing is further increased when advanced data analysis methods such as hierarchical clustering are applied to the resulting dataset, as any remaining strong artifacts will dominate the clustering.

4. Two common feature normalization methods that use negative controls as the reference are Percent Control and Z score. We commonly chose Z score because of its statistic implication. If there are positive controls with a strong treatment effect, the features can be normalized by Percent Control. The positive and negative wells are considered 0% and 100% respectively if studying inhibition, or 100% and 0% if studying stimulation. And the features are scaled accordingly. The challenge of such 2-point normalization is that positive controls are only controls for certain features but not for all. In our case, the positive/inhibition control is a benchmark for spot intensity or area, but not for nuclear morphology. It does not make biological sense to normalize nuclear morphology features using these controls. Choosing such 2-point normalization method would have limited our selection of features.

5. To remove artifacts introduced by pipetting and other systematic errors, we used Genedata's Screener software because it provides a user-friendly interface for easy artifact detection and removal. If a commercial tool is not available, one can use published algorithms such as median polish procedure [17], B-score transformation [18], or well correction [19].

6. Prior to correcting for artifacts, it may be necessary to rule out the patterns caused by a biased arrangement of the compound

library, examine multiple plates for similar shaped irregularities (i.e., the nonrandom patterns caused by local clusters of structurally related actives are unlikely to be recurring patterns).

7. It is difficult to remove every "bad" well without affecting some "good" ones, so we generally chose to use a relatively lenient threshold for automatic masking.

8. Highly correlated and non-informative features were removed in order to effectively cluster our dataset. For the purposes of this exercise, visual examination of the heat map was sufficient for identifying measurements that were either redundant or non-informative. Automated feature selection methods are actively discussed in the machine-learning field, and are out of the scope of this chapter.

9. A fundamental difference between the clustering and the line similarity approach we took is that we used the Euclidean distance as the metric to measure similarity during the clustering, striving to group compounds that have the closest degree of effect; and we used correlation as the metric in the line similarity search, striving to find compounds that have the closest feature distribution trend.

Acknowledgments

We gratefully acknowledge Rex Parker, Daniel Meyers, Debra Nickischer, and Andrea Weston for providing the HCS dataset. We would like to thank the scientists at Genedata in particular James Kristie and Jon Tupy, for their guidance and support with the Screener software. We would also like to thank Michael Lenard, Normand Cloutier, and Carlos Rios for their assistance with the data handling, their expert IT support and for providing the 64-bit computers used in the data analysis and Donald Jackson for his assistance with Spotfire and 64-bit computing. All the work presented in this chapter was performed while the authors were at Bristol-Myers Squibb.

References

1. Swinney DC, Anthony J (2011) How were new medicines discovered? Nat Rev Drug Discov 10:507–519

2. Swinney DC (2013) Phenotypic vs. target-based drug discovery for first-in-class medicines. Clin Pharmacol Ther 93:299–301

3. Kummel A, Selzer P, Beibel M et al (2011) Comparison of multivariate data analysis strategies for high-content screening. J Biomol Screen 16:338–347

4. Ong SE, Li X, Schenone M et al (2012) Identifying cellular targets of small-molecule probes and drugs with biochemical enrichment and SILAC. Methods Mol Biol 803:129–140

5. Kummel A, Gabriel D, Parker CN et al (2009) Computational methods to support high-content screening: from compound selection and data analysis to postulating target hypotheses. Expert Opin Drug Discovery 4:5–13

6. Eisen MB, Spellman PT, Brown PO et al (1998) Cluster analysis and display of genome-wide expression patterns. Proc Natl Acad Sci U S A **95**:14863–14868

7. Alon U, Barkai N, Notterman DA et al (1999) Broad patterns of gene expression revealed by clustering analysis of tumor and normal colon tissues probed by oligonucleotide arrays. Proc Natl Acad Sci U S A **96**:6745–6750

8. Durr O, Duval F, Nichols A et al (2007) Robust hit identification by quality assurance and multivariate data analysis of a high-content, cell-based assay. J Biomol Screen **12**:1042–1049

9. Giuliano KA, Chen YT, Taylor DL (2004) High-content screening with siRNA optimizes a cell biological approach to drug discovery: defining the role of P53 activation in the cellular response to anticancer drugs. J Biomol Screen **9**:557–568

10. Collinet C, Stoter M, Bradshaw CR et al (2010) Systems survey of endocytosis by multiparametric image analysis. Nature **464**:243–249

11. Golub TR, Slonim DK, Tamayo P et al (1999) Molecular classification of cancer: class discovery and class prediction by gene expression monitoring. Science 286:531–537

12. Shuguang H (2008) Classification of cell subpopulations using multiple cellular parameters from high-content imaging studies. J Biomol Screen **13**:941–952

13. Loo LH, Wu LF, Altschuler SJ (2007) Image-based multivariate profiling of drug responses from single cells. Nat Methods **4**:445–453

14. Caie PD, Walls RE, Ingleston-Orme A et al (2010) High-content phenotypic profiling of drug response signatures across distinct cancer cells. Mol Cancer Ther **9**:1913–1926

15. Jackson D, Lenard M, Zelensky A et al (2010) HCS road: an enterprise system for integrated HCS data management and analysis. J Biomol Screen **15**:882–891

16. Ogier A, Dorval T (2012) HCS-Analyzer: open source software for high-content screening data correction and analysis. Bioinformatics **28**:1945–1946

17. Tukey J (1977) Exploratory Data Analysis. Addison-Wesley, Cambridge, MA

18. Brideau C, Gunter B, Pikounis B et al (2003) Improved statistical methods for hit selection in high-throughput screening. J Biomol Screen **8**:634–647

19. Makarenkov V, Zentilli P, Kevorkov D et al (2007) An efficient method for the detection and elimination of systematic error in high-throughput screening. Bioinformatics **23**:1648–1657

Part IV

Phenotypic Models: Primary Neurons, Neural Stem/ Progenitor Cells, Induced-Pluripotent Stem Cells, Multicellular Tumor Spheroids, Endothelial/Mesenchymal Stem Cell Co-cultures, and Zebrafish Embryos

Chapter 17

High Content Screening of Mammalian Primary Cortical Neurons

Dario Motti, Murray Blackmore, John L. Bixby, and Vance P. Lemmon

Abstract

High Content Screening (HCS) can be used to analyze the morphology of neuronal primary cultures on a large scale. When used in the field of neuronal regeneration this approach allows the screening of hundreds or thousands of perturbagens, such as miRNAs, cDNAs, or compounds, for their ability to induce neuronal growth. One of the most important steps while designing these kinds of experiments is the choice of the correct neuronal model. Testing the correct neuronal type is critical to obtain results that are biologically significant and that can later be translated to a clinical setting. For example, if the goal is identifying possible therapies for Spinal Cord Injury (SCI), a challenging target is the neuronal projection from the motor cortex to the spinal cord, the corticospinal tract. Here, we describe the experimental protocols that can be used to produce primary cortical culture from young rat cortices, electroporate the neurons to study the effect of altered gene expression on neurite growth, and immunostain to measure neurite growth parameters.

Key words Phenotypic screening, Primary cortical neuron, Rat, Electroporation

1 Introduction

HCS can provide quantitative measures of cellular features from large numbers of neurons in vitro, providing neuroscientists with a high-throughput tool to measure neurite number, branching, and length [1]. In the field of nervous system regeneration, this approach provides an opportunity to screen in a relatively short amount of time different perturbagens that modulate neuronal growth. In our lab we have successfully applied this concept to test developmentally regulated genes [2–4], genes differentially expressed in the peripheral nervous system compared to the central nervous system [5–7], and kinases and phosphatases [8, 9] to identify genes and pathways regulating neuron morphology. A similar assay was used to test chemical compounds that can promote the regrowth of axons [9, 10]. Our final goal is identifying

Paul A. Johnston and Oscar J. Trask (eds.), *High Content Screening: A Powerful Approach to Systems Cell Biology and Phenotypic Drug Discovery*, Methods in Molecular Biology, vol. 1683, DOI 10.1007/978-1-4939-7357-6_17,
© Springer Science+Business Media LLC 2018

targets or drugs/compounds that can serve as robust therapeutics after Spinal Cord Injury.

A major concern for our studies is the choice of neuron type for screening. We originally started using Cerebellar Granule Cells (CGNs) because they had been widely used in neurite outgrowth assays, especially on inhibitory substrates, and because we could get millions of neurons easily from postnatal animals [5, 10, 11]. However, the relevance of CGNs to axon regeneration of projection neurons such as Corticospinal Tract (CST) neurons or rubrospinal neurons is difficult to know with certainty. We have also screened primary hippocampal neurons [8, 9]. Hippocampal neurons are a well-established culture model, and they extend readily quantifiable axons and dendrites. Because they are typically prepared from embryonic brain, however, they may not be the best choice for a regeneration model. Interestingly, postnatal cortical cultures turned out to be an excellent screening choice [3, 7]. We were interested in using post-natal cortical neurons because they are relevant to SCI and post-natal neurons undergo changes in gene expression that reduce their ability to extend or regenerate axons. Even though the cultures have a number of different neuronal cell types (based on morphology and cell type markers), as well as glia, our screen led to the discovery of KLF7 as a promoter of axon regeneration for CST neurons [4]. Finally, when testing different perturbagens, either genes or chemical compounds, we have found that different primary neurons can react differently to the same perturbagen. An example is two kinase inhibitors that we have tested across several neuronal types (Fig. 1). ML-7 strongly promotes (+200%) neurite growth from hippocampal neurons, promotes growth from cortical neurons (+50%), yet is just toxic for RGCs. The IKK inhibitor VII promotes growth of hippocampal neurons (+70%), has an insignificant effect on cortical neurons (+10%), and is toxic for RGCs (note that other compounds do

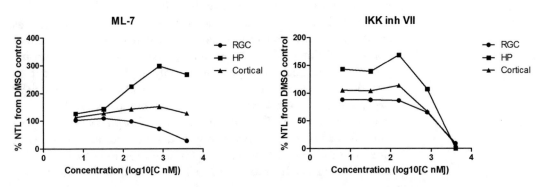

Fig. 1 The graphs in this figure are showing changes in Neurite Total Length (NTL) in different neuronal primary cultures when treated with different concentrations of ML-7 (on the *left*) or IKK inhibitor VII (on the *right*). Four different concentrations of compounds were used on three different cell types: retinal ganglion cells (RGC), hippocampal cells (HP), and cortical cells (cortical) Figure 1 adapted from [12]

promote RGC neurite growth; e.g., Go6976, data not shown) and novel triazines [10] that do not promote growth of hippocampal neurons. For these reasons, we conclude that it is important to screen perturbagens on a neuronal cell type that is related to the in vivo therapeutic target.

In this chapter, we introduce the method we use to obtain cortical neuron cultures in our laboratory, to transfect them with cDNA vectors for the overexpression of genes, or shRNAs for downregulation of specific targets and how to stain them for assessing neurite length. While postnatal neurons are a potentially useful choice for screening for regeneration-related genes or compounds, preparing them and culturing them is much more difficult than culturing embryonic neurons and a number of aspects of our procedure were introduced with this in mind. This procedure yields approximately 30 million cells from three rat pups with a transfection efficiency ranging between 30 and 50%.

2   Materials

2.1   Cell Dissociation

1. Animals: four rat pups of age P3 (*see* **Note 1**).

2. Dissection tools: tweezers, fine scissors, scalpels (*see* **Note 2**).

3. Hibernate E.

4. SM1 (*see* **Note 3**).

5. Papain.

6. Sterile glass pipettes. Fire-polish the tip of the pipettes before using.

7. Dissolve DNAse in sterile water at a concentration of 30 mg/mL. Store at $-20\ ^{\circ}C$ in 100 μL aliquots.

8. 2.5% Trypsin (*see* **Note 4**).

9. 0.22 μm filters.

2.2   Cell Transfection

1. Stock solutions for Intra Neuronal Buffer. All solutions can be stored at room temperature (RT).

   (a) 1 M KCl: weigh 75 g of KCl and transfer to a cylinder containing 80 mL of water. Bring to 100 mL always with water.

   (b) 1 M Hepes: weigh 23.83 g of Hepes and transfer to a cylinder containing 80 mL of water. Adjust pH to 7.5 with KOH. Bring volume to 100 mL with water.

   (c) 0.5 M EGTA: weigh 19 g of EGTA and transfer to a cylinder containing 80 mL of water. Adjust pH to 8.0 with NaOH. Bring volume to 100 mL with water.

(d) 0.1 M $CaCl_2$: weigh 14.7 g of $CaCl_2$ dihydrate and transfer to a cylinder containing 80 mL of water. Bring to 100 mL always with water.

(e) 1 M $MgCl_2$: weigh 20.3 g of $MgCl_2$ and transfer to a cylinder containing 80 mL of water. Bring to 100 mL always with water.

2. ECM 830 High-Throughput Electroporation System.

3. BTX 96 well-plate.

4. PNGM™ Primary Neuron Growth Medium. We add L-Glutamine and GA-1000 to PNBM® Primary Neuron Basal Medium according to factory protocol, and store the media at 4 °C. The day before plating the cells we isolate the needed amount of media we add NSF-1 to it, and we filter it with 0.22 μm filters.

5. PDL: Poly-D-lysine. Dissolve powder at 1 mg/mL in HBSS—Hank's Balanced Salt Solution. Aliquot as needed and store at −20 °C. Coat plates with PDL by covering wells with the correct volume and incubate ON at RT. The day before the preparation the plates for cell culture should be pretreated with PDL (*see* **Note 5**).

2.3 Cell Fixation and Staining

1. Paraformaldehyde (PFA) 4%. Prepare 16% stock solution of PFA by adding 16 g of paraformaldehyde to 60 mL of PBS. Stir the solution slowly, heating it at 60–65 °C, and add NaOH until the milky solution turns clear. Adjust the final volume to 100 mL. Aliquot the 16% PFA and store aliquots at −20 °C. When needed we dilute PFA 16% to 4% with PBS and warm it in a water bath at 37 °C.

2. 10× PBS.

3. 10% Bovine Serum Albumin (BSA) Stock: dissolve 10 g of BSA in 100 mL of distilled H_2O. Gently rock the solution. Do not stir. Prepare 10 mL aliquots and store at −20 °C.

4. Goat Serum. Aliquot and store at 4 °C.

5. Triton X-100.

6. Hoechst 3342, Trihydrochloride, Trihydrate—FluoroPure™ Grade. Dissolve in Dimethyl sulfoxide (DMSO) at a concentration of 10 mg/mL. Aliquot and store at 4 °C.

7. Primary Antibody: monoclonal Mouse-α-Rat-β3-Tubulin produced at the University of Miami monoclonal antibody core facility [7].

8. Secondary antibody: Alexa Fluor® 647 Goat-α-Mouse. Prepare 10 μL aliquots and store at −80 °C.

3   Methods

All the procedures should be carried out in a biological safety cabinet following basic sterility rules unless otherwise specified.

3.1   Cell Dissociation

The goal of this procedure is to very gently dissociate chunks of cortex first with papain and then trypsin.

1. Separately mix 49 mL of Hibernate E with 1 mL of SM1 in a 50 mL tube and maintain it at room temperature.

2. Prepare the solution for papain digestion by mixing 10 mL of Hibernate E with 200 U of the enzyme in a 15 mL tube to a final concentration of 20 U/mL. Place the solution in a water bath at 37 °C until needed.

3. Extract rat brains (*see* **Note 6**) and scoop them into a Petri dish containing cold Hibernate E. Remove meninges from the brain and use a scalpel to prepare a thick transverse section of the brain, as illustrated in Fig. 2. Separate the cortex from midbrain. Repeat for each brain.

4. Remove cortices from Hibernate E, move them on a flat sterile surface, like a Petri dish lid, and mince into pieces with a scalpel.

5. Remove the Papain solution from the water bath, add 50 μL of DNAse, move cortex pieces to the papain-DNAse solution using a fire-polished glass pipette and a suction bulb. Use some of the papain solution to suction the pieces, trying to avoid bubbles.

6. Incubate the tube containing the papain-DNAse solution and the minced cortices at 37 °C with shaking for 30′.

7. Remove the tube from the incubator and centrifuge cortices for 1′ at $10 \times g$.

8. Remove and discard the supernatant. Replace with 5 mL of the Hibernate E + SM1 solution.

9. Centrifuge cortices for 1′ at $10 \times g$.

10. Remove and discard the supernatant. Replace with 1.5 mL of the Hibernate E + SM1 solution. Add 5 μL of the DNAse solution.

11. Use a fire-polished Pasteur pipette attached to the suction bulb to gently triturate the cortices three times.

12. Spin down cortices for 1′ at $10 \times g$.

13. Discard the supernatant and add 10 mL of the Hibernate E.

14. Spin down cortices for 1′ at $10 \times g$.

15. Remove the supernatant and replace with 9 mL of the Hibernate E. Add 1 mL of the Trypsin solution and 50 μL of DNAse solution.

288

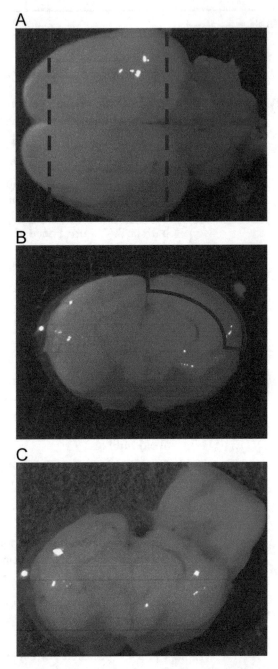

Fig. 2 This figure reports pictures of the rat brain during dissociation of the cortex. In (a) a top view of the full brain, cut along the level of the *red dotted lines* to follow the protocol. In (b) a caudal view of a coronal section of the brain. The *red box* indicates the area of interest. (c) Cortex once separated from the rest of the brain

16. Incubate the tube containing the Trypsin-DNAse solution and the minced cortices at 37 °C shaking for 30′.

17. Spin down cortices for 1′ at $10 \times g$.

18. Discard the supernatant. Replace with 5 mL of the Hibernate E + SM1 solution.

19. Spin down cortices for 1′ at $10 \times g$.

20. Discard the supernatant and replace with 1.5 mL of the Hibernate E + SM1 solution.

21. Add 5 µL of DNAse solution.

22. Use a fire-polished Pasteur pipette attached to the suction bulb to gently triturate the cortices three times.

23. Let cortices settle for 2′.

24. Move the supernatant to a new 15 mL tube.

25. Repeat **steps 20–24**, about five or six times, pooling the supernatants.

26. Determine cell concentration by counting trypan blue excluding cells with a hemocytometer.

3.2 Cell Transfection

1. Prepare INB by mixing in a tube for 50 mL:
   H_2O: 42 mL.

   1 M KCl: 6.75 mL.

   1 M HEPES: 0.5 mL.

   0.5 M EGTA: 0.5 mL.

   $CaCl_2$: 0.1 mL.

   1 M $MgCl_2$: 0.1 mL.

   Filter with 0.22 µm filters.

2. Rinse the BTX plate wells with INB three times, and leave the plate in the hood for drying.

3. Prepare the plasmids by mixing for each transfection 6 µg of DNA with INB to a total volume of 25 µL per BTX well (*see* **Note 7**). Make sure that the volume of plasmid is below 12.5 µL.

4. Move the volume of media containing needed cells (50K per transfection) to a new tube.

5. Spin down cells at $80 \times g$ for 5′.

6. Resuspend the cells in 25 µL per transfection of INB (*see* **Note 8**).

7. Transfer 25 µL of neuronal cells to each BTX plate well and mix it a couple of times with the plasmid solution.

8. Electroporate cells with a single pulse using the ECM 830 High Throughput Electroporation System at 350 V for 300 µs (*see* **Note 9**).

9. Immediately recover the cells by adding 98 μL of Hibernate E + SN1 per BTX plate well.

10. Plate the cells at the determined optimal density (*see* **Note 10**) in a glial conditioned PNGM™ Medium (*see* **Note 11**).

3.3 Cell Fixation and Staining

Sterility is not a requirement for this step of the protocol.

1. Prepare Blocking Buffer (*see* **Note 12**):

   Dilute 10 × PBS to 1× with water.

   Add 0.5–1% BSA (from 10% stock).

   Add 5% Goat Serum.

   0.03% Triton X-100.

2. After 3 days in culture (*see* **Note 13**) remove media from cells.

3. Cover cells with 4% PFA solution and incubate at RT for 25′.

4. Remove PFA and wash cells three times with PBS.

5. Cover cells in 0.5% Triton X-100 diluted in PBS for 10′ at RT.

6. Wash cells three times with PBS.

7. Cover cells with Blocking Buffer solution and incubate for 1 h at RT.

8. Add Primary Ab Mouse-α-Rat-β3-Tubulin (*see* **Note 14**) at the optimal concentration in blocking buffer and incubate for 1 h at RT.

9. Wash cells three times with PBS.

10. Add Secondary antibody Alexa Fluor® 647 Goat-α-Mo (*see* **Note 15**) at the optimal concentration in blocking buffer together with Hoechst (*see* **Note 16**) at a 1:1000 dilution (from stock 10 mg/mL aliquots) and incubate for 1 h at RT.

11. Wash cells three times with PBS.

12. Figure 3 illustrates cortical cells after staining.

4 Notes

1. The protocol can be modified for different animal ages. In our lab the procedure has been applied with small modifications to P0 and P18 animals. As animals age we found it increasingly difficult to dissociate the cells. We normally use Sprague-Dawley rats, but it can be successfully applied to any strain. Also, the protocol has been optimized for four cortices but can be used for different amounts.

2. Tools should be autoclaved sterilized before usage, and keep them in sterile condition (in a biological safety cabinet) during the entire procedure.

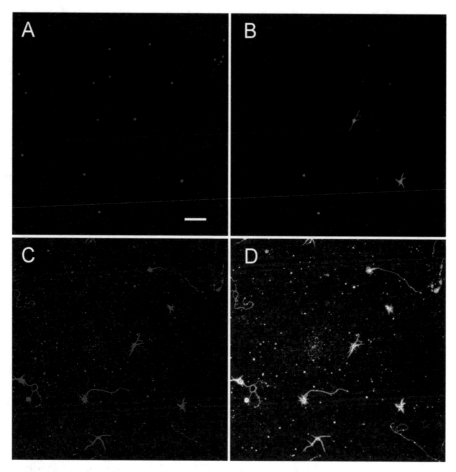

Fig. 3 This figure shows pictures of cortical cells in culture. In (**a**) is the Hoechst nuclei staining. In (**b**) the GFP channel for transfected positive cells and in (**c**) the β3-Tubulin channel staining neuronal microtubule. (**d**) It is an example of the neurite tracing performed by the high content analysis microscope

3. For storage purposes we defrost SM1 bottles and aliquot in 1 mL aliquots. SM1 is stored at −20 °C.

4. For storage purposes we aliquot Trypsin in 1 mL aliquots and store it at −20 °C.

5. Plate coating depends on the goals of the experiment. For our purpose we coat the plates only with 1 mg/mL PDL. We also experienced good results by coating with PDL at 100 μg/mL (diluted in HBSS) ON at RT followed by coating with the growth promoting substrate Laminin at 10 μg/mL (diluted from 1 mg/mL aliquots in HBSS).

6. Postnatal brain is much more substantial than embryonic brain, yet the more mature neurons are more susceptible to injury. To compensate for this our procedure uses a series of enzymatic digestions and very gentle triturations. After each trituration, chunks are allowed to settle and dissociated cells are removed

so as to minimize injury to the dissociated cells during the course of the procedure.

7. We recommend preliminary experiments to determine the conditions for obtaining the highest transfection efficiency, in terms of amount of plasmid and cells per transfection. In the protocol we used the optimal quantity ratio as determined for our system: 50K cells transfected with 6 μg of plasmid usually results in 30–50% transfection efficiency depending on the plasmid.

8. The Intra Neuronal Buffer (INB) is toxic for neurons if they are left in it for long periods; therefore, we recommend that the procedure be performed as fast as possible from the moment cells are resuspended in INB until cells are again suspended in Hibernate E media.

9. The BTX plates need to be primed before use. When we use a well for the first time we wash it twice with INB, then we add 148 μL of INB and give one single electric pulse of the same voltage and width as the actual transfection. We have also found that one well (or row of wells) can be reused for up to 5–6 rounds of transfections. After each use the well is washed twice with water to remove salts, and then once with ethanol. Plates can be stored in sterile conditions in between uses. If the efficiency of transfection drops over time, consider replacing the pins in the top of the BTX electroporation device. We do this approximately every 6 months.

10. Cell density has a profound influence of neurite growth, with higher densities typically producing more neurite growth. Most image analysis systems cannot associate particular neurites with the somas they originate from if neurites from nearby cells touch. So lower cell densities are needed. But very low cell densities have a negative impact on cell health and also increase the number of fields that need to be imaged to analyze sufficient numbers of cells to obtain statistically reliable data (in our hands, around 300 neurons per treatment). We determined that to reach the optimal density to be able to measure neurite length 3 days after plating. We need to plate 22.5 cortical neurons/mm^2. Correlation between the number of cells plated and the final density depends on cell survival, which is affected by several technical issues some of which are operator dependent. The most important appears to be the duration of the cell preparation procedure. Therefore, preliminary experiments to determine the number of cells needed to be plated per well are strongly recommended. If a substantial screening campaign is planned then standard validation procedures comparing positive and negative controls across plates to determine Z' factor

coefficients and across days to demonstrate reproducibility are essential.

11. The survival of postnatal neurons is enhanced by using a complex, growth factor rich media that includes glial conditioned media. Half of the needed media prepared the day before the preparation is overnight conditioned in glial cultures. Glial cultures and conditioning have been described before [3]. The day after the media is removed from the glial culture, mix 1:1 with the unconditioned and filter again. It is then aliquoted in the multiwall plates and stored at 37 °C, 0.5% CO_2 in incubators for a couple of hours for pH equilibration.

12. Blocking buffer should be prepared according to usage, we usually prepare 500 mL of buffer and use it for 1 or 2 weeks, storing it at 4 °C.

13. For our purposes cortical cells are fixed three days after plating. Depending on the kind of measures of interest neurons can be cultured for longer or shorter times.

14. We suggest defining the optimal antibody concentration beforehand for both primary and secondary antibodies. We used anti-β-III Tubulin antibody alone to stain neuronal processes regardless of whether they are axons or dendrites. Other kinds of antibodies, such as anti-MAP 2 or anti-Tau1, can be used to visualize specific processes or features.

15. Secondary antibody choice should be determined to avoid overlapping with reporter gene fluorescence (in the case of transfection with a reporter). Secondary antibody optimal concentration should be assessed beforehand.

16. We use Hoechst as a nuclear staining for counting cells.

Acknowledgments

We thank current and former members of the LemBix laboratory for developing our HCS approaches and thank Hassan Al-Ali for providing Fig. 1. V.P.L. holds the Walter G. Ross Distinguished Chair in Developmental Neuroscience. This work was supported by grants from the James and Esther King Biomedical Research Program JEK09KW-05 (J.L.B.), the US Army W81XWH-05-1-0061 (V.P.L., J.L.B.), the NIH [HD057521 (V.P.L.), NS059866 (J.L.B.)], and the Buoniconti Fund.

References

1. Daub A, Sharma P, Finkbeiner S (2009) High-content screening of primary neurons: ready for prime time. Curr Opin Neurobiol 19:537–543

2. Moore DL, Blackmore MG, Hu Y et al (2009) KLF family members regulate intrinsic axon regeneration ability. Science 326:298–301

3. Blackmore MG, Moore DL, Smith RP et al (2010) High content screening of cortical neurons identifies novel regulators of axon growth. Mol Cell Neurosci 44:43–54

4. Blackmore MG, Wang Z, Lerch J et al (2012) KLF7 engineered for transcriptional activation promotes axon regeneration in the adult corticospinal tract. Proc Natl Acad Sci U S A 109:7517–7522

5. Buchser WJ, Smith RP, Pardinas J et al (2012) Peripheral nervous system genes expressed in central neurons induce growth on inhibitory substrates. PLoS One 7:e38101

6. Lerch JK, Kuo F, Motti D et al (2012) Isoform diversity and regulation in peripheral and central neurons revealed through RNA-Seq. PLoS One 7:e30417

7. Lerch JK, Martinez-Ondaro YR, Bixby JL et al (2014) cJun promotes CNS axon growth. Mol Cell Neurosci 59:97–105

8. Al-Ali H, Schurer SC, Lemmon VP et al (2013) Chemical interrogation of the neuronal kinome using a primary cell-based screening assay. ACS Chem Biol 8:1027–1036

9. Buchser WJ, Slepak TI, Gutierrez-Arenas O et al (2010) Kinase/phosphatase overexpression reveals pathways regulating hippocampal neuron morphology. Mol Syst Biol 6:391

10. Usher LC, Johnstone A, Ertürk A et al (2010) A chemical screen identifies novel compounds that overcome glial-mediated inhibition of neuronal regeneration. J Neurosci 30:4693–4706

11. Smith RP, Lerch-Haner JK, Pardinas JR et al (2011) Transcriptional profiling of intrinsic PNS factors in the postnatal mouse. Mol Cell Neurosci 46:32–44

12 Al-Ali H, Blackmore M, Bixby JL, et al (2014) High Content Screening with Primary Neurons. Assay Guidance Manual [Internet]. Bethesda (MD): Eli Lilly & Company and the National Center for Advancing Translational Sciences

Chapter 18

Human-Derived Neurons and Neural Progenitor Cells in High Content Imaging Applications

Joshua A. Harrill

Abstract

Due to advances in the fields of stem cell biology and cellular engineering, a variety of commercially available human-derived neurons and neural progenitor cells (NPCs) are now available for use in research applications, including small molecule efficacy or toxicity screening. The use of human-derived neural cells is anticipated to address some of the uncertainties associated with the use of nonhuman culture models or transformed cell lines derived from human tissues. Many of the human-derived neurons and NPCs currently available from commercial sources recapitulate critical process of nervous system development including NPC proliferation, neurite outgrowth, synaptogenesis, and calcium signaling, each of which can be evaluated using high content image analysis (HCA). Human-derived neurons and NPCs are also amenable to culture in multiwell plate formats and thus may be adapted for use in HCA-based screening applications. This article reviews various types of HCA-based assays that have been used in conjunction with human-derived neurons and NPC cultures. This article also highlights instances where lower throughput analysis of neurodevelopmental processes has been performed and which demonstrate a potential for adaptation to higher-throughout imaging methods. Finally, a generic protocol for evaluating neurite outgrowth in human-derived neurons using a combination of immunocytochemistry and HCA is presented. The information provided in this article is intended to serve as a resource for cell model and assay selection for those interested in evaluating neurodevelopmental processes in human-derived cells.

Key words Human-derived neuron, Human-derived neural progenitor, Apical cellular processes, High content imaging

1 Introduction

Advances in the fields of stem cell biology and cellular engineering have resulted in the development of a variety of human stem cell-derived neural models that are appropriate for use in high content image analysis (HCA) (Tables 1 and 2). Prior to the development of these technologies, in vitro studies of neural cell types were limited to the use of cancer cell lines from humans (e.g., SH-SY5Y) or rodents (e.g., PC12 pheochromocytoma), primary neural cultures from rodents (e.g., cortical, hippocampal, cerebellar granule cells),

Paul A. Johnston and Oscar J. Trask (eds.), *High Content Screening: A Powerful Approach to Systems Cell Biology and Phenotypic Drug Discovery*, Methods in Molecular Biology, vol. 1683, DOI 10.1007/978-1-4939-7357-6_18, © Springer Science+Business Media LLC 2018

Table 1
Human stem cell-derived neural models that are commercially available[a]

Company (Web-site)	Cell product	Derivation	Description
Neural stem/progenitor cells			
Aruna Biomedical (http://arunabiomedical.com)	hNP1	ESC	Derived from WA09 hESC line. Differentiated to a neural stem/progenitor phenotype
Life Technologies (http://lifetechnologies.com)	Gibco® hNSCs	ESC	Derived from WA09 hESC line. Differentiated to a neural stem/progenitor cell phenotype
	StemPro® hNSCs	NSC	Cryopreserved primary NSCs isolated from human fetal brain
Lonza (http://lonza.com)	Poietics™ Normal Human Neural Progenitors (NHNP)	NSC	Cryopreserved neurospheres isolated from human brain cortex
EMD Millipore (http://emdmillipore.com)	ReNcell CX	NSC	c-Myc transformed neural stem/progenitor cell line from human fetal cortex
	ReNcell VM	NSC	c-Myc transformed neural stem/progenitor cell line from human fetal ventral mesencephalon
Takara/ClonTech (http://www.clontech.com)	Human Neural Cortex	NSC	Cryopreserved NSCs derived from human fetal cortex
	Human Neural Hindbrain	NSC	Cryopreserved NSCs derived from human fetal hindbrain
	Human Mid Forebrain	NSC	Cryopreserved NSCs derived from human fetal mid forebrain
	Human Neural Temporal Lobe	NSC	Cryopreserved NSCs-derived human fetal temporal lobe
	Human Neural Spinal Cord	NSC	Cryopreserved NSCs-derived human fetal spinal cord
PheonixSongs Biologicals (http://phoenixsongsbio.com)	Cortical NSCs[b]	NSC	Cryopreserved primary NSCs isolated from human cortex
	Hippocampal NSCs[b]	NSC	Cryopreserved primary NSCs isolated from human hippocampus.
	Midbrain/Hindbrain NSCs[b]	NSC	Cryopreserved primary NSCs isolated from human midbrain/hindbrain
Global Stem (http://globalstem.com)	HIP™ NSCs	iPSC	Human iPSC-derived neural stem/progenitor cell
Neuron			
Aruna Biomedical (http://arunabiomedical.com)	hN2™	ESC	Derived from WA09 hESC line. Differentiated to a neuronal phenotype

(continued)

Table 1
(continued)

Company (Web-site)	Cell product	Derivation	Description
PheonixSongs Biologicals (http://phoenixsongsbio.com)	Human Cortical CryoNeurons™	NSC	Differentiated to a neuronal phenotype from NSCs isolated from human cortex
	Human Hippocampal CryoNeurons™	NSC	Differentiated to a neuronal phenotype from NSCs isolated from human hippocampus
	Human Midbrain/Hindbrain CryoNeurons™	NSC	Differentiated to a neuronal phenotype from NSCs isolated from human midbrain/hindbrain
Cellular Dynamics (http://www.cellulardynamics.com)	iCell Neurons[c]	iPSC	Human iPSC-derived glutamatergic and GABAergic neurons
	iCell DopaNeurons[c]	iPSC	Human iPSC-derived dopaminergic neurons
GlobalStem (http://globalstem.com)	HIP™ Dopaminergic Neurons	iPSC	Human iPSC-derived neurons with a dopaminergic phenotype
Glial cells			
Cellular Dynamics (http://cellulardynamics.com)	iCell Astrocytes	iPSC	Heterogeneous population of human astrocytes

[a]Based on Internet market survey conducted in Fall of 2016. Not an exhaustive list of potential commercial suppliers of human neural cell products.
[b]Available in a variety of donor lots.
[c]iCell neurons can also be custom derived, i.e., from select patient populations

tissue explant cultures, or occasionally primary human neurons [53]. These models have proven to be useful mechanistic research tools. However, results of studies performed using these models are not easily translated to humans. This is due to the uncertainties in extrapolation across species (i.e., rodent to human) and uncertainties related to the use of cancer-derived cell lines which do not accurately model human neuronal biology. Human stem cell-derived neural cells are an alternative that can potentially aid researchers in reducing these uncertainties. Human stem cell-derived neural cells are scalable, reproducible resources that can typically be produced in large numbers and cultured in 2-D dimensional, multi-well plate formats. These features make them amenable to HCA, particularly in the case of small-molecule efficacy and toxicity screening.

The increasing prevalence of human stem cell-derived neural models in the published literature has been spurred not only by technological advancements but also by the increased commercial availability of human stem cell-derived cell products [54–56]. Table 1 lists a number of current commercially available human

Table 2
Imaging studies using human stem cell-derived neural models.

Reference	Cell type	Cell product[a]	Method
Differentiation			
Donato et al. [1]	NSC-derived	ReNcell VM and ReNcell CX	Tuj1, TH, GFAP, O1, GalC
Le et al. [2][b]	NSC-derived	ReNcell VM	Tuj1
Morgan et al. [3]	NSC-derived	ReNcell VM	Tuj1, TH
Hubner et al. [4]	NSC-derived	ReNcell VM	Tuj1, HuC/D
Lange et al. [5]	NSC-derived	ReNcell VM	Tuj1
Mazemondet et al. [6]	NSC-derived	ReNcell VM	Nestin, Tuj1, GFAP, S100b
Hernandez-Benitez et al. [7]	NSC-derived	ReNcell VM	Tuj1
Pai et al. [8]	NSC-derived	ReNcell VM	Nestin, Tuj1, TH, GFAP, O1
Lemcke and Kuznetsov [9]	NSC-derived	ReNcell VM	Nestin, Tuj1, GFAP
Lemcke et al. [10]	NSC-derived	ReNcell VM	Tuj1, GFAP
Shahbazi et al. [11]	NSC-derived	ReNcell CX	Nestin, Tuj1, GFAP
Zhao et al. [12]	NSC-derived	ReNcell CX	Tuj1, GFAP
Shin et al. [13]	ES-derived NSC	hNP1	Oct4, Nestin, Musashi1, Tuj1, HuC/D, PSN-NCAM, GFAP, A2B5, O4
Dhara et al. [14]	ES-derived NSC	hNP1	Nestin, Musashi 1, Sox2, Tuj1, GFAP, MBP
Lai et al. [15]	ES-derived NSC	hNP1	Sox1, Nestin, TUC-4, Tuj1, HuC/D, Calbindin, DCX, GABA, GFAP, OX4
Dhara et al. [16]	ES-derived NSC	hNP1	Sox2, Nestin, Musashi1, Tuj1, GABA, TH
Acharya et al. [17]	ES-derived NSC	hNP1	Sox1, Nestin, Tuj1
Cheng and Kisaalita [18]	ES-derived NSC	hNP1	Nestin, Tuj1
Dodla et al. [19]	ES-derived NSC	hNP1	Oct4, SSEA4, Sox2, Nestin
Young et al. [20]	ES-derived NSC	hNP1	Nestin, Tuj1, MAP2
Mumaw et al. [21]	ES-derived NSC	hNP1	Tra 1-81, Sox2, Nestin, Tuj1, MAP2, GFAP
Iyer et al. [22]	ES-derived NSC	hNP1	Nestin, Tuj1
Brennand et al. [23]	iPSC-derived Neuron	iCell $_{PD}$	Nanog, Tra 1-60, Sox2, Nestin, Tuj1, MAP2

(continued)

Table 2
(continued)

Reference	Cell type	Cell product[a]	Method
Chamberlain et al. [24]	iPSC-derived neuron	iCell PD	Nanog, SSEA4, TRA 1–60, TRA 1–81, Tuj1, S100b
Israel et al. [25]	iPSC-derived neuron	iCell PD	Nanog, Tra 1–81, Sox2, Nestin, Tuj1, MAP2
Lee et al. [26]	iPSC-derived NPC	iCell PD	Mash1, Tuj1
Xu et al. [27]	iPSC-derived neuron	iCell	Sox2, Nestin, Foxg1, Tuj1, MAP2, GFAP
Yagi et al. [28]	iPSC-derived neuron	iCell PD	SSEA1, SSEA3, SSEA4, Tra 1–60, Tra 1–81, Tuj1, MAP2
Cheung et al. [29]	iPSC-derived neuron	iCell PD	Nanog, SSEA4, Tra 1–60, Tra 1–81, Nestin, Tuj1
Marchetto et al. [30]	iPSC-derived neuron	iCell PD	Nanog, Tra 1–81, Lin28, Sox1, Sox2, Nestin, Musashi1, Tuj1, vGLUT1, Synapsin, GABA
Mak et al. [31]	iPSC-derived neuron	iCell	Sox1, Pax6, Nestin, Tuj1, TH
Swistowska et al. [32]	ES-derived NSC	hNP1	Sox1, Nestin, Tuj1, PSA-NCAM, MAP2, TH
Young et al. [33]	ES-derived NSC	hNP1	Nestin, Tuj1, Nurr1, EN1, Pitx3, TH, vMAT2, DAT
Seibler et al. [34]	iPSC-derived neuron	iCell PD	Oct4, Nanog, Tra 1–60, SSEA4, Tuj1, TH
Nguyen et al. [35]	iPSC-derived neuron	iCell PD	Oct4, Nanog, SSEA1, SSEA3, SSEA4, Lin28, Tra 1–60, Tra 1–81, Nestin, Tuj1, TH, Foxa2, PITX3, Nurr1
Byers et al. [36]	iPSC-derived Neuron	iCell PD	Tra 1–60, Nestin, Dcx, TH, Pitx3, Girk2, EN1, Lmx1A, FoxA2
Pasca et al. [37]	iPSC-derived neuron	iCell PD	Pax6, NCAM, Map2, TH, FoxA2, GABA
Guo et al. [38]	ES-derived NSC	hNP1	Sox1, Nestin, Tuj1, Hnk1, Peripherin, Brn3a, S100, Mash1, Parvalbumin, vGLUT1, Substance P, Hnk1, p75
Wada et al. [39]	ES-derived NSC	hNP1	Tuj1, HB9, Isl1, BF1, ChAT
Ebert et al. [40]	iPSC-derived neuron	iCell PD	Nestin, Tuj1, HB9, ChAT, SMI-32, Synapsin
Mitne-Neto et al. [41]	iPSC-derived neuron	iCell PD	Oct4, Nanog, Lin28, Islet1, Map2, ChAT

(continued)

Table 2
(continued)

Reference	Cell type	Cell product[a]	Method
Proliferation			
Donato et al. [1]	ES-derived NPC	ReNcell VM and ReNcell CX	CyQuant fluorescent assay
Lee et al. [42]	iPSC-derived NPC	iCell $_{PD}$	Calcein AM
Culbreth et al. [43][b]	NSC	ReNcell CX	BrdU
Breier et al. [44][b]	NSC	ReNcell CX	BrdU
Jaeger et al. [45]	NSC-derived	ReNcell VM	Hoechst
Lange et al. [5]	NSC-derived	ReNcell VM	DAPI, nucleus counting
Lemcke and Kuznetsov [9]	NSC-derived	ReNcell VM	EdU, % positive
Lemcke et al. [10]	NSC-derived	ReNcell VM	EdU, % positive
Mazemondet et al. [6]	NSC-derived	ReNcell VM	Propidium iodide
Zhao et al. [12]	NSC-derived	ReNcell CX	BrdU
Lai et al. [15]	ES-derived NSC	hNP1	BrdU
Acharya et al. [17]	ES-derived NSC	hNP1	BrdU
Young et al. [20]	ES-derived NSC	hNP1	CellTrace CFSE Proliferation Kit (Invitrogen)
Lee et al. [42]	iPSC-derived NPC	iCell $_{PD}$	Alamar Blue/Hoechst
Xu et al. [27]	iPSC-derived neuron	iCell	BrdU; Immunocytochemistry (Cyclin D1, Cdk2, Cdk4, phos-Rb)
Viability			
Breier et al. [44][b]	NSC	ReNcell CX	Propidium iodide exclusion
Nguyen et al. [35]	iPSC-derived neuron	iCell $_{PD}$	TH+ cell counting
Li et al. [46]	NSC-derived	ReNcell CX	Hoechst 33342, propidium iodide, annexin V-FITC
Apoptosis			
Culbreth et al. [43][b]	NSC-derived	ReNcell CX	Activated caspase 3 and p53[c]
Chaudhry and Ahmed [47]	NSC-derived	ReNcell VM	Ccaspase-2, -3 and -8[c]
Jaeger et al. [45]	NSC-derived	ReNcell VM	Hoechst, nucleus condensation
Byers et al. [36]	iPSC-derived neuron	iCell $_{PD}$	Cleaved caspase 3[c]

(continued)

Table 2
(continued)

Reference	Cell type	Cell product[a]	Method
Nguyen et al. [35]	iPSC-derived neuron	iCell $_{PD}$	Cleaved caspase 3[c]
Xu et al. [27]	iPSC-derived neuron	iCell	TUNEL
Cell migration			
Lee et al. [26]	iPSC-derived NPC	iCell $_{PD}$	DAPI, scratch test
Diaz-Coranguez et al. [48]	NSC-derived	ReNcell CX	Toluidine blue, transmembrane assay
Neurite outgrowth			
Le et al. [2][b]	NSC-derived	ReNcell VM	Tuj1
Harrill et al. [49][b]	ES-derived neuron	hN2	Tuj1
Harrill et al. [50][b]	ES-derived neuron	hN2	Tuj1
Xu et al. [27][b]	iPSC-derived neuron	iCell	Tuj1
Brennand et al. [23]	iPSC-derived neuron	iCell $_{PD}$	LV-SYNP-GFP (lentivirus)
Synapse quantification/synaptogenesis			
Brennand et al. [23]	iPSC-derived neuron	iCell $_{PD}$	vGLUT/PSD95
Israel et al. [25]	iPSC-derived neuron	iCell $_{PD}$	Synapsin/MAP2
Marchetto et al. [30]	iPSC-derived neuron	iCell $_{PD}$	Map2/vGLUT, Spine Density
Calcium signaling			
Brennand et al. [23]	iPSC-derived neuron	iCell $_{PD}$	Fluo-4AM
Gill et al. [51]	iPSC-derived neuron	iCell	Fluo-4AM
Marchetto et al. [30]	iPSC-derived neuron	iCell $_{PD}$	Fluo-4AM
Zhao et al. [12]	NSC-derived	ReNcell CX	Fura-2AM
Pasca et al. [37]	iPSC-derived neuron	iCell $_{PD}$	Fura-2AM
Wu et al. [52]	ES-derived NSC	hNP1	Calcium Green™

[a]The iCell$_{PD}$ notation indicates studies that use iPSC-derived neural cells differentiated from cells isolated from specific patient populations
[b]Studies utilize HCA instrumentation and analysis techniques
[c]Imaging is based on immunocytochemical labeling of cells. Labeled proteins are noted in the table

stem cell-derived neural cell products that can be plated and cultured using standard cell culture techniques. Current commercially available human stem cell-derived neural models can be grouped into neural stem/progenitor cells (NPCs), neurons, or glial cell products. NPCs are cells that have the capability to proliferate and whose differentiation capacity is restricted to neural lineage cells, e.g., neurons, astrocytes, and oligodendrocytes. Human stem cell-derived neurons express phenotypic markers of mature neurons (i.e., Tuj1, MAP2, HuC/D), are capable of extending neurites, and have no, or greatly diminished, capacity for cell proliferation compared to NPCs [57–60]. The phenotypes and growth characteristics of human stem cell-derived neurons can closely resemble that observed in primary rodent neural cultures. In some cases, commercially available neuron models are differentiated toward a mature neuronal phenotype by the manufacturer and cryopreserved prior to distribution to the end-user. Commercial sources for human-derived glial cells are available, but, at this time, are less prevalent in the market than human-derived neural cell models.

The derivation of commercially available NPC products differs according to manufacturer. Embryonic stem cells (ESCs) are pluripotent stem cells that have the ability to self-replicate indefinitely and are multi-potent: i.e., they have the potential to differentiate into any cell type of the body given the proper differentiation stimuli [61]. Commercially available human ESC-derived NPCs are differentiated from pre-established human ESC lines (e.g., hNP1, Gibco® hNSCs). Primary NPCs isolated from human tissue across a variety of brain regions are also available as cryopreserved end-user products (e.g., StemPro® hNSCs, PheonixSongs NSCs). Cell lines derived from transformation of primary human NPCs of the cortex and ventral mesencephalon using the c-Myc oncogene are also available as cryopreserved end-user products (e.g., ReNcell CX, ReNcell VM). Lastly, NPCs are available which are derived from induced pluripotent stem cells (HIP™). Induced pluripotent stem cells (iPSCs) are adult somatic cells that have been "de-differentiated" through transfection with a compliment of transcription factors (Oct3/4, Sox2, Klf4, c-Myc), with small molecule inhibitors that trigger transcription factor activity or through a combination of these two methods [62–66]. iPSCs can then be "reprogrammed" to differentiate into various cell types of the body using soluble factors and the appropriate culture conditions. In the case of commercially available iPSC-derived NPCs, the cells have been differentiated from a multipotent state to a proliferative, lineage-restricted cell population that express Nestin and Sox1, common markers of human NPCs. Regardless of derivation, product literature for each of the NPC products listed in Table 1 indicates that NPCs are capable of further differentiation toward a more mature neuronal phenotype.

There are several examples in the scientific literature which describe scalable production schemes for human-derived NPCs and neurons [67–71]. The production schemes for commercially available human neurons derived from ESCs, NPCs, and iPSCs vary by manufacturer (*see* product literature for products listed in Table 1) but generally involve expansion of undifferentiated cells while still in a proliferative state followed by differentiation of cells toward a more mature neuronal phenotype using soluble factors and eventual harvest and cryopreservation for end-user applications. Invariably, the production of neurons involves the targeted differentiation of cells from either an endogenous pluripotent, induced pluripotent or lineage-restricted multipotent state. In some instances, manufacturers of human neurons also market a complimentary NPC product from which the more mature neurons are derived. This provides the opportunity to study the effects of various experimental conditions (i.e., genetic manipulation, addition of small molecules, toxicity) across lineage stages of closely related cell types. The variety of commercially available ESC-derived and NPC-derived neural products has been somewhat limited by the availability of human donor tissues as well as ethical and moral concerns regarding human stem cell research. However, iPSC technology offers a means which human neurons from diverse sets of donors can be studied in vitro. For instance, somatic cells from specific patient populations can be isolated, de-differentiated, and reprogrammed to a neuronal lineage into order to investigate the influence of genetic polymorphisms, abnormalities, or susceptibility factors on neuronal development and functions. To date, this approach has been used for a variety of human neurological diseases including Alzheimer's disease, Parkinson's disease, Rett syndrome, schizophrenia, and others [24–26, 29, 30, 32, 33]. Also, studies have demonstrated that iPSCs can be stimulated to differentiate into specific neuronal subtypes, such as dopaminergic midbrain neurons [34–36], sensory neurons [38], and motor neurons [39–41], therefore providing flexibility in the target tissues which can be studied using in vitro methodologies.

1.1 Apical Cellular Processes of Neurodevelopment Evaluated Using HCA

Commercially available human NPCs are appropriate for use in HCA applications focused on examining cell proliferation, apoptosis, migration, or neural differentiation in response to experimental stimuli. Commercially available human neurons display a number of phenotypic characteristics and biological processes that are consistent with the development and function of human neurons in situ. These include extension of neurites (i.e., neurite outgrowth), development of functionally distinct axons and dendrites (i.e., neurite maturation), formation of synapses, development of cellular Ca^{+2} currents, and the eventual development of synaptic networks [72–75]. Each of these biological endpoints is amenable to analysis using HCA methodologies. Table 2 provides a summary of peer-reviewed primary research publications that either: (1) use commercially available human stem cell-derived neural cell models to investigate a critical process of nervous system development or function or (2) were essential preliminary work in the development of a commercial model. Only a small number of these studies employ high content image analysis (HCA) instrumentation and methodologies. However, each publication summarized in Table 2 utilizes lower throughput imaging-based assays that may be easily adapted to HCA research applications. The intent of Table 2 is to serve as a guide to aid researchers in the selection of an appropriate human stem cell-derived neural model for specific research needs and to provide examples of critical cellular processes in neural cells which may be examined using HCA. Specific biological processes, associated quantitative endpoints, and HCA strategies related to each are discussed below.

1.1.1 Cell Differentiation

The most common biological process examined in human stem cell-derived neural models is cell differentiation. In vivo, differentiation of NPCs toward mature neural phenotypes is critical for establishment of neural networks that support integrated nervous system function. Disruption of NPC differentiation is hypothesized to play a key role in the mode-of-action of some human developmental neurotoxicants [76, 77]. Therefore, in vitro assays that measure chemical effects on neuronal differentiation represent potential screening tools for developmental neurotoxicity hazard identification and prioritization for further in vivo testing. In addition, heterologous protein overexpression or protein knockdown (siRNA/RNAi) screening strategies may be used to identify genes that promote or inhibit neuronal differentiation. HCA is an ideal tool for assessing neuronal differentiation with these various types of screens. In microscopy imaging studies, including HCA, neural differentiation is typically assessed by measuring the expression of one or more markers whose expression varies through the progression of neural differentiation. Following experimental manipulation, cells are fixed, nuclei are labeled with DNA-binding fluorescent dyes (e.g., 4′,6-diamidino-2-phenylindole, DAPI) and proteins are labeled through the use of antibody-based immunocytochemistry methods. Following image acquisition, individual cells are identified based on the nucleus label and the presence or absence of protein expression (or alternatively the intensity of fluorescent signal from the immunolabeled proteins) is used to assess the differentiation state of the cells on a cell-by-cell basis. Data are expressed as the percent of the total cell population expressing the protein of interest (i.e., % of Tuj1+ cells) [5, 7]. If multiple proteins are labeled, data may be expressed as the percent of cells expressing one and not the other, both or none of the markers of interest (i.e., % of ChaT+/Tuj1+ cells) [35, 40]. The

Boolean method of %-positive gating is similar to that employed in flow cytometry experiments.

In the studies summarized in Table 2, the large number of examples for measuring differentiation of human stem-cell-derived neural cells is partially driven by a need to initially qualify the models and confirm that cell phenotypes are consistent with a particular lineage stage or neuronal subtype following differentiation protocols. Oct4, Nanog, and Lin28 are proteins involved in the self-renewal of undifferentiated ESCs. These proteins in addition to stage-specific embryonic antigens (SSEA1, SSEA3, SSEA4) and TRA 1–60/TRA 1–81 antigens (putative identifiers of the heavily glycosylatted podocalyxin membrane protein [78]) are used as markers of ESCs and as well as de-differentiated iPSCs prior to reprogramming [23–25, 28–30, 34–36, 41]. Cells differentiated toward a neural lineage will no longer possess multipotent differentiation capability and will no longer express these markers. Sox1, Sox2, and Pax6 are transcription factors expressed in neuroepithelia and are used to evaluate the progression of cells from a multipotent stage toward the germ cell layer from which neural cell types arise [14–17, 19, 23, 25, 27, 31, 32, 37, 38]. Nestin and Musashi 1 are the most common markers used to identify NPCs [6, 8, 9, 11, 13, 14, 16–22]. Additional markers may be used to characterize the differentiation of NPCs corresponding to specific structure and regions of the developing brain such as the neural crest (Mash1, Hnk1, p75) [26, 38], forebrain (Foxg1) [27], or midbrain (Lmx1a) [36]. GFAP is used a marker of differentiation toward the astrocyte lineage. Likewise, a number of markers (O1, O4, OX4, Gal C, MBP) have been used as markers of differentiation toward oligodendrocytes [8, 13, 15]. The cytoskeletal protein Tuj1 (otherwise known as β_{III}-tubulin) is the most commonly used marker for cells committed to the neuronal lineage [1–5, 7]. Neural cell adhesion molecule (NCAM), HuC/D, Tuc-4, and microtubule-associated protein 2 (MAP2) are also used as markers of neuronal differentiation [4, 13, 15, 20, 21, 23, 25, 27, 28, 32, 37, 41]. Synapse-associated proteins such as synapsin, synaptophysin, synaptotagmin can also be used to identify neurons that have differentiated to a neuronal phenotype and have progressed toward a state of neuronal maturation [40]. Furthermore, variations in differentiation conditions can result in the generation of specific subtypes of neurons such as interneurons (GABA, calbindin) [15, 16, 30], dompaminergic midbrain neurons (TH, VMAT2, DAT, GIRK2, Nurr1, EN1, FoxA2, Pitx3) [32–36], sensory neurons (parvalbumin, Brn3a, Substance P) [38], and spinal motor neurons (BF1, ChAT, SMI-32, HB9, Isl1) [39–41]. Each of these markers has the potential to be used as immunocytochemical markers for quantification of population percentages using HCA methodologies.

Proliferation of NPCs can be measured similar to other proliferative cell types using a variety of HCA strategies. The simplest strategy is counting of labeled nuclei in fixed samples, i.e., comparing the number of nuclei in control versus treated samples after a specified period of time following experimental manipulation. DNA-binding dyes such as DAPI, DRAQ5, Hoechst stains, and propidium iodide have been used to visualize nuclei in cell proliferation applications [5]. In a similar approach, live cells can be labeled using membrane permeable molecules that become fluorescent after cleavage by intracellular enzymes. The fluorescent signal can be quantified using HCA methodologies similar to the fluorescent signals emitted from fixed, immunolabeled cells. Examples of this strategy include Calcien AM [42] and carboxyfluorescein diacetate, succinimidyl ester (CFSE) [20]. These molecules diffuse into live cells and are cleaved by intracellular esterases, thus increasing the fluorescent signal. The number of fluorescent labeled cells is compared between control and treatment conditions to yield a measure of relative cell proliferation. Cell/nucleus counting is an indirect measure of cell proliferation which is dependent upon comparisons between separate wells or conditions.

Alternatively, the incorporation of synthetic nucleotides into DNA during the process of DNA replication can be used as a more direct measure of cell proliferation. Bromodeoxyuridine (BrdU) is a synthetic nucleotide commonly used for cell proliferation assays both in vitro and in vivo. The incorporation of BrdU is detected in fixed cell through the use of BrdU-specific antibodies either directly labeled with a fluorophore or compatible with a fluorophore-labeled secondary antibody. 5-ethynyl-2-deoxyuridine (EdU) is a relatively more recent technique that does not involve the use of fluorophore-labeled antibodies. Instead, this technique relies upon a copper-catalyzed chemical reaction between alkyne and azide moieties (Click-iT® chemistry, Life Technologies) present in the EdU and in fluorescent secondary reagents, respectively, to release fluorescent signal. An advantage of EdU over BrdU is that the latter technique requires DNA denaturation for antigen unmasking, whereas the former technique does not, allowing for multiplexing opportunities. The percent of BrdU [12, 15] and EdU [9] positive cells has been used as an endpoint to measure cell proliferation in human stem cell-derived NPCs. BrdU incorporation in human NPCs has been assessed using HCA methodology [43, 44] . Lastly, immunocytochemistry with cell cycle-regulated proteins (Ccnd1, Cdk2, Cdk4, phospho-Rb) has been used as a complimentary measure of cell proliferation in conjunction with BrdU incorporation [27]. The protein abundance of cyclin proteins (e.g., Ccnd1) varies in a cyclical pattern during cell cycle progression with particular cyclins showing peak levels of expression during different phases of the cell cycle. Likewise, the phosphorylation of some proteins (e.g., Rb) varies in a cyclical pattern during progression of the cell cycle. Immunolabeling with these markers and determination of a percent positive population can yield estimates of what proportion of cells are progressing through the cell cycle. In each of the HCA strategies that involve determination of a percent positive population, the researcher must empirically determine a fluorescent intensity threshold that separates positive cells from negative cells.

1.1.3 Cell Viability and Apoptosis

Cell viability is another measure that is commonly of interest in HCA-based studies, especially in the case of small-molecule efficacy and/or toxicity screens. In some instances, luminescence-based or fluorescence-based plate reader assays have been employed in a "sister-plate" strategy to compare phenotypic responses to changes in cell viability [44]. Pairs of plate are prepared and treated in an identical manner and run on two different instrument platforms, i.e., a high content imager and a luminescent/fluorescent plate reader. Direct measurement of cell viability with imaging techniques can be performed in live cells using a combination of cell permeable and cell impermeable DNA-binding dyes, such as Hoechst 33258 and YoYo-1, Toto-3 or propidium iodide, respectively [44]. The cell permeable DNA dye labels all cells while the cell impermeable dye only labels dead or dying cells in which cell membrane integrity has been compromised. Data are typically quantified as a percent positive population. This technique can also be coupled with immunocytochemical labeling for markers known to be upregulated in dying cells.

Apoptosis or programmed cell death is also an endpoint commonly of interest in studies of human stem-cell-derived neural cells. During nervous system development, apoptosis serves to prune and shape the nervous system, removing excess neurons that have not established synaptic connections with target sites that provide trophic support [79]. Increased apoptosis in the developing brain is hypothesized to result in developmental neurotoxicity [80]. Apoptotic cell death is also hypothesized to play a role in neurodegenerative diseases [81, 82]. HCA has been used to examine apoptosis in NPC cell lines through immunocytochemical labeling of activated caspase 3 and activated p53 [43]. Caspases are cysteine-dependent aspartate-directed proteases activated during programmed cell death [79]. The p53 tumor suppressor gene is activated in response to apoptotic stimuli and regulates the activation of downstream effector proteins (e.g., Bax, caspase 9, 3, 6, and 7) [83]. An increase in the percent of activated caspase 3 or activated p53 positive cells is indicative of an increased amount of apoptosis. An alternative technique for measuring apoptosis is terminal deoxynucleotide transferase (TdT) deoxyuridine triphosphate (dUTP) nick end labeling (TUNEL). One of the intracellular processes that is associated with apoptosis is DNA fragmentation. The

TUNEL technique detects late stages of apoptotic cascade and requires the use of terminal deoxynucleotidyl transferase (TdT) enzyme to end-label the 3'-OH ends of fragmented DNA with the dUTP nucleoside analog. Labeling of the DNA ends can be detected using secondary labeling reagents and immunocytochemistry. An increase in the percent of TUNEL positive cells is indicative of an increased amount of apoptosis. Nuclear morphology is another technique that may be employed to assess apoptosis using high content image analysis. As apoptosis progresses, the nuclei of apoptotic cells condense and if labeled with DNA-binding dyes, form brightly labeled "apoptotic bodies" that are readily distinguishable from normal cell nuclei. This method has been employed to examine the effects of glycogen synthase kinase 3β inhibition on apoptosis of human NPC cell lines [45].

Annexin V is a cellular membrane protein that contains phosphatidylserines localized to the cytoplasmic surface of the membrane [84]. During early events of apoptosis cascade, the phosphatidylserine residues become externalized and can be detected using immunocytochemistry [85]. Detection of externalized Annexin V has been used extensively as a marker of apoptosis in flow cytometry applications, but has also been employed in imaging studies aimed as evaluating cell viability [32, 86]. Measures of cell viability (or cell health) have also been performed by counting the number of cells positive for a immunolabeled marker proteins (i.e., Tuj1 or TH) and comparing the number of counted cells across treatment groups [35, 49, 50]. This technique is appropriate for use in post-mitotic cells and assumes that cells in each treatment well were seeded at the same density and that a decrease in cell number across wells is representative of a decrease in cell viability or cytotocixity in response to treatment. Similar to other techniques utilizing percent positive calculations, the researcher must empirically determine a fluorescent intensity threshold that separates positive cells from negative cells.

1.1.4 Cell Migration

DNA labeling can also be used in cell migration assays to track the movement of human stem cell-derived NPCs in vitro. Lee et al. employed an in vitro model of wound healing to examine the migration of human iPSC-derived NPCs harboring a gene mutation associated with familial dysautonomia compared to cells from a control subject. Cells along the plating surface of a well were removed using a scratch technique and the movement of cells into the vacated area was monitored using a DNA-binding dye [26]. Diaz-Coranguez et al. examined the movement of human NPC cells through a layer of microvascular endothelial cells meant to recapitulate the blood–brain barrier [48]. Monolayers of endothelial cells were established in net well inserts and NPC cells were added in suspension. A conditioned media stimulus was used to stimulate the movement of the cells through the endothelial cells layers toward the bottom of the membrane. Migration of the NPC cells was quantified by imaging cells stained with toluidine blue which were present on the bottom of the net-well insert membranes. While this method is not particularly suited to image acquisition using HCA methodologies, manually acquired images could be analyzed using high content imaging analysis software algorithms.

1.1.5 Neurite Outgrowth

To date, neurite outgrowth is the biological process measured most frequently in human stem cell-derived neural cells using HCA. A majority of the high content imaging studies utilize immunocytochemical labeling of neurons with β_{III}-tubulin (Tuj1), a neuronal cytoskeletal protein that is distributed evenly throughout all neurites and the neuronal cell body [87]. Measurements typically reported in HCA neurite outgrowth studies include the total length of neurites per neuron, the average number of neurites per neuron, and the length of the longest neurite per neuron [49, 50]. In primary rodent neural cultures, neurons develop two different types of neurites, axons and dendrites, which are functionally and morphologically distinct. Axons are small-caliber neurites that grow rapidly in culture and serve to conduct electrical impulses (i.e., action potentials) away from the cell body. Dendrites are large-caliber neurites that grow more slowly in culture and receive synaptic inputs from the axons of neighboring cells [88]. Studies utilizing primary rodent neural cultures have used immunocytochemical labeling of cytoskeletal proteins that localize to axons (i.e., p-NF) and dendrites (i.e., MAP2) to measure the effects of chemicals on specific neurite subpopulations [89]. HCA-based neurite subpopulation analysis is feasible for use with human stem cell-derived neural cells given similarities in marker localization as compared to rodent primary neurons (Fig. 1). However, to date, these methods have not been applied to human-derived cells. Heterologous expression of fluorescent proteins driven by a neuronal promoter element has also been used to trace and quantify neurite outgrowth [23]. Care must be taken when employing heterologous expression techniques to ensure that the fluorescent protein is not toxic to the cells and that the protein is being trafficked throughout the length of the neurites. Otherwise, automated measurements may underestimate the amount of neurite outgrowth that occurs. Fusion proteins that combine the fluorescent protein and a neuronal protein that is actively transported throughout the length of the neurites are a potential solution to this caveat.

1.1.6 Synaptogenesis

Quantification of synapses or studies of synaptogenesis can also be performed with HCA [90, 91]. Synapses are visualized through immunocytochemical labeling of synaptic proteins which localize to the synaptic contact sites to form discrete punctate staining patterns (e.g., synapsin, synaptophysin, synaptotagmin) [92].

320 Joshua A. Harrill

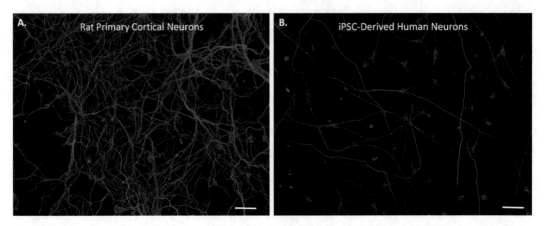

Fig. 1 Rat primary cortical neurons at 5 days in vitro (DIV) (**a**) or cellular dynamics human iPSC-derived iCell neurons at seven DIV (**b**) were labeled with antibodies against Tuj1 (*red*, Covance PRB-435P) or axonal neurofilaments (*green*, Covance SMI 312). Neurons in both culture types have clearly defined axons, a hallmark of cortical neuron differentiation. Scale bar = 50 μm

Often colabeling with a dendrite-localized cytoskeletal protein (e.g., MAP2) is used to specifically quantify synaptic puncta in close proximity to a post-synaptic target cell. The MAP2-labeled dendrites are traced, as in a neurite outgrowth assay, and the number of synaptic puncta per cell or per unit length of dendrite is quantified [91]. Different proteins can be labeled to identify and quantify specific synaptic sub-populations such as excitatory (e.g., vGLUT+) or inhibitory (e.g., vGAT+) synapses. Presynaptic contacts on postsynaptic targets can also be transient in nature. Thus, colabeling for presynaptic (e.g., synapsin) and postsynaptic (e.g., PSD95) proteins can also be used in conjunction with puncta detection overlap functions to quantify the number of structures with established pre- and postsynaptic elements [23]. Of note, imaging methods for synapse quantification using puncta immunocytochemical labeling cannot distinguish between active and inactive synapses. Measurement of changes in membrane potential (e.g., patch clamp) or synchronous electrical activity (e.g., microelectrode array recording) can be used to compliment HCA methods and provide information on functional synaptic activity in response to experimental conditions.

1.1.7 Calcium Signaling

Calcium plays a significant role in the function of mature neuronal networks. Release of neurotransmitters from pre-synaptic vesicles is dependent upon the entry of calcium into the neuron through voltage-sensitive calcium channels following membrane depolarization. Calcium also enters the neuron via neurotransmitter activation of post-synaptic ionotropic glutamate receptors. Post-synaptic calcium influx plays a critical role in synaptic plasticity and strengthening of synaptic connections [93]. Calcium may also be released from intracellular calcium stores (i.e., the endoplasmic reticulum) and regulate neuronal excitability, pre-synaptic exocytosis, and synaptic plasticity [94]. Excitotoxicity can result from an excess entry or release of calcium into the neuronal cytoplasm. Thus, measurement of intracellular calcium levels can be used to assess synaptic function, response to ligands and as a complementary assay for toxicity assessments. In imaging studies, intracellular calcium levels are visualized using cell-permeable molecules that bind calcium and emit higher levels of fluorescence upon excitation when bound with calcium. Examples include Fluo-4 AM, Fura-2 AM, and Calcium Green™. Given that calcium flux is a dynamic process, imaging studies using calcium indicators are commonly performed in live-cell applications [23, 30, 51]. Cells are incubated or "dye-loaded" prior to imaging. In the case of high content imaging studies, instrumentation would need to be capable of kinetic data analysis and equipped with live-cell chambers to maintain cultures under standard atmospheric conditions during image acquisition.

Overall, commercially available human stem-cell-based neural cells are amenable for use in HCA applications. Even though there is a current scarcity of studies using these cell types in combination with high content imaging methods, the information presented above indicates that a number of lower throughput methods are readily adaptable for higher throughput applications.

2   Neuroprogenitor Cell Proliferation and Apoptosis Assay

Below is a generic HCA method for evaluating proliferation and apoptosis in a human neuroprogenitor cell model. The method was adapted from Culbreth et al. [43]. Proliferation is evaluated by measuring incorporation of BrdU into the DNA of dividing cells. Apoptosis is evaluated using immunocytochemistry to detect two marker proteins: (1) activated (i.e., cleaved) caspase 3, a cysteine protease activated as a component of the apoptosis-associated proteolytic cascade [95] and (2) p53, a tumor suppressor gene that plays a role in initiating apoptosis and whose expression increases at the beginning of the apoptotic process [96]. Immunocytochemistry reagents have been updated due to discontinuation of the labeling kit used in Culbreth et al. [43]. The method is based upon the analysis of ReNcell CX human neural progenitor cells; however, this method would also be applicable to other human-derived neural progenitor cell types. The overall image acquisition and analysis workflow associated with this assay would be similar regardless of which HCA instrument is used. However, the exact terminology and instrument settings used to execute the workflow are likely to vary across HCA instruments. Therefore, a generalized workflow, absent of instrument-specific terminology, is presented below.

2.1 Materials

1. ReNcell CX Human Neural Progenitor Cells (EMD Millipore, Billerica, MA).

2. ReNcell Neural Stem Cell Maintenance Media (EMD Millipore, Billerica, MA).

3. Epidermal Growth Factor (EGF) (EMD Millipore, Billerica, MA).

4. Basic Fibroblast Growth Factor (bFGF) (EMD Millipore, Billerica, MA).

5. T75 Cell Culture Treated Polystyrene Tissue Culture Flasks (Corning Inc., Corning, NY).

6. 96-well Polystyrene Tissue Culture Plates (Corning Inc., Corning, NY).

7. Laminin (Sigma-Aldrich, St. Louis, MO).

8. Dulbecco's Phosphate-Buffered Saline (DPBS).

9. Accutase™ (EMD Millipore, Billerica, MA).

10. 5-Bromo-2′-Deoxyuridine (BrdU).

11. BrdU Primary Antibody (mouse monoclonal, IgG2a).

12. Cleaved Caspase 3 Primary Antibody (rabbit polyclonal IgG).

13. p53 primary antibody (goat polyclonal IgG).

14. Alexa Fluor 488 conjugated goat-anti mouse IgG (Thermo Scientific).

15. Alexa Fluor 546 conjugated goat-anti rabbit IgG (Thermo Scientific).

16. Alexa Fluor 647 conjugated donkey-anti goat IgG (Thermo Scientific).

17. 4′,6-diaminido-2′-phenylindole, dihydrochloride (DAPI), 1 mg/mL Solution (Thermo Scientific).

18. Permeabilization Buffer ($1 \times$ DPBS $+1\%$ Triton X-100).

19. Blocking Buffer (DPBS $+5\%$ Bovine Serum Albumin (BSA)).

20. 12% Paraformaldehyde (Electron Microscopy Sciences, Hatfield, PA).

2.2 Thawing and Expansion of Cryopreserved ReNcell CX Cells

1. Prior to thawing cells, coat the growing surface of a T75 tissue culture flask with a solution of 20 µg/mL of laminin and incubate in a 37 °C, 5% CO_2 incubator for 4 h (*see* **Note 1**).

2. Rinse the flask once with DPBS.

3. Add 10 mL of ReNcell Neural Stem Cell Maintenance Media to the flask.

4. Place the flask in a 37 °C, 5% CO_2 incubator while cells are being thawed.

5. Remove cells from liquid nitrogen storage (*see* **Note 2**) and transport to culture area on bed of dry ice.

6. Partially submerge the vial in a 37 °C water bath for 1–2 min to thaw the cells (*see* **Note 3**).

7. Once thawed, immediately transfer cells to a 15 mL conical tube.

8. In a drop-wise manner, add 9 mL of pre-warmed ReNcell Neural Stem Cell Maintenance Media (minus EGF an FGF growth factors) to the 15 mL conical tube (*see* **Note 4**).

9. Resuspend cells by gentle tituration.

10. Centrifuge the resuspended cells at 133 × g at room temperature for 5 min.

11. Aspirate residual media from cell pellet (*see* **Note 5**).

12. Resuspend the cell pellet in a volume of ReNcell Neural Stem Cell Maintenance Media +20 ng/mL EGF + 20 ng/mL bFGF (*see* **Note 6**). A resuspension volume of 10 mL is used in serial passaging of cells to T75 flasks.

13. For additional passaging of ReNcell CX cells, seed the pre-warmed laminin-coated T75 flasks with ~2 × 10^6 cells and place in a 37 °C, 5% CO_2 incubator (*see* **Note 7**).

14. Allow 24 h for attachment and perform a complete media change with Neural Stem Cell Maintenance Media + EGF + bFGF.

15. Perform additional complete media change with Neural Stem Cell Maintenance Media + EGF + bFGF every other day until a desired confluency (~80%) is reached.

2.3 Passaging and Seeding of ReNcell CX Cells into 96-Well Test Plates

1. Prior to beginning the passaging and 96-well seeding procedure, coat the growing surface of the desired number of 96-well test wells/plates with 20 μg/mL laminin solution for 4 h in a 37 °C, 5% CO_2 incubator (*see* **Note 8**).

2. Leave the plates with the laminin coating solution in the 37 °C, 5% CO_2 incubator while performing the cell passaging procedure (*see* **Note 9**).

3. Remove flask of ReNcell CX cells from the incubator and place in a cell culture cabinet.

4. Aspirate the Neural Stem Cell Maintenance Media and rinse cells once with 20 mL of warm DPBS.

5. Add 5 mL of Accutase to the flask.

6. Place cells in a 37 °C, 5% CO_2 incubator for 4 min.

7. Remove the flask from the incubator and add 10 mL of Neural Stem Cell Maintenance Media (minus growth factors).

8. Rinse the growing surface 2–3 times with the media that was just added to the flask.

9. Transfer the contents of the tissue culture flask to a fresh 50 mL conical tube.

10. Rinse the T75 flask with an additional 10 mL of Neural Stem Cell Maintenance Media (minus growth factors) to capture residual cells.

11. Transfer rinse to the 50 mL conical tube.

12. Centrifuge cells at $133 \times g$ for 5 min at room temperature.

13. Aspirate residual media from the cell pellet (*see* **Note 5**).

14. Resuspend the cell pellet in a volume of Neural Stem Cell Maintenance Media +20 ng/mL EGF + 20 ng/mL bFGF appropriate for downstream application and counting of cells (*see* **Note 6**).

15. If dilution of the cell suspension is required to achieve a target concentration of cells in suspension (i.e., 1.11×10^5 cells/mL for 10,000 cells/well in a 96-well plate) (*see* **Note 10**), add an additional volume of Neural Stem Cell Maintenance Media +20 ng/mL EGF + 20 ng/mL bFGF.

16. Remove laminin-coated 96-well plates from the incubator and place in a cell culture cabinet.

17. Aspirate laminin coating solution and rinse each well with 100 μL of sterile DPBS.

18. Dispense entire volume of cell suspension into a sterile, plastic cell culture basin (*see* **Note 11**).

19. Dispense 90 μL of cell suspension per well in a 96-well culture format (*see* **Note 11**).

20. Place cells in a 37 °C, 5% CO_2 incubator for at least 24 h to allow attachment to the growing surface.

2.4 Treatment of ReNcell CX Cells with Test Chemicals

1. Prepare test chemicals in 100% DMSO at a 1000× the desired test concentration (*see* **Notes 12** and **13**).

2. Dispense 990 μL of Neural Stem Cell Maintenance Media + FGF + bFGF into one well of a 96-well "deep well" plate.

3. Dispense 10 μL of 1000× chemical stock into the 990 μL of media (1:100 dilution). Mix with a pipette two to three times.

4. Repeat **steps 2** and **3** for each chemical/concentration to be tested (*see* **Note 14**).

5. Remove the 96-well test plate from the incubator and place in cell culture hood.

6. Dispense 10 μL of 10× intermediate dosing stock into the 90 μL of media present in the test well (1:10 dilution) (*see* **Note 15**).

7. Place the test plates in a 37 °C, 5% CO_2 incubator for 20 h (*see* **Note 16**).

2.5 BrdU Labeling

1. Prepare a solution of 150 μM BrdU in Neural Stem Cell Maintenance Media. The BrdU solution should be of sufficient volume to allow 50 μL to be dispensed into each test well.

2. Remove test plate(s) from the incubator and place in a cell culture cabinet.

3. Add 50 μL of 150 μM BrdU solution to each test well. The final concentration of BrdU in test wells is 50 μM.

4. Place test plates in a 37 °C, 5% CO_2 incubator for an additional 4 h (see **Note 16**).

2.6 Immunocyto-chemistry

1. Remove test plate(s) from the incubator and place in a cell culture cabinet.

2. Gently aspirate media from each well (see **Note 17**).

3. Add 100 μL of warm (37 °C) 12% paraformaldehyde into each well (see **Note 18**).

4. Incubate the plates for 1 h at 37 °C.

5. Aspirate fixative.

6. Rinse cells twice with 100 μL of 1× DBPS.

7. Add 100 μL of Permeabilization Buffer to each well and incubate for 15 min at 37 °C.

8. Aspirate Permeabilization Buffer.

9. Add 100 μL of Blocking Buffer to each well and incubate for 1 h at 37 °C.

10. During the Blocking Buffer incubation, prepare Primary Antibody Solution by diluting BrdU, Cleaved Caspase 3 and p53 Primary antibodies at 1:100–1:250 dilution in Blocking Buffer (see **Note 19**).

11. Aspirate Blocking Buffer.

12. Add 50 μL of Primary Antibody Solution to each test well and incubate at room temperature for 1 h at 37 °C (see **Note 20**).

13. During primary antibody incubation, prepare Secondary Antibody Solution #1 by diluting Alexa Fluor 647 conjugated donkey-anti goat IgG secondary antibody at 1:500 in Blocking Buffer.

14. Aspirate Primary Antibody Solution.

15. Rinse cells twice with 100 μL of Blocking Buffer.

16. Add 50 μL of Secondary Antibody Solution #1 to each test well and incubate for 1 h at 37 °C, protected from light.

17. During Secondary Antibody Solution #1 Incubation, prepare Secondary Antibody Solution #2 by diluting Alexa Fluor 488 conjugated goat-anti mouse IgG and Alexa Fluor 546 conjugated goat-anti rabbit IgG at 1:500 in Blocking Buffer.

18. Add 1 µL of 1000 µg/mL DAPI solution per mL of Secondary Antibody Solution #2. The final DAPI concentration in the Secondary Antibody Solution #2 is 1 µg/mL.

19. Aspirate Secondary Antibody Solution #1 from test wells.

20. Rinse the cells twice with 100 µL of Blocking Buffer.

21. Add 50 µL of Secondary Antibody Solution #2 to each test well and incubate for 1 h at 37 °C, protected from light.

22. Aspirate Secondary Antibody Solution #2.

23. Rinse cells twice with DPBS.

24. Dispense 100 µL of DPBS into each well as a storage buffer.

25. Seal the plate with an optical plate seal.

26. Wrap in tin foil and store at 4 °C until ready for image acquisition.

2.7 High Content Image Analysis

The steps listed below describe a generic workflow for image acquisition and analysis of proliferation and apoptosis in ReNcell CX cells using a HCA instrument. The exact steps and terminology for this workflow will likely vary from one HCA instrument to another. As opposed to using terminology specific to a particular platform, the steps below are intended to aid the researcher in adapting the workflow to the platform they wish to use. It is recommended that researchers new to HCA consult with field application scientists or other professionals familiar with the HCA instrument they intend to use for help in setting up the assay.

2.7.1 Image Acquisition

1. The ReNcell CX proliferation/apoptosis assay uses four fluorescent channels. Assign fluorescent channels 1 through 4 to DAPI, BrdU, Caspase 3, and p53. Channel 1 (i.e., the fluorescent channel used for autofocusing the instrument) should be assigned to DAPI. The remaining Channels 2 through 4 can be assigned to each of the other fluorescent labels at the discretion of the researcher.

2. Assign appropriate excitation/emission filter settings to each of the fluorescent channels (*see* **Note 21**).

3. Select imaging objective to be used for image acquisition. In Culbreth et al. [43], a 10×/0.3 NA imaging objective was used for image acquisition (*see* **Note 22**).

4. Determine camera settings (i.e., standard, low resolution, high resolution) for image acquisition and autofocusing (*see* **Note 23**).

5. Set integration (i.e., exposure) times for each fluorescent channel (*see* **Note 24**).

6. Define scan limits such as the number of unique fields-of-view to be sampled per well or the number of cells to be sampled per well (*see* **Note 25**).

2.7.2 Background Correction and Object Segmentation

The Channel 1 (Nucleus) Image will appear as bright objects on a dark background. The goal of the background correction and object segmentation steps in Channel 1 is to separate foreground from background, accurately identify individual nuclei in the image, and define object masks that will then be transposed onto subsequent fluorescent channels (i.e., 2–4) to facilitate quantification of BrdU, caspase 3, and p53 labeling (*see* **Note 26**).

1. Determine the need for background correction and assign a background correction method and threshold (*see* **Note 27**).

2. Select an object identification thresholding method and select an object identification threshold (*see* **Note 28**).

3. Select an object segmentation methods and select an object segmentation value (if available as an option) (*see* **Note 29**).

4. Set object selection parameters (*see* **Note 30**).

2.7.3 Quantification of BrdU, Caspase 3, and p53 Fluorescent Intensity

The nucleus masks established in Channel 1 are superimposed on subsequent fluorescent channels (i.e., 2, 3, and 4) and used to define the area of each cell where fluorescent intensity of the various markers will be quantified (*see* **Note 31**). Each marker is evaluated in a separate fluorescent channel. Measures of fluorescent labeling such as total fluorescent intensity, average fluorescent intensity and variation in fluorescent intensity are measured on a cell-by-cell basis. The resulting cell level data can then be used to calculate well-level averages, or alternatively set "responder gates" that produced a dichotomous data output for each well (i.e., % Responder). The chemical concentration-response data in Culbreth et al. [43] was presented as average fluorescent intensity normalized to DMSO controls. Responder gating methods have also been used to evaluate proliferation of neuronal lineage cells [97]. In the responder gating approach for proliferation, the average fluorescent intensity of BrdU labeling in a non-proliferating cell was quantified and a gate for BrdU-positive cells was established as a multiplier (3×) of that value. Chemical concentration-response data was presented as percent responder normalized to DMSO controls. A similar responder gating approach may be used to quantify apoptosis using the caspase 3 and p53 labels in ReNcell CX cells (*see* **Note 32**). Evaluations of response variability in vehicle and positive control-treated samples and quantitative evaluation of assay performance using the different quantification methods will help guide researcher's regarding which data analysis method to use.

3 Notes

1. The growing surface of the tissue culture flask should be evenly covered with the laminin solution during the coating process. Check whether the shelves in the incubator are level to help

avoid irregularities in coating of the growing surface. The manufacturer's protocol for ReNcell CX cells recommends coating with 20 µg/mL laminin for at least 4 h. However, lower concentrations of laminin or shorter incubation times may be adequate for providing even laminin coating and supporting cell growth. Optimization experiments for laminin concentration in the coating solution and incubation time may be performed in an effort to reduce cost or speed up the overall workflow.

2. Upon purchase from vendor, ReNcell CX cells can be expanded and cryopreserved. Consult with manufacturer for cryopreservation protocol. Typically, cells are expanded to a target passage number and multiple aliquots of cells (i.e., 2×10^6 cells/vial) are frozen to create a cryogenic stock. During experiments, individual vials of cells are thawed and taken through a uniform expansion and passaging procedure prior to seeding in test plates. Using this approach, experimental replicates can be performed using a uniform passage number. ReNcell CX cells expanded to passage 10 were used in Culbreth et al. [43].

3. Take care not to completely submerge the vial to prevent possible contamination of cells with water from the water bath. Grasping the top of the vial with a hemostat and gently swirling in the water bath may help decrease thawing time. Spray the vial thoroughly with 70% ethanol and wipe with a low lent ChemWipe before placing in the cell culture cabinet.

4. Dropwise addition of Neural Stem Cell Maintenance Media is recommended by the manufacturer and is intended to reduce osmotic shock to the cells. It is important to minimize the amount of time between thawing of cells and dilution in media in order to reduce the impact of residual DMSO in the cryopreservation media on cell viability.

5. Take care to avoid aspirating cells along with the residual media. Tilting the conical tube in the direction of the aspiration pipette may help decant the cell pellet and minimize the chance of aspirating the cells.

6. Once cells are resuspended, cells are typically counted by removing a small aliquot (i.e., 10–20 µL), diluting in 0.4% trypan blue, and counting manually on a hemocytometer or using an automated cell counting device (i.e., ThermoFisher Countess™ II, Millipore Sceptre™). The cell suspension can then be adjusted to the desired cell concentration by further addition of media. The researcher should take into account the type of culture vessel that will be receiving the cells (i.e., flask or plate) as well as the desired media volume and number of cells seeded (i.e., seeding density) per unit of growing surface for determining target cell concentration in solution. If seeding a

96-well plate, it is advisable to overestimate the amount of cell suspension needed to complete the experimental design. The overestimation factor will be laboratory specific and dependent upon the accuracy and configuration of the plating instruments (i.e., hand pipetting from a cell culture basin or dead-space in an automated dispenser).

7. One or more laminin-coated flasks containing 10 mL of Neural Stem Cell Maintenance Media are placed in a 37 °C, 5% CO_2 incubator prior to thawing cells. The flasks are seeded by adding a volume of cell suspension (10 mL) containing the total number of cells to be seeded to the 10 mL of media present in the flask. After the addition of cell suspension, invert the flask cap up, swirl gently, and return the flask to a horizontal position to ensure even seeding of cells. Avoid letting the media enter the neck of the culture flask.

8. As above, optimization experiments for laminin concentration in the coating solution and incubation time may be performed in an effort to reduce cost or speed up the overall workflow. Typically, the minimum recommended working volume of a 96-well is used as a starting point to determine the amount of laminin coating solution to add to each well. The minimal amount of coating solution needed may be less than the minimal working volume of the well, but needs to be sufficient to evenly coat the entire growing surface of the well.

9. With experience, the laminin coating procedure and cell passaging procedure can be synchronized so that when cells are passaged, a sufficient time has passed for the plate coating to be effective. The plates can be quickly rinsed with DPBS and ready to receive the cells.

10. In Culbreth et al. [43], ReNcell CX cells were plated at a concentration of 10,000 cells/well in 90 μL of media. Researchers may want to optimize the seeding density to meet the particular needs of their experiment. For the HCA assay performed in Culbreth et al. [43], it was determined that a seeding density of 10,000 cells/well was sufficient for the assessment of cell proliferation and apoptosis using a 10× imaging objective on a ThermoFisher Cellomics ArrayScan V^{TI} HCA with the goal of collecting data on at least 500 cells per well. Using 10,000 cells/well, the spacing of the cells was sufficient to allow for accurate identification and segmentation of cell nuclei and to allow for the analysis of 500 cells over a few (2–4) individual fields of view in untreated cultures.

11. **Steps 18** and **19** refer to a manual plating procedure using a handheld multichannel pipette. Automated cell dispensers may also be used to seed cells and will not be discussed further here. It is recommended the researcher consult the instrument

manufacturer for specifications, operating instructions, and assistance in developing automated plating protocols for ReNcell CX cells.

12. The procedure in Subheading 2.4 is for DMSO solubilized chemicals assuming a 1000× total dilution from chemical stocks to the final test well concentration and a final DMSO concentration in test wells of 0.1%. The dilution scheme would also be applicable for aqueous solubilized chemicals (*see* Culbreth et al. [43]). The procedure also assumes dosing solutions will be prepared manually using handheld laboratory pipettes. Researchers may wish to devise an alternative dilution scheme based on their experimental needs. Researchers may wish to evaluate DMSO tolerance of ReNcell CX cells in their laboratory to ensure that vehicle effects do not confound treatment-related observations.

13. For concentration-response studies, or other types of studies involving multiple test chemicals/conditions, intermediate dose plates can be prepared from DMSO-solubilized stocks. Tissue culture plates (96- or 384-well format) or "deep well" plates (96-well format) are typically used as intermediate dose plates for these types of studies. The researcher should consider the working volume of the intermediate dose plate, the volume of intermediate dilution that needs to be prepared, the number of technical replicate wells within an experiment, and the volume of intermediate dosing solution to be transferred to the test plates when devising a dilution scheme.

14. For concentration-response studies, a dilution series for each test chemical can be prepared in pure DMSO at a 1000× concentration above the final test concentration. Each concentration in the dilution series is then subsequently diluted via the intermediate dose plate and finally by dispensing into the test plate.

15. Intermediate dosing solutions should be applied to test wells in a uniform manner, taking care not to disturb the attached monolayer of cells. An automated handheld pipette (single or multichannel) many reduce variability associated with manual plunger pipettes.

16. Culbreth et al. [43] used a 24 h chemical exposure period. BrdU is added 4 h prior to the end of the exposure period (i.e., at 20 h). Researchers may choose to vary the timing of the chemical exposure and/or the timing of BrdU addition to suite experimental needs.

17. Aspiration may be performed effectively using a multichannel stainless-steel vacuum manifold. Take care not to touch the growing surface of the plates during aspiration to prevent shearing of cells from the growing surface. Holding the test

plates in slightly tilted vertical orientation may help decant liquid from the growing surface during aspiration.

18. Paraformaldehyde additions should be performed within a chemical safety cabinet or an externally vented cell culture cabinet to reduce possible exposure to PFA vapors.

19. The procedure in Culbreth et al. [43] was performed using a Cellomics Multiplex Mitosis-Apoptosis Hit Kit. This kit is no longer commercially available in the US, but analogous primary antibody reagents are available from other vendors. A recommended primary antibody dilution range of 1:100–1:250 is recommended as a starting point for testing antibody reagents and optimizing antibody dilutions for immunocytochemistry.

20. A 1 h incubation is recommended based on Culbreth et al. [43]. However, optimization of primary antibody incubation time is recommended. Longer incubation times (i.e., 24 h overnight at 4 °C) may increase the observed fluorescent signal intensity for some reagents.

21. The researcher should consult available resources, including technical manuals, product inserts, field application specialists, and/or imaging experts to determine which excitation/emission filters installed on the HCA instrument are appropriate for use with the fluorophore combination used to label the ReNcell CX cells. It is advisable to confirm that appropriate excitation/emission filters are installed on the instrument prior to labeling cells or performing experiments.

22. Typically, a HCA instrument is equipped with more than one imaging objective ranging from 5× (or lower) to 20× (or higher). It is possible to quantify BrdU, caspase 3, and p53 labeling using these different objectives. The researcher should determine which imaging objective is appropriate to meet the goals of their experiment. Lower magnification objectives capture more cells in a single field of view, but may limit the ability of the researcher to evaluate fine details of the fluorescent labeling (i.e., texture measurements). Higher magnification objectives allow for greater resolution of cell features, but capture less cells in a single field of view, thereby potentially increasing the number of images which need to be collected per well and subsequently the total time needed for scanning a plate.

23. Typically, HCA instruments allow the researcher to capture images at varying levels of image resolution. For example, on the ThermoFisher Cellomics ArrayScan VTI, "Standard" image acquisition uses 2 × 2 pixel binning for image acquisition. "Low resolution" image acquisition uses 4 × 4 pixel binning while "High resolution" image acquisition does not

utilize pixel binning. Autofocus features typically use lower resolution settings to speed up the autofocus process. Higher resolution images can then be captured for downstream analysis. Image file size increases with increasing levels of resolution. If the size of the database used for image storage is a limiting factor (such as on a common use instrument), then lower resolution images may be used to minimize required storage space and still be of sufficient quality for accurate quantification of fluorescent intensity from BrdU, caspase 3, or p53 labeled cells. The images analyzed in Culbreth et al. [43] were collected on a ThermoFisher Cellomics ArrayScan VTI using a standard (2×2 pixel binning) image acquisition setting.

24. The ReNcell CX proliferation/apoptosis assay is essentially a gain-of-signal assay. Cells that have undergone a mitotic division will have higher BrdU incorporation and thus higher fluorescent signal than quiescent cells. Likewise, cells undergoing apoptosis will have higher expression of cleaved caspase 3 and p53 and thus higher fluorescent signal than cells not undergoing apoptosis. When determining integration times, the researcher should take into account the dynamic range of the CCD camera and avoid integration times that result in saturation of pixel intensity. For setting the BrdU exposure time, the researcher should use a representative sample (i.e., test well) which contains BrdU positive cells. Likewise, for cleaved caspase 3 and p53, it is recommended that the research use representative samples that have been treated with positive control compounds (i.e., stuarosporine, paclitaxel, etc.) known to increase apoptosis. Defining an integration time that positions the pixel intensity histogram below 50–70% of the dynamic range of the CCD camera should ensure that pixel saturation is not occurring and provide an adequate working range to detect treatment-related increases in marker expression.

25. Some HCA instruments analyze images during image acquisition and provide an option for defining the minimum number of cells which need to be analyzed before moving to the next test well. Such instruments may also provide options for flagging "sparse fields" or "sparse wells"; that is, fields or wells that contain less cells than a user-defined threshold. Sparse fields or wells may be indicative of plating or labeling artifacts or overt cytotoxicity. The researcher should tailor the scan limits to meet the requirements of their experiment.

26. Optimization of an image analysis algorithm is an iterative process. Representative images are analyzed under an initial set of image analysis algorithm settings. Parameters of the algorithm are then adjusted in an effort to increase the accuracy of object identification and masking. The accuracy of object

identification and masking is typically a subjective analysis by the researcher. Once an optimized algorithm is developed, the same algorithm should be applied to all image sets in an experiment.

27. There are a number of background correction methods (*see* Buchser et al. [98] for an introductory review). The availability and nuances of these methods will vary from system to system. Background correction is used to "reduce noise" in the image, eliminate out-of-focus artifacts, and increase contrast between foreground objects and background toward the goal of more reproducible and accurate object segmentation and tracing. The researcher may have the option of setting the size of a background correction pixel radius (i.e., kernel); that is, the dimensions of a shape used to raster an image and remove local background from each pixel in the image. This is also referred to as a smoothing function or "rolling ball" background correction method. It is recommended that the researcher evaluate different background correction methods or background correction pixel radii and assess impacts on pixel intensity within objects-of-interest and in areas of background using a set of representative images. If the background correction kernel is set smaller than the objects of interest, there is a potential subtract too much background and lose the ability to detect objects of interest.

28. There are a number of thresholding methods for object identification (*see* Buchser et al. [98] and Ljosa and Carpenter [99] for an introductory review). The availability and nuances of these methods will vary from system to system. Thresholding is the process of distinguishing foreground from background. It is accomplished by selecting a cut-off value or cut-off function that is applied to the pixel intensity histogram and separates bright (i.e., foreground) from dim (i.e., background) pixels. It is recommended that the researcher evaluate different thresholding methods and assess impacts on accuracy of nucleus tracing using a set of representative images. Setting the threshold too high may reduce tracing accuracy by retracting the tracing mask within the true edges of the nucleus, resulting in over-segmentation of nuclei (i.e., division of a nucleus into several artifactual objects), or result in an inability to identify nuclei at all. Setting the threshold too low may result in mis-identification of areas of background as foreground objects (i.e., false objects).

29. There are a number of methods for object segmentation (*see* Buchser et al. [98] and Ljosa and Carpenter [99]) for an introductory review). The availability and nuances of these methods will vary from system to system. Object segmentation is used to separate adjacent objects into separate objects. Local

intensity peaks, object shape features, or watershed methods are used to determine the boundaries of adjacent objects. It is recommended that the researcher evaluate different object segmentation methods and assess impacts on accuracy of nucleus tracing using a set of representative images.

30. Once background correction, object identification and object segmentation have been optimized, most HCA software provides a list of settings which can be used to select or reject traced objects. Object selection settings use location, size, shape, and pixel intensity measurements to identify a subpopulation of traced objects of interest to the researcher. For example, the researcher may want to reject traced objects that intersect the border of an image (i.e., border objects) due to the fact the entirety of the object is not represented in the image. Size and shape parameters can be used reject objects (potential artifacts) which do not conform to the expected morphology of the cell type of interest. Size and shape selection parameters should be used judiciously to avoid rejecting cells whose morphology has changed due to a biological response. For example, apoptotic cells will become rounded and compact with condensed, bright nuclei. It would be important not to reject such objects when evaluating a test chemical's potential to induce apoptosis.

31. Background correction in the BrdU, caspase 3, and p53 fluorescent channels may not be necessary since the geometry of the nucleus masks was determined in Channel 1. Some HCA software packages allow the researcher to dilate or contract the nucleus mask in subsequent channels to quantify cytoplasmic or perinuclear fluorescent signal. The need for background correction and nucleus mask adjustment should be evaluated by the researcher.

32. Some HCA software packages allow the researcher to designate control wells that can be used to automatically calculate responder gates for the calculation of well-level measurements. Alternatively, cell level data may be exported.

Acknowledgments

The author would like to thank Dr. William Mundy and Dr. Timothy Shafer for their advice and technical assistance regarding this work. The author would also like to thank the editors, O. Joseph Trask and Peter O'Brien, for their insightful comments on earlier versions of this work.

References

1. Donato R et al (2007) Differential development of neuronal physiological responsiveness in two human neural stem cell lines. BMC Neurosci 8:36

2. Le MT et al (2009) MicroRNA-125b promotes neuronal differentiation in human cells by repressing multiple targets. Mol Cell Biol 29 (19):5290–5305

3. Morgan PJ et al (2009) Protection of neurons derived from human neural progenitor cells by veratridine. Neuroreport 20(13):1225–1229

4. Hubner R et al (2010) Differentiation of human neural progenitor cells regulated by Wnt-3a. Biochem Biophys Res Commun 400 (3):358–362

5. Lange C et al (2011) Small molecule GSK-3 inhibitors increase neurogenesis of human neural progenitor cells. Neurosci Lett 488 (1):36–40

6. Mazemondet O et al (2011) Quantitative and kinetic profile of Wnt/beta-catenin signaling components during human neural progenitor cell differentiation. Cell Mol Biol Lett 16 (4):515–538

7. Hernandez-Benitez R et al (2013) Taurine enhances the growth of neural precursors derived from fetal human brain and promotes neuronal specification. Dev Neurosci 35 (1):40–49

8. Pai S et al (2012) Dynamic mass redistribution assay decodes differentiation of a neural progenitor stem cell. J Biomol Screen 17 (9):1180–1191

9. Lemcke H, Kuznetsov SA (2013) Involvement of connexin43 in the EGF/EGFR signalling during self-renewal and differentiation of neural progenitor cells. Cell Signal 25 (12):2676–2684

10. Lemcke H et al (2013) Neuronal differentiation requires a biphasic modulation of gap junctional intercellular communication caused by dynamic changes of connexin43 expression. Eur J Neurosci 38(2):2218–2228

11. Shahbazi M et al (2013) Inhibitory effects of neural stem cells derived from human embryonic stem cells on differentiation and function of monocyte-derived dendritic cells. J Neurol Sci 330(1–2):85–93

12. Zhao X et al (2013) Dual effects of isoflurane on proliferation, differentiation, and survival in human neuroprogenitor cells. Anesthesiology 118(3):537–549

13. Shin S et al (2006) Long-term proliferation of human embryonic stem cell-derived neuroepithelial cells using defined adherent culture conditions. Stem Cells 24(1):125–138

14. Dhara SK et al (2008) Human neural progenitor cells derived from embryonic stem cells in feeder-free cultures. Differentiation 76 (5):454–464

15. Lai B et al (2008) Endothelium-induced proliferation and electrophysiological differentiation of human embryonic stem cell-derived neuronal precursors. Stem Cells Dev 17 (3):565–572

16. Dhara SK et al (2009) Genetic manipulation of neural progenitors derived from human embryonic stem cells. Tissue Eng Part A 15 (11):3621–3634

17. Acharya MM et al (2010) Consequences of ionizing radiation-induced damage in human neural stem cells. Free Radic Biol Med 49 (12):1846–1855

18. Cheng K, Kisaalita WS (2010) Exploring cellular adhesion and differentiation in a micro−/nano-hybrid polymer scaffold. Biotechnol Prog 26(3):838–846

19. Dodla MC et al (2011) Differing lectin binding profiles among human embryonic stem cells and derivatives aid in the isolation of neural progenitor cells. PLoS One 6(8):e23266

20. Young A et al (2011) Ion channels and ionotropic receptors in human embryonic stem cell derived neural progenitors. Neuroscience 192:793–805

21. Mumaw JL et al (2010) Neural differentiation of human embryonic stem cells at the ultrastructural level. Microsc Microanal 16 (1):80–90

22. Iyer S et al (2012) Mitochondrial gene replacement in human pluripotent stem cell-derived neural progenitors. Gene Ther 19(5):469–475

23. Brennand KJ, Gage FH (2012) Modeling psychiatric disorders through reprogramming. Dis Model Mech 5(1):26–32

24. Chamberlain SJ et al (2010) Induced pluripotent stem cell models of the genomic imprinting disorders Angelman and Prader-Willi syndromes. Proc Natl Acad Sci U S A 107 (41):17668–17673

25. Israel MA et al (2012) Probing sporadic and familial Alzheimer's disease using induced pluripotent stem cells. Nature 482 (7384):216–220

26. Lee G et al (2009) Modelling pathogenesis and treatment of familial dysautonomia using patient-specific iPSCs. Nature 461 (7262):402–406

27. Xu X et al (2013) Prevention of beta-amyloid induced toxicity in human iPS cell-derived neurons by inhibition of Cyclin-dependent kinases

and associated cell cycle events. Stem Cell Res 10(2):213–227

28. Yagi T et al (2012) Establishment of induced pluripotent stem cells from centenarians for neurodegenerative disease research. PLoS One 7(7):e41572

29. Cheung AY et al (2011) Isolation of MECP2-null Rett syndrome patient hiPS cells and isogenic controls through X-chromosome inactivation. Hum Mol Genet 20(11):2103–2115

30. Marchetto MC et al (2010) A model for neural development and treatment of Rett syndrome using human induced pluripotent stem cells. Cell 143(4):527–539

31. Mak SK et al (2012) Small molecules greatly improve conversion of human-induced pluripotent stem cells to the neuronal lineage. Stem Cells Int 2012:140427

32. Swistowska AM et al (2010) Stage-specific role for shh in dopaminergic differentiation of human embryonic stem cells induced by stromal cells. Stem Cells Dev 19(1):71–82

33. Young A et al (2010) Glial cell line-derived neurotrophic factor enhances in vitro differentiation of mid−/hindbrain neural progenitor cells to dopaminergic-like neurons. J Neurosci Res 88(15):3222–3232

34. Seibler P et al (2011) Mitochondrial Parkin recruitment is impaired in neurons derived from mutant PINK1 induced pluripotent stem cells. J Neurosci 31(16):5970–5976

35. Nguyen HN et al (2011) LRRK2 mutant iPSC-derived DA neurons demonstrate increased susceptibility to oxidative stress. Cell Stem Cell 8(3):267–280

36. Byers B et al (2011) SNCA triplication Parkinson's patient's iPSC-derived DA neurons accumulate alpha-synuclein and are susceptible to oxidative stress. PLoS One 6(11):e26159

37. Pasca SP et al (2011) Using iPSC-derived neurons to uncover cellular phenotypes associated with Timothy syndrome. Nat Med 17(12):1657–1662

38. Guo X et al (2013) Derivation of sensory neurons and neural crest stem cells from human neural progenitor hNP1. Biomaterials 34(18):4418–4427

39. Wada T et al (2009) Highly efficient differentiation and enrichment of spinal motor neurons derived from human and monkey embryonic stem cells. PLoS One 4(8):e6722

40. Ebert AD et al (2009) Induced pluripotent stem cells from a spinal muscular atrophy patient. Nature 457(7227):277–280

41. Mitne-Neto M et al (2011) Downregulation of VAPB expression in motor neurons derived from induced pluripotent stem cells of ALS8 patients. Hum Mol Genet 20(18):3642–3652

42. Lee G et al (2012) Large-scale screening using familial dysautonomia induced pluripotent stem cells identifies compounds that rescue IKBKAP expression. Nat Biotechnol 30(12):1244–1248

43. Culbreth ME et al (2012) Comparison of chemical-induced changes in proliferation and apoptosis in human and mouse neuroprogenitor cells. Neurotoxicology 33(6):1499–1510

44. Breier JM et al (2008) Development of a high-throughput screening assay for chemical effects on proliferation and viability of immortalized human neural progenitor cells. Toxicol Sci 105(1):119–133

45. Jaeger A et al (2013) Glycogen synthase kinase-3beta regulates differentiation-induced apoptosis of human neural progenitor cells. Int J Dev Neurosci 31(1):61–68

46. Li N et al (2010) Prosaposin in the secretome of marrow stroma-derived neural progenitor cells protects neural cells from apoptotic death. J Neurochem 112(6):1527–1538

47. Chaudhry ZL, Ahmed BY (2013) Caspase-2 and caspase-8 trigger caspase-3 activation following 6-OHDA-induced stress in human dopaminergic neurons differentiated from ReNVM stem cells. Neurol Res 35(4):435–440

48. Diaz-Coranguez M et al (2013) Transmigration of neural stem cells across the blood brain barrier induced by glioma cells. PLoS One 8(4):e60655

49. Harrill JA et al (2010) Quantitative assessment of neurite outgrowth in human embryonic stem cell-derived hN2 cells using automated high-content image analysis. Neurotoxicology 31(3):277–290

50. Harrill JA et al (2011) Comparative sensitivity of human and rat neural cultures to chemical-induced inhibition of neurite outgrowth. Toxicol Appl Pharmacol 256(3):268–280

51. Gill JK et al (2013) Contrasting properties of alpha7-selective orthosteric and allosteric agonists examined on native nicotinic acetylcholine receptors. PLoS One 8(1):e55047

52. Wu ZZ et al (2010) Effects of topography on the functional development of human neural progenitor cells. Biotechnol Bioeng 106(4):649–659

53. Harry GJ, Tiffany-Castiglioni E (2005) Evaluation of neurotoxic potential by use of in vitro systems. Expert Opin Drug Metab Toxicol 1(4):701–713

54. Comley J 2013 Stem cells rapidly gaining traction in research and drug discovery. Accessed

21 Dec 2016 http://www.ddw-online.com/ther apeutics/p213497-stem-cells-rapidly-gaining-traction-in-research-and-drugdiscoverysummer-13.html

55. BioInformant Worldwide, LLC (2016) Strategic development of neural stem and progenitor cell products. Biotechnology 2013 November 2013. Accessed 21 Dec 2016 https://www.bioinformant.com/launching-march-2016-strategic-development-of-neural-stem-and-pro genitor-cellproducts/

56. Dage J, K Merchant (2012) The application of iPS cells and differentiated neuronal cells to advance drug discovery. Accessed 21 Dec 2016 http://www.ddw-online.com/therapeu tics/p149535-the-application-of-ips-cells-and-differentiated-neuronal-cellsto-advance-drug-discovery-summer-12.html

57. Goldman SA, Sim F (2005) Neural progenitor cells of the adult brain. Novartis Found Symp 265:66–80. discussion 82–97

58. Mokry J, Karbanova J, Filip S (2005) Differentiation potential of murine neural stem cells in vitro and after transplantation. Transplant Proc 37(1):268–272

59. Deleyrolle LP, Reynolds BA (2009) Isolation, expansion, and differentiation of adult mammalian neural stem and progenitor cells using the neurosphere assay. Methods Mol Biol 549:91–101

60. Reynolds BA, Weiss S (1992) Generation of neurons and astrocytes from isolated cells of the adult mammalian central nervous system. Science 255(5052):1707–1710

61. Thomson JA et al (1998) Embryonic stem cell lines derived from human blastocysts. Science 282(5391):1145–1147

62. Shi Y et al (2008) Induction of pluripotent stem cells from mouse embryonic fibroblasts by Oct4 and Klf4 with small-molecule compounds. Cell Stem Cell 3(5):568–574

63. Yu J et al (2007) Induced pluripotent stem cell lines derived from human somatic cells. Science 318(5858):1917–1920

64. Takahashi K, Yamanaka S (2006) Induction of pluripotent stem cells from mouse embryonic and adult fibroblast cultures by defined factors. Cell 126(4):663–676

65. Huangfu D et al (2008) Induction of pluripotent stem cells by defined factors is greatly improved by small-molecule compounds. Nat Biotechnol 26(7):795–797

66. Hou P et al (2013) Pluripotent stem cells induced from mouse somatic cells by small-molecule compounds. Science 341 (6146):651–654

67. Badja C et al (2014) Efficient and cost-effective generation of mature neurons from human induced pluripotent stem cells. Stem Cells Transl Med 3(12):1467–1472

68. Kim DS et al (2012) Highly pure and expandable PSA-NCAM-positive neural precursors from human ESC and iPSC-derived neural rosettes. PLoS One 7(7):e39715

69. Yan Y et al (2013) Efficient and rapid derivation of primitive neural stem cells and generation of brain subtype neurons from human pluripotent stem cells. Stem Cells Transl Med 2 (11):862–870

70. Stover AE et al (2013) Process-based expansion and neural differentiation of human pluripotent stem cells for transplantation and disease modeling. J Neurosci Res 91 (10):1247–1262

71. D'Aiuto L et al (2014) Large-scale generation of human iPSC-derived neural stem cells/early neural progenitor cells and their neuronal differentiation. Organogenesis 10(4):365–377

72. Dotti CG, Sullivan CA, Banker GA (1988) The establishment of polarity by hippocampal neurons in culture. J Neurosci 8(4):1454–1468

73. de Lima AD, Merten MD, Voigt T (1997) Neuritic differentiation and synaptogenesis in serum-free neuronal cultures of the rat cerebral cortex. J Comp Neurol 382(2):230–246

74. Lowenstein PR et al (1995) Synaptogenesis and distribution of presynaptic axonal varicosities in low density primary cultures of neocortex: an immunocytochemical study utilizing synaptic vesicle-specific antibodies, and an electrophysiological examination utilizing whole cell recording. J Neurocytol 24 (4):301–317

75. Basarsky TA, Parpura V, Haydon PG (1994) Hippocampal synaptogenesis in cell culture: developmental time course of synapse formation, calcium influx, and synaptic protein distribution. J Neurosci 14(11 Pt 1):6402–6411

76. Latchney SE et al (2011) Neural precursor cell proliferation is disrupted through activation of the aryl hydrocarbon receptor by 2,3,7,8-tetrachlorodibenzo-p-dioxin. Stem Cells Dev 20 (2):313–326

77. Schreiber T et al (2010) Polybrominated diphenyl ethers induce developmental neurotoxicity in a human in vitro model: evidence for endocrine disruption. Environ Health Perspect 118(4):572–578

78. Schopperle WM, DeWolf WC (2007) The TRA-1-60 and TRA-1-81 human pluripotent stem cell markers are expressed on podocalyxin in embryonal carcinoma. Stem Cells 25 (3):723–730

79. Yuan J, Yankner BA (2000) Apoptosis in the nervous system. Nature 407(6805):802–809

80. Lei X, Guo Q, Zhang J (2012) Mechanistic insights into neurotoxicity induced by anesthetics in the developing brain. Int J Mol Sci 13(6):6772–6799

81. Behl C (2000) Apoptosis and Alzheimer's disease. J Neural Transm (Vienna) 107 (11):1325–1344

82. Shimohama S (2000) Apoptosis in Alzheimer's disease—an update. Apoptosis 5(1):9–16

83. Shen Y, White E (2001) p53-dependent apoptosis pathways. Adv Cancer Res 82:55–84

84. Vermes I et al (1995) A novel assay for apoptosis. Flow cytometric detection of phosphatidylserine expression on early apoptotic cells using fluorescein labelled Annexin V. J Immunol Methods 184(1):39–51

85. Walton M et al (1997) Annexin V labels apoptotic neurons following hypoxia-ischemia. Neuroreport 8(18):3871–3875

86. Martin HL et al (2014) High-content, high-throughput screening for the identification of cytotoxic compounds based on cell morphology and cell proliferation markers. PLoS One 9 (2):e88338

87. Caceres A, Banker GA, Binder L (1986) Immunocytochemical localization of tubulin and microtubule-associated protein 2 during the development of hippocampal neurons in culture. J Neurosci 6(3):714–722

88. Barnes AP, Polleux F (2009) Establishment of axon-dendrite polarity in developing neurons. Annu Rev Neurosci 32:347–381

89. Harrill JA et al (2013) Use of high content image analyses to detect chemical-mediated effects on neurite sub-populations in primary rat cortical neurons. Neurotoxicology 34:61–73

90. Harrill JA et al (2015) Ontogeny of biochemical, morphological and functional parameters of synaptogenesis in primary cultures of rat hippocampal and cortical neurons. Mol Brain 8:10

91. Harrill JA, Robinette BL, Mundy WR (2011) Use of high content image analysis to detect chemical-induced changes in synaptogenesis in vitro. Toxicol In Vitro 25(1):368–387

92. Fletcher TL et al (1991) The distribution of synapsin I and synaptophysin in hippocampal neurons developing in culture. J Neurosci 11 (6):1617–1626

93. Kawamoto EM, Vivar C, Camandola S (2012) Physiology and pathology of calcium signaling in the brain. Front Pharmacol 3:61

94. Berridge MJ (1998) Neuronal calcium signaling. Neuron 21(1):13–26

95. Nicholson DW (1999) Caspase structure, proteolytic substrates, and function during apoptotic cell death. Cell Death Differ 6 (11):1028–1042

96. Fridman JS, Lowe SW (2003) Control of apoptosis by p53. Oncogene 22(56):9030–9040

97. Mundy WR, Radio NM, Freudenrich TM (2010) Neuronal models for evaluation of proliferation in vitro using high content screening. Toxicology 270(2–3):121–130

98. Buchser W et al (2004) Assay development guidelines for image-based high content screening, high content analysis and high content imaging. In: Sittampalam GS et al (eds) Assay guidance manual. Eli Lilly & Company and the National Center for Advancing Translational Sciences, Bethesda, MD

99. Ljosa V, Carpenter AE (2009) Introduction to the quantitative analysis of two-dimensional fluorescence microscopy images for cell-based screening. PLoS Comput Biol 5(12): e1000603

Chapter 19

Determination of Hepatotoxicity in iPSC-Derived Hepatocytes by Multiplexed High Content Assays

Oksana Sirenko and Evan F. Cromwell

Abstract

We present here methods for assessing hepatotoxicity by high content imaging and image analysis. The assays focus on the characterization of toxic effects using a variety of phenotypic endpoint readouts. Multi-parametric automated image analysis is used in the protocols to increase assay sensitivity and provide important information about possible in vitro toxicity mechanisms. iPSC-derived hepatocytes were used as a model for the hepatotoxicity assays, but the methods would also be suitable for other liver toxicity cell models. The methods contain detailed step-by-step descriptions of the cell treatment, staining, image acquisition, and image analysis.

Key words Screening, High content, Imaging, Hepatotoxicity, Stem cell models, Hepatocytes

1 Introduction

High content imaging based in vitro toxicity assays show great promise for safety and efficacy testing because they can be automated to provide sufficient capacity to screen many pharmaceutical compounds or environmental chemical hazards in high throughput [1]. High content imaging has been used with primary hepatocytes [2] and immortalized cell lines [3–5]. We developed imaging acquisition and analysis methods using iPSC-derived hepatocytes [6–8]. Human iPSC-derived hepatocytes are considered a useful model for toxicity evaluations because they exhibit a primary tissue-like phenotype, have consistent and unlimited availability, and have the potential to establish genotype-specific cells from different individuals. iPSC-derived tissue-specific cells provide relevant human biology in vitro and are increasingly being studied for their potential to accurately predict drug-induced toxicity to multiple organs [8, 9]. The HCS assay model provides tools for multi-parametric characterization of various toxicity phenotypes that include: characterization of cell size and shape, cell adhesion and spreading,

Paul A. Johnston and Oscar J. Trask (eds.), *High Content Screening: A Powerful Approach to Systems Cell Biology and Phenotypic Drug Discovery*, Methods in Molecular Biology, vol. 1683, DOI 10.1007/978-1-4939-7357-6_19,
© Springer Science+Business Media LLC 2018

number of viable cells, nuclear condensation, lipid accumulation, cytoskeleton organization, as well as short-term and long-term changes in mitochondria potential. The combination of these readouts has potential to increase assay sensitivity and provide important information about possible insights into toxicity mechanisms.

Two assay methods were developed for both live cells and fixed cells. The first assay, developed to provide general cellular toxicity readouts, uses a simplified staining protocol compatible with screening for environmental hazards. It employs a panel of three dyes, Calcein AM, MitoTracker Orange, and Hoechst, that provide information on cell viability [10, 11], mitochondria membrane potential [12], and nuclear condensation [13] respectively (Figs. 1–3). The advantages of this method include simplicity, cost, and ease-of-workflow. Cell staining is achieved by the co-addition of the dyes in a

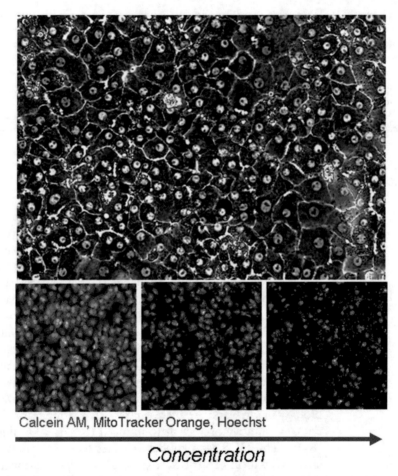

Calcein AM, MitoTracker Orange, Hoechst

Concentration

Fig. 1 *Top*: Morphological characterization of iCell Hepatocytes. Transmitted light image of iCell Hepatocytes after 4 days in culture. Bi-nucleation and bile canaliculi (thick cell-cell borders) can be seen in the image. *Bottom*: iCell Hepatocytes treated with increasing concentrations of amitriptyline for 48 h (*left* to *right*: 0, 3, & 10 µM). Images shown are composites of the FITC, TRITC, and DAPI channels and were taken with a 10× objective

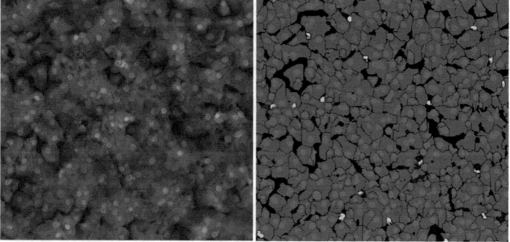

Fig. 2 Live-cell imaging for toxicity assessment. *Top*: Representative settings for analysis using the MWCS application module. *Bottom left*: Composite image of a control well of iCell Hepatocytes stained with Calcein AM, Hoechst and MitoTracker Orange. *Bottom right*: Image analysis results after processing with the MWCS module. Live cells are indicated by Green masks while dead cells (Hoechst stain only) are indicated by Gray masks

single step, without the need to fix cells or wash repeatedly. The second method has been adapted for fixed cells by the inclusion of phalloidin staining to enable an assessment of cytoskeletal organization. Although the fixed cell protocol requires additional steps, assay plates may be stored and re-imaged as required. We also describe optimized protocols and methods of analysis for several other commercially available toxicity assays including the JC-10 mitochondria potential assay, the Cyto-ID Autophagy Detection assay, and the HCS LipidTOX™ Phospholipidosis and Steatosis Detection assay (Fig. 4). We describe the typical settings for all the assays including

Multi-Parametric Analysis Using Custom Module Editor

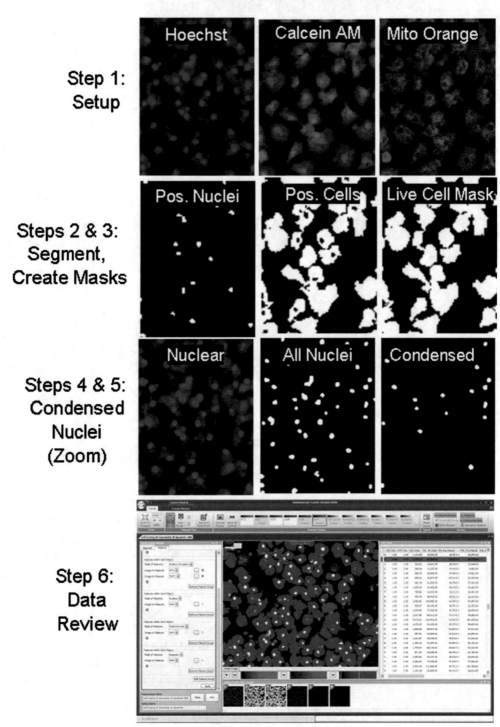

Fig. 3 Images from various steps in the CME analysis module used for the multi-parametric live cell hepatotoxicity assay. Details of the analysis performed in each step are provided in the text. Briefly, images are selected (*step 1*); then segmentation is performed to form various masks (*steps 2 and 3*). Object filtering can be used to further classify cells (*steps 4 and 5*) into those, for example, with condensed nuclei. Finally (*step 6*) data can be reviewed with linkage between images, masks, and data outputs on a cell-by-cell basis

JC-10   Autophagy   Phospho-lipidosis

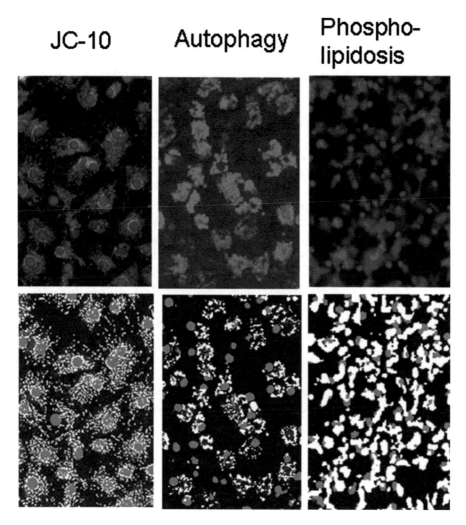

Fig. 4 Composite images and analysis results from JC-10 (*left*), autophagy (middle), and phospholipidosis (*right*) assays. The JC-10 image was acquired from untreated (control) cells stained with JC-10 reagent. The autophagy image was acquired from cells treated with rapamycin for 24 h and stained with Cyto-ID™ Autophagy detection kit. The phospholipidosis image was acquired from cells treated with 30 μM of propranolol for 48 h. All the images were acquired with a 10× objective and analyzed using the Granularity module. Analysis results are shown beneath of each image

image acquisition and analysis, and recommend the most valuable output parameters for the evaluation of hepatotoxicity.

The JC-10 assay readout is complimentary to MitoTracker Orange and provides information on the immediate direct effects of compounds on mitochondria potential. JC-10 is selectively taken up by mitochondria, and changes color from green to orange as the mitochondrial membrane potential increases. Mitochondrial toxicity can be evaluated by quantifying the number of JC-10 aggregates using granularity image analysis algorithms. Autophagy is a process of selective degradation of intracellular targets such as mis-folded proteins and damaged organelles. The Cyto-ID stain forms granules during autophagy and this can also be quantified using granularity image analysis algorithms. Phospholipidosis, or excessive accumulation of phosphilipids in tissues, is an important concern for drug development as drug-induced phospholipidosis represents a safety risk. The LipidTOX reagent allows for evaluation of phospholipidosis and steatosis by the classification of neutral lipid and phospholipid granules using the granularity image analysis algorithms.

2 Materials

2.1 Cell Culture and Plating

1. The human iPSC-derived hepatocytes used were iCell® Hepatocytes (Cellular Dynamics International, Madison WI) (CDI). Cells are received fresh and require processing before use according to the protocol provided and recommended by CDI (*see* **Note 1**). Media formulations and cell-handling procedures were performed according to the protocols provided by CDI.

2. Cells are seeded at a density of 60 K/well (96-well plate) or 15 K/well (384-well plate) in black-wall clear-bottom collagen-coated microplate (*see* **Note 2**) and incubated for at least for 2 days at 37 °C and 5% CO_2 in the appropriate growth media containing glucose (Media D from CDI is recommended) prior to compound exposure. Cells should exhibit a uniform confluent monolayer with cobblestone-like morphology as shown in Fig. 1 (*see* **Note 3**).

3. For compound exposure, growth media is typically switched to a glucose-free, galactose-containing solution for 24 h at 37 °C and 5% CO_2. For the media change, simply remove glucose media, wash 1× with galactose-containing media, and replace with galactose media.

2.2 Multi-parametric Hepatotoxicity Assay Stains and Reagents

1. Calcein AM.
2. Alexa Fluor® 488 (AF488) conjugated Phalloidin.
3. Hoechst 33258.
4. MitoTracker Orange CM-H2TMROS.
5. Phosphate-buffered Saline (PBS).
6. Blocking buffer: PBS + 0.1% of Fetal Bovine Serum (FBS).
7. Formaldehyde.
8. Permeabilization solution: 0.02% saponin plus 2% FBS.

2.3 Assay Kits

1. JC-10 (ultra pure) Mitochondria Dye (AAT Bioquest, Inc., Sunnyvale, CA).
2. Cyto-ID Autophagy Detection Kit (Enzo Life Sciences, Farmingdale, NY).
3. HCS LipidTOX Phospholipidosis and Steatosis Detection Kit (Life Technologies, Carlsbad, CA).

2.4 Automated Imaging Setup

1. The methods described herein were developed on the ImageXpress® Micro 4 or ImageXpress® Micro XLS high content imaging system (Molecular Devices LLC, Sunnyvale, CA), but could be adapted to any automated high content imaging platform.

2. MetaXpress® 6 or MetaXpress® 5 high content image acquisition and analysis software or equivalent.

3. 4× Plan Apo objective, Numerical Aperture (NA) = 0.2 (Nikon).

4. 10× Plan Fluor objective, NA = 0.3 (Nikon).

5. 20× ELWD Plan Fluor objective, NA = 0.45 (Nikon).

6. DAPI Filter Cube, Excitation Filter (Ex) = 377/50, Emission Filter (Em) = 447/60 (Center wavelength/Bandpass width in nm).

7. FITC Filter Cube, Ex = 482/35, Em = 536/40.

8. TRITC Filter Cube. Ex = 543/22, Em = 593/40.

3 Methods

3.1 Compound Treatment

1. Prepare serial dilutions of individual compounds from DMSO stock solutions. Final DMSO concentration in the media should not exceed 1%, since toxicity can occur at higher concentrations (data not shown). We recommend testing compounds in triplicates or quadruplicates in a half-log dilution series over a range of 10 nM–1000 μM. This provides a balance between statistical relevancy and assay throughput. Other dilution ranges can be selected based on compound potency. To increase throughput, compound libraries can be tested in duplicates, with appropriate positive and negative controls included in greater numbers of replicates.

2. Incubate cells with compounds for an appropriate period of time at 37 °C and 5% CO_2. Suggested times are provided in Table 1, although optimal exposure times should be empirically determined for each compound. Compounds were added once at the beginning of the incubation period.

Table 1
Suggested compound incubation times for assays

	Mutli-tox	JC-10	Autophagy	Phospholipidosis
Compound treatment (h)	72	1	24	48

3.2 Multi-parametric Hepatotoxicity Assay—Live and Fixed-Cell Staining

1. Staining solution: Prepare working concentrations of staining reagents in PBS buffer: 6 mL per plate (final concentrations are given in **step 3**).

2. Carefully remove cell culture media containing compounds by manually pipetting media out of wells. Removal of media by this method is important; flipping or aspiration methods can lead to loss of damaged or dead cells (*see* **Note 4**).

3. Live Cell Staining Multi-Tox Assay: Remove the media and incubate cells with a mix of Calcein AM (1 μM), Hoechst 33,258 (2 μg/mL), and MitoTracker Orange (200 nM) (*see* **Note 5**). Gently add staining solution into wells (60 μL or more for 96-well plates, 15 μL or more for 384-well plates). Incubate for 30 min at 37 °C and 5% CO_2.

4. Remove staining solution by manually pipetting. Add PBS buffer: 100 μL for 96-well plates, 50 μL for 384-well plates (for better stability of cells include 0.1% of serum into PBS). Do not wash cells. Proceed to imaging (*see* **step 5**).

5. Acquire images using the ImageXpress Micro XLS system with 20×, 10×, or 4× objective. A 4× or 10× objective is recommended for cell scoring analysis where resolving subcellular features is not critical. A 20× objective provides better visualization of features such as mitochondria particles (*see* **Note 6**). Excitation light is from a solid-state white light engine with emission from 380 to 680 nm.

6. Use a FITC filter cube for Calcein AM or AF488-phalloidin staining (typical exposure time 10–50 ms for Calcein AM, and 100–300 ms for phalloidin staining). Use a TRITC filter cube for MitoTracker Orange (typical exposure times 30–100 ms). Use a DAPI filter cube for Hoechst nuclear stain (typical exposure times 20–100 ms).

7. Number of Sites per Well: 1–2 images per well for 384-well plates and 2–4 images per well for 96-well plates were taken when using the 10× or 20× objectives. 1 image per well was taken when using the 4× objective (*see* **Note 7**). For image analysis use MetaXpress 5 or MetaXpress 6 software with the Multi-Wavelength Cell Scoring (MWSC) application module. The MWSC algorithm is preferred over the use of single-color scoring because of the increased flexibility to classify cells with multiple readouts. Images are first segmented into individual cells using the nuclear stain to identify nuclear objects by both staining intensity over background, and morphology. Then standard cell segmentation algorithms are used to define and associate specific regions of interest within each cell such as cytoplasm, nucleus, or whole cell. Typical settings for image analysis are given in Table 2 (*see* **Note 8**). *See* Fig. 2 for a typical software dialog setup screen.

Table 2
Image analysis setup parameters for multi-parametric hepatotoxicity assay

I.D	Channel	Min width (μM)	Max width (μM)	Int. above background (gray level)	Min area (μM²)	Notes
W1	DAPI	6	20	100		Hoechst/nuclei
W2	FITC	12	30	2000	100	Calcein AM
W2	FITC	12	30	300	100	488 Phalloidin
W3	TRITC	12	30	2000	100	Mito-tracker orange

8. A variety of parameters output by the MWCS module can be used to evaluate toxicity. The ones best suited for the hepatotoxicity assay are listed below (*see* **Note 9**). The measurements are used to quantitate the total cell number, number of viable cells (Positive W3, *see* Table 2 for definitions), and number of cells with intact mitochondria (Positive W3). The measurements are also used to measure cell spreading or coverage (Positive W2 stain area) and intensity of viability or actin staining (average or integrated intensities of the positive W2 signal). Nuclear mean area or average intensity from W1 is used to evaluate nuclei and nuclear condensation.

   - Total Cells (number of all cells, nuclear count).
   - Positive W2 (number of Calcein AM, or AF488-phalloidin positive cells).
   - % Positive W2 (% of cells that are viable).
   - Positive W3 (number of MitoTracker Orange positive cells).
   - Positive W2 Mean Stain Area (mean area of viable cells).
   - Positive W2 Mean Stain Integrated Intensity (intensity of positive stain).
   - Positive W2 Mean Stain Average Intensity (intensity of positive stain).
   - All Nuclei Mean Area (mean nuclear area).
   - All Nuclei Mean Average Intensity (mean nuclear stain intensity).

9. Custom Module Editor: A more advanced image analysis data set can be generated using the MetaXpress Custom Module Editor (CME). The CME is used to combine cell scoring algorithms along with other analyses and provide additional output parameters to characterize phenotypic changes associated with toxicity. The CME also contains features like a "click-to-find" tool where clicking on selected cells or nuclei

Table 3
CME setup parameters for multi-parametric hepatotoxicity assay

Step 2	Image	Min width (μM)	Max width (μM)	Int. above background (gray level)	Min area (μM^2)	Output mask
	DAPI	7.8	22	50		Positive or negative nuclei
	FITC	13	41	22		Positive cytoplasm
Step 4	Image	Min width (μM)	Max width (μM)	Int. above background (gray level)	Min area (μM^2)	Output mask
	DAPI	7.5	20	50		Nucleus
Step 5	Image	Mask	Max area (μM^2)	Min avg intensity (gray scale)		Output mask
	DAPI	Nucleus	105	90		Condensed nuclei (apoptotic)

automatically defines the minimum and maximum size as well as a contrast threshold. The CME protocol used for the Live Cell multi-parametric hepatotoxicity assay (*see* Fig. 3) is outlined below. The specific parameters used in each step are listed in Table 3.

- Step 1—Setup: Select Images to analyze (denoted by Filter Cubes: DAPI, FITC, and TRITC).

- Step 2—Cell Scoring Objects: Segment live nuclei and live cells from background (*see* **Notes 10** and **11**) using width and intensity and label output parameters, Positive Nuclei and Positive Cytoplasm respectively. A viable (Live) cell is identified by the presence of both Hoechst and Calcein AM staining.

- Step 3—Logical Operation: Create Total Live Cell mask by the addition of Positive Cytoplasm and Positive Nuclei masks.

- Step 4—Find Round Objects: Find all nuclei in the DAPI channel that satisfy width and intensity measurements (*see* Table 3).

- Step 5—Filter Mask: Identify small, bright nuclei using area and intensity filters as indication of nuclear condensation (*see* **Note 12**).

- Step 6—Measure: Calculate outputs for each object using masks defined during the previous steps. Output parameters include Calcein AM and MitoTracker Orange intensities in Positive Cells, percent Positive Cells, and percent cells with Condensed Nuclei.

3.3 Multi-parametric Hepatotoxicity Assay—Fixed Cell

1. The Multi-parametric Hepatotoxicity Assay can also be performed with fixed cells. This provides flexibility if a delay between plate preparation and imaging is desirable.

2. Remove the media and incubate cells first with the mix of Hoechst 33258 (2 μg/mL) and MitoTracker Orange (200 nM) for 30 min at 37 °C and 5% CO_2, remove the dye, then fix cells using formaldehyde (4% final concentration) in PBS for 15 min at room temperature.

3. Wash once with PBS, then once with permeabilization solution, and then permeabilize/block cells for 1 h using permeabilization solution.

4. After permeabilization step stain cells with AF488-phalloidin (50 nM) for 2 h. Note that this step may be combined with permeabilization step if desired. Remove staining solution and wash cells 1× or 2× with PBS before acquiring images.

5. Follow protocols for image acquisition and analysis as detailed in Subheading 3.2.

3.4 Mitochondria Membrane Potential Assay

1. Expose cells to selected compounds for 30 min at 37 °C and 5% CO_2.

2. Monitor mitochondria membrane potential by the addition of the mitochondrial active dye JC-10 following the recommended protocol and appropriate concentrations of included reagents (*see* **Note 13**).

3. A nuclear stain such as Hoechst can be included if desired (*see* **Note 14**).

4. Add 6× concentrated solution of JC-10 reagent directly to the cell culture media without the removal of compounds.

5. Incubate with JC-10 reagent for additional 30 min at 37 °C (total time for compound treatment is 60 min).

6. Acquire images using ImageXpress Micro XLS system with 10× or 20× objective.

7. Use FITC and TRITC filter cubes for JC-10 emissions (typical exposure 20–100 ms). The cells can be defined using the green emission of the JC-10 monomers in mitochondria. One can also use Hoechst as a nuclear stain and then define/count cells using the DAPI channel. Typically, one image per well was taken for 384-well plates and 2–4 images per well for 96-well plates were taken using 10× or 20× objectives. Multiple sites can be acquired with the 20× objective to obtain improved cell statistics.

8. For the image analysis use MetaXpress 6 or MetaXpress 5 software with the Granularity application module. Image analysis setup parameters are given in Table 4.

Table 4
Image analysis parameters for mitochondria membrane potential assay

I.D	Channel	Min width (μM)	Max width (μM)	Int. above background (gray level)	Min area (μM^2)	Notes
Granule	TRITC	0.6	4	2300		J-aggregates
Nuclear	FITC	10	30	2300		JC-10 cytoplasm
Nuclear	DAPI	6	20	100	100	Hoechst stain (if used)

Table 5
Image analysis setup parameters for autophagy assay

I.D	Channel	Min width (μM)	Max width (μM)	Int. above background (gray level)	Min area (μM^2)	Notes
Granule	FITC	0.3	9	800		Cyto-ID dye
Nuclear	DAPI	6	20	100		Hoechst stain

9. Measurements for Granularity:
   - Granules (granularity, total number of granules per image).
   - Total Granule Area (area covered by granules, per image).
   - Integrated Granule Intensity (sum of the intensities of all granules per image).
   - Mean Granular Area or Granule Area per Cell (mean area covered by granules, per cell).
   - Granules per Cell (number of granules per cell).
   - Total Cells (count of all cells in the field).

3.5 Autophagy Assay

1. Autophagy can be assessed by the Cyto-ID Autophagy Detection Kit after treatment of cells with compounds for 24 h at 37 °C and 5% CO_2.

2. Acquire images using the ImageXpress Micro XL system with 20× or 10× objectives.

3. Use FITC filter cube for Cyto-ID Autophagy Detection assay.

4. For the image analysis use MetaXpress® 6 or MetaXpress® 5 software with the Granularity application module. Typical settings for image analysis for the Autophagy assay are given in Table 5.

5. Measurements from Granularity module used to characterize autophagy are listed below:
   - Granules (granularity, total number of granules per image).
   - Total Granule Area (area covered by granules, per image).

- Integrated Granule Intensity (sum of the intensities of all granules per image).
- Mean Granular Area or Granule Area Per Cell (mean area covered by granules, per cell).
- Granules Per Cell (number of granules per cell).
- Total Cells (count of all cells).

3.6   Phospholipidosis Assay

1. Phospholipidosis was measured using the HCS LipidTOX Phospholipidosis and Steatosis Detection Kit for neutral lipids and phospholipids after treatment of cells with compounds for 48 h at 37 °C and 5% CO_2. The staining protocol followed the product recommended reagent dilutions and buffers.

2. Acquire images using the ImageXpress Micro XLS system with $10\times$ or $20\times$ objective.

3. Use FITC filter cube for LipidTox staining of neutral lipids, and TRITC filter cube for LipidTox staining of phospholipids.

4. For the image analysis use MetaXpress® 6 or MetaXpress® 5 software with the Granularity application module. Typical settings for image analysis for the phospholipidosis assay are given in Table 6. The neutral lipids and phospholipids need to be analyzed for separately.

5. Measurements from the Granularity module used to characterize phospholipidosis are listed below. Separate Granule parameters will be determined for neutral lipids and phospholipids.

- Granules (granularity, total number of granules per image).
- Total Granule Area (area covered by granules, per image).
- Integrated Granule Intensity (sum of the intensities of all granules per image).
- Mean Granular Area or Granule Area Per Cell (mean area covered by granules, per cell).
- Granules Per Cell (number of granules per cell).
- Total Cells (count of all cells).

Table 6
Image analysis setup parameters for phospholipidosis assay

I.D	Channel	Min width (μM)	Max width (μM)	Int. above background (gray level)	Min area (μM^2)	Notes
Granule	FITC	3	10	1000		Neutral lipids
Granule	TRITC	3	10	1000		Phospholipids
Nuclear	DAPI	6	20	100		Hoechst stain

3.7 Data Processing and Further Analysis

AcuityXpress™ high content informatics software (Molecular Devices) or other analysis packages can be used for statistical data analysis, curve fitting (e.g., generating EC_{50} or IC_{50} values), obtaining Z'-values, generating heat-maps, or other types of analysis. Detailed description of such an analysis is beyond the scope of this chapter.

4   Notes

1. iPSC-derived hepatocytes are received on cold pack, and need disaggregation by trypsin followed by purification by a Ficoll density gradient. Detailed protocols are provided by CDI.

2. For imaging applications black-walled with clear-bottom plates are strongly recommended. Good results were achieved using collagen-coated plates from BD. Plates from other manufacturers (e.g., Costar or Greiner) can also be used. It is important to get plates recommended for imaging as they use higher quality plastic and result in better images.

3. It is essential that cells have the right morphology. They should form an adherent monolayer, have round nucleus, distinct nucleoli, high cytoplasmic/nuclear ratio, and show evidence of bi-nucleation and bile canaliculi.

4. It is essential to remove media with compounds prior to the staining step. Addition of staining solution directly into media results in a nonuniform staining. The number of washes/manipulations is minimized during processing in order not to lose loosely attached or damaged cells, which can contribute to experimental error. All additions of solutions or washes must be done at low speed with maximal care.

5. Some compounds may fluoresce at the same wavelengths as the reporter dyes and can cause artifacts in measurement parameters such as mean fluorescence intensity.

6. The multi-parameter toxicity assay can be done with either the $10\times$ or $4\times$ objectives. The $10\times$ objective provides higher resolution of cells and subcellular structures while allowing the capture of relatively large cell numbers (2000–3000) in a single image because of the large field of view captured by the instrument (~ 2 mm^2). The system captures an entire well area from a 384-well plate when using a $4\times$ objective providing typically >10,000 cells/image with sufficient resolution to detect cells and nuclei. However, detection of nuclear size or cell size is less accurate using the $4\times$ objective than with the $10\times$ objective. Other assays described here (mitochondria membrane potential, autophagy, phospholipidosis) require at least the $10\times$ objective. The $20\times$ objective provides even better

resolution but typically requires taking multiple sites per well for better cell statistics.

7. Hepatocytes form confluent cultures and with accurate cell seeding provide consistent cell numbers in all wells. If the cells are equally distributed within a well one image per 384 well is sufficient for a representative sample. However, if the cells are non-confluent or unequally distributed, then taking additional sites can improve the assay performance. Also, counting of cell numbers in samples affected with compounds (where fewer cells are present) will be more precise when multiple images per well are taken. Taking more images does increase the time of acquisition and analysis and needs to be considered.

8. Indicated numbers for Min and Max width may vary depending on staining intensity and plating density. The numbers for intensity over the local background (gray levels) must be adjusted for each specific experiment, since those will significantly vary depending on staining intensity and exposure time during acquisition.

9. All the listed output parameters may be considered for data analysis, however some might not give significant response or have good Z'-factor coefficient values. This will depend on specifics of the experiment and response of the compounds selected for the assay. For the multi-parameter toxicity assay the number of Calcein AM or actin-positive cells, as well as mean cell area for positive cells, typically provides the largest assay window and highest Z'-factor coefficients between positive and negative samples (Z'-factor coefficient > 0.5). For the mitochondria potential assays, the number of granules per cell, total granule area, or granule area per cell typically provide the best outputs.

10. The Min and Max values can be entered manually, or selected using the Click-To-Find tool. In the latter case, the user selects objects with the pointer and size and intensity values are calculated based on that set of objects. This is often a good starting point and then values can be optimized manually by observing the analysis results.

11. There are two algorithm types available to the user: Fast and Standard. The Fast algorithm can be up to $7\times$ faster in processing speed, but may not be as accurate in segmenting and splitting cells and nuclei. The Standard algorithm is a more involved calculation that has been the basis of segmentation and splitting for previous MetaXpress software and MetaMorph® software image analysis packages.

12. The Apoptotic classification is based on the observation of condensed nuclei. To positively determine that cells are in apoptosis a secondary measurement, such as Caspase 3 assay, may be considered.

13. The JC-10 protocol was used as recommended by the manufacturer (AAT Bioquest, Sunnyvale, CA) with the following modifications. JC-10 dye-loading solution was prepared by diluting 100× JC-10 (Component A) solution in provided buffer (Assay Buffer A, Component B). Cells were incubated with 50 μL/well (96-well plate) or 12.5 μL/well (384-well plate) of JC-10 dye-loading solution for 30 min. Assay Buffer B (Component C) was not used.

14. The mitochondria membrane potential JC-10 assay does not include nuclear stain. When run with just the JC-10 dye, the granularity algorithm uses the green channel (FITC) to define the cell mask and count individual cells. In this case cell dimensions should be used instead of nuclear dimensions (included in the example) for the Nuclear channel. Alternatively, one can improve the accuracy of the cell count by using Hoechst nuclear stain. The nuclear stain does not interfere with the JC-10 staining protocol (data not shown). In this case nuclear stain/dimensions should be used for the Nuclear channel in the analysis protocol.

References

1. Thomas N (2010) High-content screening: a decade of evolution. J Biomol Screen 15:1–9

2. Gomez-Lechon MJ, Castell JV, Donato MT (2010) The use of hepatocytes to investigate drug toxicity. Methods Mol Biol 640:389–415

3. Rodrigues RM, Bouhifd M, Bories G et al (2013) Assessment of an automated in vitro basal cytotoxicity test system based on metabolically-competent cells. Toxicol In Vitro 27:760–767

4. Mennecozzi M, Landesmannn B, Harrix GA (2012) Hepatotoxicity scfreening taking a mode-of-action approach using HepaRG cells and HCA. ALTEX Proc 1:193–204

5. O'Brien PJ, Irwin W, Diaz D et al (2006) High concordance of drug-induced human hepatotoxicity with in vitro cytotoxicity measured in a novel cell-based model using high content screening. Arch Toxicol 80:580–604

6. Anson BD, Kolaja KL, Kamp TJ (2011) Opportunities for use of human iPS cells in predictive toxicology. Clin Pharmacol Ther 89:754–758

7. Mann DA, Einhorn S, Block K (2013) Human iPSC-derived hepatocytes: functional model tissue for in vitro predictive metabolism, toxicity, and disease modeling. Gen Eng Biotech News 33:28–29

8. Medine CN, Lucendo-Villarin B, Storck C et al (2013) Developing high-fidelity hepatotoxicity models from pluripotent stem cells. Stem Cells Transl Med 2:505–509

9. Schweikart K, Guo L, Shuler Z et al (2013) The effects of jaspamide on human cardiomyocyte function and cardiac ion channel activity. Toxicol In Vitro 27:745–751

10. Lichtenfels R, Biddison WE, Schulz H et al (1994) CARE-LASS (calcein-release-assay), an improved fluorescence-based test system to measure cytotoxic T lymphocyte activity. J Immunol Methods 172:227–239

11. Papadopoulos NG, Dedoussis GV, Spanakos G et al (1994) An improved fluorescence assay for the determination of lymphocyte-mediated cytotoxicity using flow cytometry. J Immunol Methods 177:101–111

12. Presley AD, Fuller KM, Arriaga EA (2003) MitoTracker green labeling of mitochondrial proteins and their subsequent analysis by capillary electrophoresis with laser-induced fluorescence detection. J Chromatogr B Analyt Technol Biomed Life Sci 793:141–150

13. Latt SA, Stetten G, Juergens LA et al (1975) Recent developments in the detection of deoxyribonucleic acid synthesis by 33258 Hoechst fluorescence. J Histochem Cytochem 23:493–505

Chapter 20

The Generation of Three-Dimensional Head and Neck Cancer Models for Drug Discovery in 384-Well Ultra-Low Attachment Microplates

David A. Close, Daniel P. Camarco, Feng Shan, Stanton J. Kochanek, and Paul A. Johnston

Abstract

The poor success rate of cancer drug discovery has prompted efforts to develop more physiologically relevant cellular models for early preclinical cancer lead discovery assays. For solid tumors, this would dictate the implementation of three-dimensional (3D) tumor models that more accurately recapitulate human solid tumor architecture and biology. A number of anchorage-dependent and anchorage-independent in vitro 3D cancer models have been developed together with homogeneous assay methods and high content imaging approaches to assess tumor spheroid morphology, growth, and viability. However, several significant technical challenges have restricted the implementation of some 3D models in HTS. We describe a method that uses 384-well U-bottomed ultra-low attachment (ULA) microplates to produce head and neck tumor spheroids for cancer drug discovery assays. The production of multicellular head and neck cancer spheroids in 384-well ULA-plates occurs in situ, does not impose an inordinate tissue culture burden for HTS, is readily compatible with automation and homogeneous assay detection methods, and produces high-quality uniform-sized spheroids that can be utilized in cancer drug cytotoxicity assays within days rather than weeks.

Key words Three-dimensional models, Head and neck cancer, High content screening, Ultra-low attachment plates, Cancer drug discovery

1 Introduction

New cancer drug leads are most commonly identified in high-throughput growth inhibition screening campaigns conducted in tumor cell lines that have been maintained and assayed in traditional two-dimensional (2D) cell culture models [1–3]. Molecularly targeted agents that disrupt specific oncogenes and cytotoxic compounds that inhibit tumor cell line growth advance to anti-tumor efficacy studies in mice, and mechanism of action studies is initiated for compounds that display in vivo efficacy [2, 4, 5].

Paul A. Johnston and Oscar J. Trask (eds.), *High Content Screening: A Powerful Approach to Systems Cell Biology and Phenotypic Drug Discovery*, Methods in Molecular Biology, vol. 1683, DOI 10.1007/978-1-4939-7357-6_20, © Springer Science+Business Media LLC 2018

However, new cancer drug approval rates are $\leq 5\%$, more than twofold lower than for other therapeutic areas [2, 4–6]. The poor success rate of cancer drug discovery has raised concerns that in vitro 2D tumor cell line growth inhibition assays do not faithfully represent the biological complexity and heterogeneity of human cancers, prompting efforts to develop more physiologically relevant cellular models that can be utilized in early preclinical cancer lead discovery [2, 7–10]. For solid tumors, this would mandate the implementation of three-dimensional (3D) tumor models that more accurately recapitulate human solid tumor architecture and biology [7–10].

Solid tumors exist in a highly interactive 3D microenvironment comprised of tumor cells, stromal cells (vascular, immune, and fibroblast cells), and extracellular matrix (ECM) components where cell-cell interactions, cell-ECM interactions and local gradients of nutrients, growth factors, secreted factors and oxygen control cell function and behavior [11–15]. Tumor cells cultured in 3D microenvironments experience dramatically different adhesive, topographical, and mechanical forces and the altered cellular cues received in 3D cultures modify responses to stimuli and drugs [11–23]. In addition, the increased cell-cell and cell-ECM interactions of cells in solid tumors represent a permeability barrier through which therapeutic agents must penetrate [11–24].

3D tumor spheroids are self-assembled cultures of tumor cells formed in conditions where cell-cell interactions predominate over cell-substrate interactions [12–16, 25, 26]. Multicellular tumor spheroids resemble avascular tumor nodules, micro-metastases, or the intervascular regions of large solid tumors with respect to their morphological features, microenvironment, volume growth kinetics and gradients of nutrient distribution, oxygen concentration, cell proliferation, and drug access [12–16, 25, 26]. A variety of anchorage-dependent and anchorage-independent in vitro 3D cancer models have been developed; tissue slices or explants, gel or matrix embedded cultures and cocultures, scaffold-based supports, micro-patterning and cell printing approaches, microfluidic systems, and multicellular spheroids [12–17, 19, 21–23, 25–32]. Although a number of homogeneous assay methods and high content imaging approaches have been developed to assess tumor spheroid morphology, growth, and viability [31], a variety of significant technical challenges have restricted the implementation of some 3D models in HTS and/or HCS [16, 18–20, 22, 26, 27]: protracted 2–3 week cell culture periods can be both rate-limiting and resource intensive; models developed in petri dishes or 6-well, 12-well, or 24-well microtiter plates limit throughput and capacity; natural gel matrices such as matrigel contain an ill-defined myriad of endogenous components and growth factors that may exhibit considerable batch-to-batch variability; reproducible calibration of the mechanical and biochemical properties of naturally derived hydrogels can also be difficult; matrices such as soft agar may need to be preheated and mixed with tumor cells without compromising viability; complex multi-step procedures can be challenging to automate and temperature must be tightly controlled during automated liquid transfers to prevent premature gelation of matrices; hydrogels may limit the penetration of compounds and detection reagents; some models produce spheroids with large variations in morphologies and size; variable Z-plane locations and heterogeneity in 3D colony size and shape together with the opacity of gels or matrices may interfere with signal capture; and many 3D models are labor intensive requiring experimentalists with extensive expertise and experience for their implementation.

We describe herein a method for the use of 384-well U-bottomed ultra-low attachment (ULA) microplates to produce head and neck tumor spheroids for cancer drug discovery assays. A variety of human tumor cells lines, but not all, have been shown to spontaneously form tight spheroids, compact aggregates, or loose aggregates when seeded into U-bottomed 96-well ULA-plates treated with a hydrophilic, neutrally charged coating covalently bound to the polystyrene or co-polymer well surface [26]. Multi-cellular tumor spheroids formed in 96-well U-bottomed ULA-plates can be maintained and grown in culture, exhibited similar morphologies and immuno-histochemistry staining to spheroids grown in agar, and could be used for tumor cell migration and invasion assays [26]. Eight lung cancer cell lines underwent self-assembly to form viable 3D tumor spheroids in 96-well ULA-plates [16]. 3D lung cell line spheroids exhibited altered epidermal growth factor receptor (EGFR) and cMET expression and signal transduction when compared to 2D cultures [16]. The ULA lung cancer cell line spheroid culture microenvironment altered the responses to drugs and growth factors that mimicked the natural lung tumors better than 2D cultures [16]. Non-transformed epithelial cells and breast cancer epithelial cells cultured in 96-well ULA-plates, either alone or as cocultures with human fibroblasts or HUVECs, formed 3D spheroids [33]. Microtubule disrupting agents and EGFR inhibitors exhibited preferential cytotoxicity against breast cancer cells over normal cells when they were cultured as 3D spheroids, but not in 2D cultures [33]. Similar selective breast cancer drug sensitivity was also observed in 3D cocultures when compared to 2D cocultures [33]. H-RAS-transformed fibroblasts were screened against 633 compounds from the NIH clinical collection and a small kinase focused library in growth inhibition assays conducted in 2D monolayers and 3D cultures grown in 96-well ULA-plates [34]. Four drugs exclusively inhibited H-RAS fibroblast growth in 3D cultures, and five preferentially inhibited the growth of 3D cultures over 2D monolayers [34].

Several head and neck squamous cell carcinoma (HNSCC) cell lines have also been shown to generate spheroids after 4 days of culture in 96-well ULA-plates [16, 26]. The Cal33 HNSCC cell line can be used for mouse xenograft models and we have previously used it for cancer drug discovery high content screening (HCS) campaigns [35–37]. To investigate whether Cal33 cells would form 3D multicellular tumor spheroids in 384-well ULA-plates, we seeded wells with 625, 1250, 2500, and 5000 cells and then placed the plates in an incubator at 37 °C, 5% CO_2 and 95% humidity for 24 h. After 24 h in culture, live (Calcein AM, CAM) and dead (Ethidium homodimer, EtHD) reagents were added to the wells and single best focus images per well were acquired with a $10\times$ objective in each of three channels (Ch) on the ImageXpress Micro (IXM) automated HCS platform; transmitted light (Ch1, TL), FITC (Ch2, CAM) and Texas Red (Ch3, EtHD) (Fig. 1a). To acquire fluorescent images of spheroids we used an automated image-based focus algorithm to acquire both a coarse focus (large µm steps) set of CAM stained spheroid images for the first spheroid to be imaged, followed by a fine (small µm steps) set of images to select the best focus image Z-plane. The same best focus Z-plane was then used to acquire images in all other fluorescent channels. For all subsequent wells and spheroids to be imaged only a fine focus set of CAM images were acquired. Cal33 HNSCC cells formed compact 3D spheroids after 24 h of culture in ULA-plates and the size of the spheroids increased with the number of cells added to the well. Cal33 tumor spheroids exhibited strong CAM staining in the periphery and throughout the spheroid with only sparse EtHD staining in the central core, indicating that the overwhelming majority of the cells in the spheroids were viable (Fig. 1a). We used the multi-wavelength cell scoring (MWCS) image analysis module to quantify the integrated fluorescent intensities of the CAM (Ch2) and EtHD (Ch3) signals in the digital images of the Cal33 spheroids acquired on the IXM (Fig. 1b). For Cal33 spheroids we defined the approximate minimum width of the CAM stained spheroid in Ch2 (FITC) to be 160 µm and the approximate maximum width to be 700 µm, and the threshold intensity above local background to be 50, and the MWCS module image segmentation created total spheroid masks that were applied to Ch2 and Ch3. The spheroid masks were then used to quantify the mean integrated fluorescence intensity of CAM and EtHD within the Cal33 spheroid (Fig. 1b). Consistent with the images presented in Fig. 1a, Cal33 spheroids produced much higher mean integrated fluorescent intensity values (>100-fold) for CAM than EtHD, indicating that most of the cells were viable (Fig. 1b). The CAM integrated fluorescent intensity signal increased linearly with respect to the number of Cal33 cells seeded per well of the 384-well ULA-plate and the corresponding larger sizes of the spheroids formed (Fig. 1b). We used the line-scan tool of the MetaXpress image analysis software to measure the diameters of the Cal33 spheroid using the transmitted light images (Fig. 1a). The

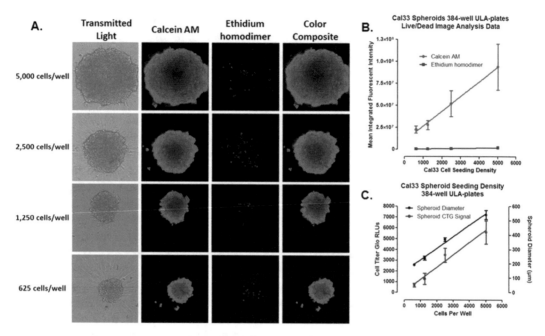

Fig. 1 Production and analysis of Cal33 head and neck squamous cell carcinoma (HNSCC) spheroids in U-bottomed 384-well ultra-low attachment plates. (**a**) Transmitted light and fluorescent images of spontaneously formed Cal33 HNSCC spheroids 24 h after seeding into U-bottomed 384-well ULA-plates. Cal33 cells were seeded into the wells of 384-well ULA-plates at 625, 1250, 2500, and 5000 cells per well and then incubated at 37 °C, 5% CO_2, and 95% humidity for 24 h. After 24 h in culture, 25 μM Calcein AM (CAM) live and 50 μM Ethidium homodimer (EtHD) dead reagents were added to the wells for an additional 45 min incubation, and then single best focus images per well were acquired with a $10\times$ objective in each of 3 Ch on the IXM HCS platform; Transmitted Light (Ch1, TL), FITC (Ch2, CAM), and Texas Red (Ch3, EtHD). Representative images from multiple independent experiments are shown. (**b**) Quantification of the mean integrated fluorescent intensities of the CAM (live) and EtHD (dead) signals in the digital images of the Cal33 spheroids acquired on the IXM. We used the multi-wavelength cell scoring (MWCS) image analysis module to quantify the integrated fluorescent intensities of the CAM and EtHD signals in the digital images of the Cal33 spheroids acquired on the IXM. We defined the approximate minimum width of the CAM stained Cal33 spheroid in Ch2 (FITC) to be 160 μm and the approximate maximum width to be 700 μm, and the threshold intensity above local background to be 50, and the MWCS image segmentation created total spheroid masks that were applied to Ch2 and Ch3. The spheroid masks were then used to quantify the mean integrated fluorescence intensity of CAM and EtHD within the spheroid images. The mean $\pm$ SD ($n = 18$) mean integrated fluorescence intensity of CAM (●) and EtHD (■) from 18 wells for each seeding density are presented. Representative experimental data from three independent experiments are shown. (**c**) Measuring the diameters of Cal33 spheroids and the correlation between Cell Titer Glo® RLU signals and Cal33 spheroid cell number and size. Cal33 HNSCC cells were seeded at seeding densities ranging from 625 to 5000 cells per well into 384-well ULA-plates and placed in an incubator for 24 h. We used the line-scan tool of the MetaXpress image analysis software to measure the diameters of the Cal33 spheroids captured in the transmitted light images. The mean $\pm$ SD ($n = 18$) diameters (μm) from 18 wells for each seeding density are presented on the right Y axis (●). After 24 h in culture, the Cell Titer Glo™ (Promega, Madison, WI) reagent was added to the wells and the RLU signals were captured on the M5e microtiter plate reader (Molecular Devices LLC, Sunnyvale, CA). The mean $\pm$ SD ($n = 9$) RLU signals from nine wells for each seeding density are presented on the left Y axis (●). Representative experimental data from multiple independent experiments are shown

diameters of the Cal33 spheroids ranged between 193 and 540 µM, and increased linearly with respect to the number of Cal33 cells seeded into the wells of the 384-well ULA-plate (Fig. 1c). The homogenous Cell Titer Glo™ (Promega, Madison, WI) (CTG) reagent measures the luciferase luminescent signal (RLUs) produced by cellular ATP levels and is frequently used as an indicator of cell number, viability and for proliferation/growth inhibition assays. We have previously developed 384-well CTG growth inhibition assays in 2D Cal33 cultures to determine compound cytotoxicity [36, 37], and others have used the reagent to measure cell viability, cell number, and cancer drug effects in multicellular tumor spheroid cultures generated in 96-well ULA-plates [16, 26, 33, 34]. When added to Cal33 spheroids formed 24 h after seeding into 384-well ULA plates, the CTG RLU signal increased linearly with respect to the number of viable Cal33 cells seeded per well and the sizes of the resulting spheroids (Fig. 1c). In summary, all three quantitative indicators of spheroid size and cell viability exhibited a linear relationship with respect to the number of Cal33 cells seeded into the wells of 384-well ULA plates (Fig. 1b, c).

To illustrate that uniformly sized 3D HNSCC spheroids can be readily produced in 384-well ULA-plates at a scale compatible with HTS/HCS, we seeded wells with 5000 Cal33 cells and maintained the plates in an incubator at 37 °C, 5% CO_2, and 95% humidity for 24, 48, 72, and 96 h (Fig. 2). After each 24 h period in culture, CAM and EtHD reagents were added to wells and single best focus images per well were acquired with a $4\times$ objective in each of 3 Ch on the IXM HCS platform; TL, FITC, and Texas Red. Images acquired using the $4\times$ objective capture the whole well of the 384-well plate thereby minimizing the chances of missing spheroids. The montage of 384 TL images acquired 24 h after the Cal33 cells were seeded into the 384-well ULA-plate shows that spheroids had formed in every well, and except for a small number of wells (<4%), the size, shape, and central location of the spheroids were remarkably uniform (Fig. 2a). As described above, we used the linescan tool to measure the diameters of the Cal33 spheroids captured in the TL images (Figs. 2b, c) and the MWCS image analysis module to quantify the integrated fluorescent intensities of the CAM and EtHD signals in the FITC and Texas Red images (Fig. 2d, e). The daily time sequence TL images of the Cal33 spheroid generated in a single well of the 384-well ULA-plate suggest that although spheroids formed within 24 h of cell seeding, spheroids appear to contract and become more compact and acquire a more defined boundary over the next 72 h in culture (Fig. 2b). On average, 5000 Cal33 cells seeded into 384-well ULA-plates formed spheroids with a diameter of ~460 µM, and although the mean spheroid diameter did not appear to change significantly with time in culture, diameters were trending lower (~448 µm) at

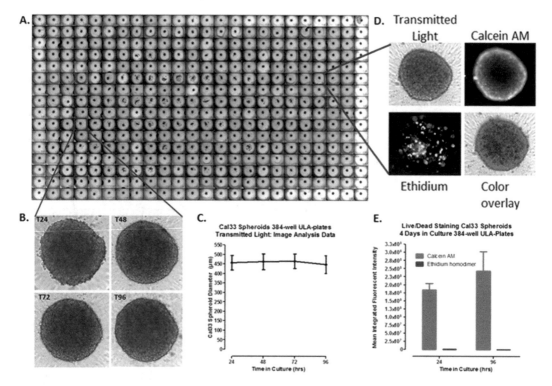

Fig. 2 Cal33 head and neck squamous cell carcinoma (HNSCC) spheroids produced in U-bottomed 384-well ultra-low attachment plates are compatible with HTS/HCS. (**a**) Montage of transmitted light images of Cal33 spheroids formed after 24 h in ULA-plates. 5000 Cal33 cells were seeded into 384-well ULA-plates and maintained in an incubator at 37 °C, 5% CO_2, and 95% humidity for 24 h. (**b**) Time sequence of transmitted light images of a Cal33 spheroid generated in a single well of the 384-well ULA-plate after 24, 48, 72, and 96 h in culture. 5000 Cal33 cells were seeded into 384-well ULA-plates and maintained in an incubator at 37 °C, 5% CO_2, and 95% humidity for 96 h. Single best focus TL images per well were acquired with a 4× objective after 24, 48, 72, and 96 h in culture. The daily time sequence of TL images of the Cal33 spheroid generated in a single well are presented. (**c**) Time sequence of Cal33 spheroid diameters extracted from transmitted light images of the Cal33 spheroids generated in the wells of 384-well ULA-plate after 24, 48, 72, and 96 h in culture. 5000 Cal33 cells were seeded into 384-well ULA-plates and maintained in an incubator at 37 °C, 5% CO_2, and 95% humidity for 96 h. Single best focus TL images per well were acquired with a 4× objective after 24, 48, 72, and 96 h in culture. We used the line-scan tool to measure the diameters of the Cal33 spheroids captured in the TL images and the daily time sequence for mean ± SD ($n = 73$) of the Cal33 spheroid diameters (μM) in 73 wells is presented (●). (**d**) Transmitted light and fluorescent images of a Cal33 HNSCC spheroid 96 h after seeding into a U-bottomed 384-well ULA-plate. Cal33 cells were seeded into the wells of 384-well ULA-plates wells at 5000 cells per well and then incubated at 37 °C, 5% CO_2, and 95% humidity for 96 h. After 96 h in culture, Calcein AM (live, *green*) and Ethidium homodimer (dead, *red*) reagents were added to the wells at a final concentration of 2.5 μM and 5 μM respectively, and the plates were incubated for an additional 1 h before single best focus images per well were acquired with a 4× objective in each of 3 Ch on the IXM HCS platform; transmitted light (Ch1, TL), FITC (Ch2, CAM), and Texas Red (Ch3, EtHD). (**e**) Mean integrated fluorescent intensities of the CAM (live) and EtHD (dead) signals in the digital images of the Cal33 spheroids acquired on the IXM after 24 and 96 h in culture. As described above, we used the multi-wavelength cell scoring (MWCS) image analysis module to quantify the integrated fluorescent intensities of the CAM and EtHD signals in the digital images of the Cal33 spheroids acquired on the IXM. The mean ± SD of the mean integrated fluorescence intensities of CAM (■) and EtHD (■) from multiple wells at 24 h ($n = 8$) and 96 h ($n = 15$) are presented. Representative experimental data from three independent experiments are shown

96 h (Fig. 2c). Fluorescent images of 96 h Cal33 tumor spheroids stained with CAM and EtHD exhibited strong CAM staining in the periphery and throughout the spheroid, with some punctate EtHD staining in the central core (Fig. 2d). A comparison of the mean integrated fluorescent intensity data from 24 to 96 h Cal33 spheroids stained with CAM and EtHD indicated that most of the cells in the spheroid were viable at both time points, and that while the CAM signal increased by ~33% during the 72 h in culture, the EtHD signal decreased (Fig. 2e).

We used Cal33 spheroids produced in 384-well ULA-plates to evaluate the potency and efficacy of four molecularly targeted anti-cancer drugs; the HSP90 inhibitor STA9090 (Ganetspib), the receptor tyrosine kinase inhibitor (PDGF-R/VEGF-R) Sunitinib, the mTOR inhibitor Everolimus, and the proteasome inhibitor Bortezomib (Fig. 3). Cal33 spheroids formed 24 h after seeding 5000 cells per well into ULA-plates that were then treated with the cancer drugs at the indicated concentrations and returned to the incubator for an additional 72 h of culture. Representative TL, Hoechst, CAM, and EtHD images of Cal33 spheroids exposed to DMSO or selected concentrations of the four drugs illustrate the effects of compound exposure on spheroid size, morphology, and viability (Fig. 3a). At 5 μM STA9090, the top concentration tested, 72 h exposure to the HSP90 inhibitor significantly reduced the size of the Cal33 spheroids relative to DMSO controls, dramatically reducing both the CAM (live) and Hoechst (DNA) staining while significantly increasing the relative proportion of EtHD staining of dead cells (Fig. 3a). At the top concentrations of Bortezomib and Everolimus tested, 0.1 and 25 μM respectively, 72 h exposure to the proteasome and mTOR inhibitors reduced the sizes and apparent densities of the Cal33 spheroids relative to DMSO controls, reduced Hoechst staining, and dramatically reduced CAM staining while significantly increasing the relative proportion of EtHD staining of dead cells (Fig. 3a). At 5.56 μM Sunitinib, a concentration slightly lower than its IC_{50} in our 3D model (Fig. 3b), 72 h exposure to the receptor tyrosine kinase inhibitor only partially reduced the size of the Cal33 spheroids relative to DMSO controls, but produced a more pronounced reduction in Hoechst and CAM staining while increasing the relative proportion of EtHD staining of dead cells (Fig. 3a). For the Cal33 spheroid growth inhibition assays we used DMSO control wells ($n = 64$) to represent 100% cell viability and to normalize the data from the compound treated wells as % of DMSO controls (Fig. 3b). The four molecularly targeted anti-cancer drug inhibitors inhibited the viability of Cal33 spheroids in a concentration-dependent manner and exhibited IC_{50}s of 0.015, 0.034, 7.27, and 9.25 μM for STA9090, Bortezomib, Sunitinib, and Everolimus respectively.

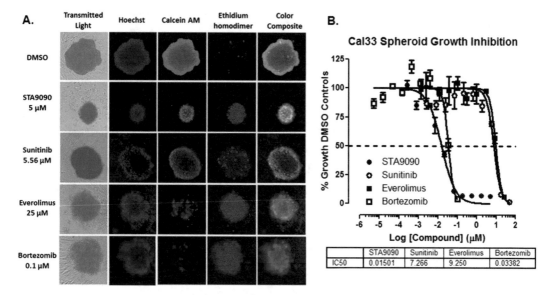

Fig. 3 Disruption of Cal33 head and neck squamous cell carcinoma (HNSCC) spheroid morphology and viability in U-bottomed 384-well ultra-low attachment plates by molecularly targeted anti-cancer drugs. (**a**) Transmitted light, Hoechst, CAM, and EtHD images of Cal33 spheroids exposed to cancer drugs for 72 h. Cal33 HNSCC cells were seeded at 5000 cells per well into 384-well ULA-plates and after 24 h in culture they were exposed to the indicated concentrations of four molecularly targeted anti-cancer drugs for an additional 72 h of culture; the HSP90 inhibitor STA9090 (Ganetspib), the receptor tyrosine kinase inhibitor (PDGF-R/VEGF-R) Sunitinib, the mTOR inhibitor Everolimus, and the proteasome inhibitor Bortezomib. Calcein AM (live, *green*) and Ethidium homodimer (dead, *red*) reagents were added to the wells at a final concentration of 2.5 µM and 5 µM respectively, and the plates were incubated for an additional 1 h before single best focus images per well were acquired with a 4× objective in each of 4 channels on the IXM HCS platform; transmitted light, Hoechst, FITC and Texas red. Representative images from three independent experiments are shown. (**b**) Cytotoxicity toward Cal33 spheroids exposed to cancer drugs for 72 h. Cal33 HNSCC cells were seeded at 5000 cells per well into 384-well ULA-plates and after 24 h in culture the indicated concentrations of the four molecularly targeted anti-cancer drugs were added to the spheroids and the plates were returned to the incubator for an additional 72 h of culture. The molecularly targeted anti-cancer drugs included the HSP90 inhibitor STA9090 (Ganetspib) (●), the receptor tyrosine kinase inhibitor (PDGF-R/VEGF-R) Sunitinib (○), the mTOR inhibitor Everolimus (■), and the proteasome inhibitor Bortezomib (□). After 72 h in culture, the CTG reagent was added to the wells and the RLU signals were captured on the M5e microtiter plate reader. For the Cal33 spheroid growth inhibition assays, we used DMSO control wells ($n = 64$) to represent 100% cell viability and to normalize the data from the compound treated wells as % of DMSO controls. We used GraphPad Prism 5 software to plot and fit the normalized data to curves using the Sigmoidal log [inhibitor] versus normalized response variable slope equation: $Y = 100/(1 + 10^{\wedge}((LogIC50-X) \times HillSlope))$. The mean ± SD ($n = 3$) growth inhibition data from triplicate wells for each compound concentration are presented as the % of the DMSO plate controls. Representative experimental data from three independent experiments are shown

The production of tumor spheroids in 384-well in ULA-plates occurs in situ, does not impose an inordinate tissue culture burden for HTS, is readily compatible with automation and homogeneous assay detection methods, and produces high-quality uniform-sized spheroids that can be assayed within days rather than weeks.

2   Materials

1. Head and neck squamous cell carcinoma (HNSCC) cell lines such as the Cal33 and FaDu tumor cell lines.

2. Dulbecco's Modified Eagle Medium (DMEM) with 2 mM L-glutamine supplemented with 10% fetal bovine serum,100 μM nonessential amino acids, and 100 U/mL penicillin and streptomycin.

3. Trypsin 0.25%, 1 g/L EDTA solution (trypsin-EDTA).

4. Dulbecco's Mg^{2+} and Ca^{2+} free phosphate-buffered saline (PBS).

5. Hoechst 33342.

6. Ultra-Low Attachment (ULA) black walled, clear round-bottom 384-well spheroid microtiter plates.

7. Calcein AM (CAM).

8. Ethidium Homodimer-1 (EtHD).

9. Promega CellTiter-Glo® Luminescent Cell Viability Assay (CTG).

10. An automated imaging high content screening (HCS) platform with integrated image analysis software capable of acquiring and analyzing transmitted light and florescent images.

11. A microtiter plate reader platform capable of detecting luminescence.

3   Methods

1. Rehydrate the ULA plate by dispensing 50 μL of serum-free medium (SFM) into each well. Centrifuge the plate at $50 \times g$ for 1 min, and then incubate the plate at room temperature for 15–45 min. Remove the media from the wells by flicking the plate over a sink or an appropriate container. Immediately add 15 μL of DMEM tissue culture medium to each well, and then centrifuge the plate at $50 \times g$ for 1 min (*see* **Note 1**). Set the plate aside until ready for cell seeding.

2. Aspirate spent tissue culture medium from HNSCC cells in tissue culture flasks that are <70% confluent (*see* **Note 2**), wash cell monolayers 1× with PBS, and expose cells to trypsin-EDTA until they detach from the surface of the tissue culture flasks. Add the serum-containing tissue culture medium to neutralize the trypsin. Transfer the cell suspension to a 50 mL capped sterile centrifuge tube and centrifuge at $500 \times g$ for 5 min to pellet the cells. Resuspend cells in the serum-containing tissue culture medium and count the number of trypan blue excluding viable cells using a hemocytometer.

3. Use a multichannel automated pipet to seed HNSCC cells at 5000 cells per well in 30 μL of DMEM tissue culture media into a previously rehydrated 384-well ULA assay plate (45 μL final). Briefly centrifuge the plate at $50 \times g$ for 30 s. Culture cells overnight at 37 °C, 5% CO_2, and 95% humidity allowing spheroids to form for 24 h (*see* **Note 3**).

4. After the multi-cellular HNSCC tumor spheroids have formed, typically 24 h for most HNSCC cell lines, use an automated liquid-handling platform to transfer 5 μL of compounds or controls pre-diluted in SFM to the wells of the ULA assay plate (0.2% DMSO final), centrifuge the plate at $50 \times g$ for 1 min, and incubate at 37 °C, 5% CO_2, and 95% humidity for 72 h (*see* **Note 4**).

5. The self-aggregation of HNSCC cells to form spheroids and other assembled cell morphologies can be monitored sequentially and noninvasively by acquiring transmitted light images on an automated imaging HCS platform. We describe the use of the ImageXpress Micro (IXM) automated field-based HCS platform (Molecular Devices LLC, Sunnyvale, CA) that has been integrated with the MetaXpress imaging and analysis software (*see* **Note 5**). The IXM optical drive includes a 300 W Xenon lamp broad spectrum white light source and 2/3″ chip Cooled CCD Camera and optical train for standard field of view fluorescence imaging and an IXM transmitted light option with phase contrast. We utilized either a 4× Plan Apo 0.20 NA objective or a 10× Plan Fluor 0.3 NA objective to acquire transmitted light and/or fluorescent images of HNSCC spheroids in 348-well ULA assay plates. For fluorescent imaging, the IXM is equipped with the following ZPS filter sets; DAPI, FITC/ALEXA 488, TRITC/CY3, CY5, and Texas Red. To acquire transmitted light images of spheroids we used an automated image-based focus algorithm to acquire both a coarse focus (large μm steps) set of images for the first spheroid to be imaged, followed by a fine (small μm steps) set of images to select the best focus image. For all subsequent wells and spheroids to be imaged only a fine focus set of images were acquired to select the best focus Z-plane.

6. Calcein AM/Ethidium Homodimer-1/Hoechst 33342 staining. To each well, dispense 5 μL of a 10× cocktail containing 25 μM CAM, 50 μM EtHD, and 20 μg/mL Hoechst in SFM. Centrifuge at $100 \times g$ for 1 min, then incubate at 37 °C, 5% CO_2, and 95% humidity for 1 h. Acquire fluorescent images in three channels (Hoechst—DAPI, CAM—FITC, EtHD—Texas Red) on an automated imaging platform with a 4× or 10x objective. To acquire fluorescent images of spheroids we used the automated image-based focus algorithm to acquire both a coarse focus (large μm steps) set of DAPI (Hoechst

stained nuclei) images for the first spheroid to be imaged, followed by a fine (small μm steps) set of images to select the best focus image Z-plane. The same best focus Z-plane was then used to acquire images in all other fluorescent channels. For all subsequent wells and spheroids to be imaged only a fine focus set of DAPI images were acquired. Transmitted light images can also be acquired now (*see* **Note 6**).

7. For the Promega CellTiter-Glo® Luminescent Cell Viability Assay: Remove the 384-well ULA assay plate from the incubator and allow cooling to room temperature for 30 min. To each well, add 20 μL of CellTiter-Glo® Luminescent Cell Viability reagent. Centrifuge at $100 \times g$ for 1 min, and then shake on a microplate shaker for 20 min at 500 rpm. Measure the luminescence using a microplate luminometer (*see* **Note 7**).

4   Notes

1. While some consider rehydration of the ULA plate to be an optional step, we have found that rehydrating the plate prior to seeding cells diminishes the effects of any fissures or cracks in the ULA coating, as well as washes away any loose coating particles. This results in spheroids that are more uniform in shape. After the SFM is removed, it is important to immediately add back complete growth media or cell suspension to each well so that the coating is not allowed to dry. Adding a small volume of complete growth media (e.g., 15 μL) eliminates the danger of the wells drying prior to seeding of the cell suspension. This is particularly helpful for more complex experiments where multiple seeding densities or cells lines are being tested.

2. Typically, better responses are obtained when Cal33 cells are harvested from tissue culture flasks that are <70% confluent. Although we have described the formation of multicellular tumor spheroids by Cal33 HNSCC cells seeded into 384-well ULA-plates, we have observed similar spheroid formation with several other HNSCC cell lines.

3. The appropriate seeding density will depend on the experimental design. We have successfully formed HNSCC spheroids from 500 to 20,000 cells. Care should be taken during cell seeding to avoid contact with the bottom of the ULA plate with the pipet. If the plate coating is damaged, the resulting spheroids may be malformed. The centrifugation step is optional, but we have found that we see more uniform spheroids if this step in included. The objective is to gently corral all the cells into the same space to promote spheroid formation, and centrifuging too fast or for too long should be avoided.

4. Due to the three dimensionality and potentially fragile nature of the spheroids, an appropriate dispense height and dispense speed on the liquid-handling platform is critical in preventing unintentional damage to the spheroids. Compounds are dispensed 3 mm above the plate bottom at a rate of 1 µL/s to maintain the integrity of the spheroids.

5. We have described the use of the IXM automated field-based HCS platform and the MetaXpress imaging and analysis software to acquire and analyze transmitted light and fluorescent images of HNSCC spheroids in 384-well ULA assay plates. Any commercially available automated HCS platform with comparable transmitted light and fluorescent imaging capabilities and an integrated image analysis software package should be able to achieve similar results.

6. Another option would be to acquire images of HNSCC spheroids on the IXM that involves the use of the infrared (IR) laser autofocus to detect the bottom of the plate and well. A series of 10–20 Z-stack images are then acquired each separated by a step size of 20 µm in a range equally distributed above and below a set Z-position. A journal is then used to either select the best focus image from the Z-stack, or to collapse the images in the Z-stack into a single maximum projection image. Typically, however, this method takes much longer to acquire a 384-well plate with scan times of several hours compared to ~90 min using the optical focus method. In addition, the IR laser-based autofocus is not always successful at detecting the bottom of the plate and wells of the U-bottom ULA-plates.

7. The Promega CellTiter-Glo® Luminescent Cell Viability Assay is a homogeneous "mix-and-read" format that lyses cells to release intracellular ATP. This free ATP, which is directly proportional to the number of cells present in the wells, then drives the generation of a stable "glow-type" luminescent signal by a proprietary thermostable luciferase (Ultra-Glo™ Recombinant Luciferase). Since the luciferase signal is proportional to the amount of free ATP in the well, there is a direct correlation between cell number and luminescent output.

Acknowledgments

This project was supported in part by funds from a Development Research Project award (Johnston, PI) from the Head and Neck Spore P50 award (CA097190) of the University of Pittsburgh Cancer Institute.

References

1. Al-Lazikani B, Banerji U, Workman P (2012) Combinatorial drug therapy for cancer in the post-genomic era. Nat Biotechnol 30:679–692

2. Ocana A, Pandiella A, Siu LL, Tannock IF (2011) Preclinical development of molecular-targeted agents for cancer. Nat Rev Clin Oncol 8:200–209

3. Ocaña A, Pandiella A (2010) Personalized therapies in the cancer "omics" era. Mol Cancer 9:202–214

4. Hait W (2010) Anticancer drug development: the grand challenges. Nat Rev Drug Discov 9:253–254

5. Hutchinson L, Kirk R (2011) High drug attrition rates–where are we going wrong? Nat Rev Clin Oncol 8:189–190

6. Kamb A, Wee S, Lengauer C (2007) Why is cancer drug discovery so difficult? Nat Rev Drug Discov 6:115–120

7. Abbot A (2003) Biology's new dimension. Nature 424:870–872

8. Pampaloni F, Reynaud EG, Stelzer EH (2007) The third dimension bridges the gap between cell culture and live tissue. Nat Rev Mol Cell Biol 8:839–845

9. Ryan S, Baird AM, Vaz G, Urquhart AJ, Senge M, Richard DJ, O'Byrne KJ, Davies AM (2016) Drug discovery approaches utilizing three-dimensional cell culture. Assay Drug Dev Technol 14:19–28

10. Zips D, Thames HD, Baumann M (2005) New anticancer agents: in vitro and in vivo evaluation. In Vivo 19:1–7

11. Baker B, Chen CS (2012) Deconstructing the third dimension: how 3D culture microenvironments alter cellular cues. J Cell Sci 125:3015–3024

12. Friedrich J, Seidel C, Ebner R, Kunz-Schughart LA (2009) Spheroid-based drug screen: considerations and practical approach. Nat Protoc 4:309–324

13. Lovitt C, Shelper TB, Avery VM (2013) Miniaturized three-dimensional cancer model for drug evaluation. Assay Drug Dev Technol 11:435–448

14. Lovitt C, Shelper TB, Avery VM (2014) Advanced cell culture techniques for cancer drug discovery. Biology 3:345–367

15. Wang C, Tang Z, Zhao Y, Yao R, Li L, Sun W (2014) Three-dimensional in vitro cancer models: a short review. Biofabrication 6:022001

16. Ekert J, Johnson K, Strake B, Pardinas J, Jarantow S, Perkinson R, Colter DC (2014) Three-dimensional lung tumor microenvironment modulates therapeutic compound responsiveness in vitro–implication for drug development. PLoS One 9:e92248

17. Fischbach C, Chen R, Matsumoto T, Schmelzle T, Brugge JS, Polverini PJ, Mooney DJ (2007) Engineering tumors with 3D scaffolds. Nat Methods 4:855–860

18. Härmä V, Virtanen J, Mäkelä R, Happonen A, Mpindi JP, Knuuttila M, Kohonen P, Lötjönen J, Kallioniemi O, Nees M (2010) A comprehensive panel of three-dimensional models for studies of prostate cancer growth, invasion and drug responses. PLoS One 5:e10431

19. Hongisto V, Jernström S, Fey V, Mpindi JP, Kleivi Sahlberg K, Kallioniemi O, Perälä M (2013) High-throughput 3D screening reveals differences in drug sensitivities between culture models of JIMT1 breast cancer cells. PLoS One 8:e77232

20. Horman S, Toja J, Orth AP (2013) An HTS-compatible 3D colony formation assay to identify tumor specific chemotherapeutics. J Biomol Screen 18:1298–1308

21. Shin C, Kwak B, Han B, Park K (2013) Development of an in vitro 3D tumor model to study therapeutic efficiency of an anticancer drug. Mol Pharm 10:2167–2175

22. Wenzel C, Riefke B, Gründemann S, Krebs A, Christian S, Prinz F, Osterland M, Golfier S, Räse S, Ansari N, Esner M, Bickle M, Pampaloni F, Mattheyer C, Stelzer EH, Parczyk K, Prechtl S, Steigemann P (2014) 3D high-content screening for the identification of compounds that target cells in dormant tumor spheroid regions. Exp Cell Res 323:131–143

23. Yip D, Cho CH (2013) A multicellular 3D heterospheroid model of liver tumor and stromal cells in collagen gel for anti-cancer drug testing. Biochem Biophys Res Commun 433:327–332

24. Minchinton A, Tannock IF (2006) Drug penetration in solid tumours. Nat Rev Cancer 6:583–592

25. Foty R (2011) A simple hanging drop cell culture protocol for generation of 3D spheroids. J Vis Exp 51:2720

26. Vinci M, Gowan S, Boxall F, Patterson L, Zimmermann M, Court W, Lomas C, Mendiola M, Hardisson D, Eccles SA (2012) Advances in establishment and analysis of three-dimensional tumor spheroid-based functional assays for target validation and drug evaluation. BMC Biol 10:29–49

27. Krausz E, de Hoogt R, Gustin E, Cornelissen F, Grand-Perret T, Janssen L, Vloemans N,

Wuyts D, Frans S, Axel A, Peeters PJ, Hall B, Cik M (2013) Translation of a tumor microenvironment mimicking 3D tumor growth co-culture assay platform to high-content screening. J Biomol Screen 18:54–66

28. Li X, Zhang X, Zhao S, Wang J, Liu G, Du Y (2014) Micro-scaffold array chip for upgrading cell-based high-throughput drug testing to 3D using benchtop equipment. Lab Chip 14:471–481

29. Rimann M, Angres B, Patocchi-Tenzer I, Braum S, Graf-Hausner U (2014) Automation of 3D cell culture using chemically defined hydrogels. J Lab Autom 19:191–197

30. Tibbitt M, Anseth KS (2009) Hydrogels as extracellular matrix mimics for 3D cell culture. Biotechnol Bioeng 103:655–663

31. Singh M, Close DA, Mukundan S, Johnston PA, Sant S (2015) Production of uniform 3D microtumors in hydrogel microwell arrays for measurement of viability, morphology, and signaling pathway activation. Assay Drug Dev Technol 13:570–583

32. Singh M, Mukundan S, Jaramillo M, Oesterreich S, Sant S (2016) Three-dimensional breast cancer models mimic hallmarks of size-induced tumor progression. Cancer Res 76:3732–3743

33. Howes A, Richardson RD, Finlay D, Vuori K (2014) 3-dimensional culture systems for anti-cancer compound profiling and high-throughput screening reveal increases in EGFR inhibitor-mediated cytotoxicity compared to monolayer culture systems. PLoS One 9:e108283

34. Rotem A, Janzer A, Izar B, Ji Z, Doench JG, Garraway LA, Struhl K (2015) Alternative to the soft-agar assay that permits high-throughput drug and genetic screens for cellular transformation. Proc Natl Acad Sci U S A 112:5708–5713

35. Lin C, Grandis JR, Carey TE, Gollin SM, Whiteside TL, Koch WM, Ferris RL, Lai SY (2007) Head and neck squamous cell carcinoma cell lines: established models and rationale for selection. Head Neck 29:163–188

36. Johnston P, Sen M, Hua Y, Camarco D, Shun TY, Lazo JS, Grandis JR (2014) High-content pSTAT3/1 imaging assays to screen for selective inhibitors of STAT3 pathway activation in head and neck cancer cell lines. Assay Drug Dev Technol 12:55–79

37. Johnston P, Sen M, Hua Y, Camarco DP, Shun TY, Lazo JS, Wilson GM, Resnick LO, LaPorte MG, Wipf P, Huryn DM, Grandis JR (2015) HCS campaign to identify selective inhibitors of IL-6-induced STAT3 pathway activation in head and neck cancer cell lines. Assay Drug Dev Technol 13:356–376

Chapter 21

An Endothelial Cell/Mesenchymal Stem Cell Coculture Cord Formation Assay to Model Vascular Biology In Vitro

Michelle Swearingen, Beverly Falcon, Sudhakar Chintharlapalli, and Mark Uhlik

Abstract

Blood vessels are crucial components for normal tissue development and homeostasis, so it is not surprising that endothelial dysfunction and dysregulation results in a variety of different pathophysiological conditions. The large number of vascular-related disorders and the emergence of angiogenesis as a major hallmark of cancer has led to significant interest in the development of drugs that target the vasculature. While several in vivo models exist to study developmental and pathological states of blood vessels, few in vitro assays have been developed that capture the significant complexity of the vascular microenvironment. Here, we describe a high content endothelial colony forming cells (ECFC)/adipose-derived stem cell (ADSC) coculture assay that captures many elements of in vivo vascular biology and is ideal for in vitro screening of compounds for pro- or anti-angiogenic activities.

Key words Adipose-derived stem cell (ADSC), Angiogenesis, CD31, Cord formation, Endothelial cell, High content assay (HCA), Pericyte, Phenotypic drug discovery, Smooth muscle actin (SMA), Vascular endothelial growth factor (VEGF)

1 Introduction

For several decades, the field of vascular biology has served as a rich source for the study and development of therapeutic agents aimed at treating cardiovascular diseases such as heart disease and hypertension. More recently, intense research efforts have delved into this space to examine the roles of endothelial cells and the vascular microenvironment in other therapeutic areas such as autoimmunity, kidney disease, and oncology. Judah Folkman's 1971 hypothesis that sustained tumor growth must always be accompanied by expansion of the tumor vascular network via neoangiogenesis [1] has been observed in a number of laboratory experiments and is now universally accepted as one of the key "hallmarks of cancer" [2, 3]. The discovery and characterization of vascular endothelial

Paul A. Johnston and Oscar J. Trask (eds.), *High Content Screening: A Powerful Approach to Systems Cell Biology and Phenotypic Drug Discovery*, Methods in Molecular Biology, vol. 1683, DOI 10.1007/978-1-4939-7357-6_21,
© Springer Science+Business Media LLC 2018

growth factor (VEGF) [4, 5] and its receptors (vascular endothelial growth factor receptors; VEGFRs) [6, 7] as the prominent signaling axis that drives tumor angiogenesis led to the development of numerous agents to target this pathway and inhibit tumor growth [8]. However, clinical responses with these antiangiogenic agents are usually short-term [9], with multiple mechanisms of "resistance" or progression on therapy having been postulated [10]. Consequently, there are ongoing efforts to discover targets, pathways, and drugs for the next generation of antiangiogenic therapy for the treatment of cancers.

Numerous in vivo models have been successfully employed to study developmental and pathological angiogenesis. These include such assays as the chicken chorioallantoic membrane (CAM) assay [11], developmental retinal angiogenesis and oxygen-induced retinopathy (OIR) models [12, 13], femoral artery ligation/ischemia reperfusion model [14], zebrafish intersegmental vessel (ISV) development [15], rodent Ad-VEGF ear/flank [16], and tumor models [17]. While no single model can recapitulate all aspects of vascular biology, each of these models displays specific strengths of the developing or pathological vasculature (such as growth factor-induced or hypoxia-driven angiogenesis) that make them particularly useful to study the effects of therapeutic agents. Most of these models incorporate all the biologically relevant cell types in their proper locations and retain the physiological or pathophysiological functions of the vasculature allowing for studies to be performed in the context of blood flow, smooth muscle/pericyte contraction of vessels, cell-cell contact/adhesion, and vessel permeability [11–17].

However, studying vascular biology in vitro is fraught with many inherent difficulties. It is challenging to replicate the immense complexity of in vitro cell types (endothelial cells, pericytes, fibroblasts, immune cells, or platelets), soluble factors (growth factors, cytokines, chemokines, hormones, metabolites, or signaling lipids), matrix components (fibrin, collagen, or laminins), and cell-cell interactions present within the in vivo vascular microenvironment [18]. Most of the "gold standard" or historical in vitro assays that have been used to explore vascular biology have centered on the use of primary endothelial cells, particularly human umbilical endothelial cells (HUVEC), in culture to examine effects on their proliferation, apoptosis, migration, or tube formation using a Matrigel matrix [19]. These vascular biological models are relatively easy to set up and are considered a mainstay to support basic research and drug discovery. However, many biological components and interactions are not present in these vascular biological systems and the linkage to in vivo biology is not always maintained.

In an effort to design a more physiologically relevant in vitro assay for vascular biology, we have developed a coculture cord formation model that utilizes endothelial progenitor cells (endothelial colony forming cells; ECFCs) and mesenchymal stem cells (adipose-derived stem cells; ADSCs) [20]. In this assay, the endothelial cells form cord-like projections that develop into complex, branched networks over a period of 3–5 days. The cord formation occurs progressively through a series of events involving endothelial adhesion, sprouting, migration, proliferation, differentiation, and maintenance of cell survival over time. Thus, this single assay captures many of the phenotypic features required for vessel formation in vivo. Many of the effects of known pharmacological agents observed in the ECFC/ADSC cord formation assay have been confirmed with in vivo vascular models to provide supporting evidence for the relevance of this vascular biological assay system [20–24].

Furthermore, this model and other similar coculture models described in the literature [25, 26] have numerous advantages over mono-culture endothelial cell tube formation assays (Fig. 1). One of the most important features of the coculture cord formation assay is the ability of endothelial cells to physically interact with the stem cells and communicate via cell-cell interactions and via proximal secreted microenvironmental cues such as growth factors, chemokines, and other "angiocrine factors" [27, 28]. In our coculture assay, the ADSCs respond to endothelial cues and undergo differentiation into smooth muscle actin (SMA)-expressing myofibroblast-like cells, similar to the perivascular cells, or pericytes, that form close associations with vessels during vascular maturation. Interestingly, the stem cells only differentiate to express SMA when in close contact with the ECFCs, suggesting that cell-cell contact may be mediating this biology. The endothelial cords that form are long-lived and stable for more than 2 weeks in culture, allowing for both "neoangiogenic" and "established" cord networks to be used for drug screening [20].

Another significant advantage of the ECFC/ADSC coculture is that it is conducted in a media devoid of serum and extraneous matrix; each of which may produce their own exogenous growth factors. Unlike many other endothelial assays [29, 30], there is no need to supply a matrix for cell adhesion in this assay: the ADSCs are plated as a feeder layer or "lawn" and they readily secrete and lay down a matrix that supports endothelial adhesion and migration. Interestingly, no additional exogenous growth factors or serum is required to sustain a basal level of cord formation in this system. The self-contained assay format serves to eliminate some of the variability often encountered with different lots of FBS, Matrigel, and growth factor preparations. We have identified that numerous soluble factors are produced by the ECFCs and ADSCs, many of which are known pro-angiogenic factors, and these likely synergize to enable the proliferation and survival of the culture. This is further exemplified by the fact that ECFCs will not survive more than a few hours in the basal media alone but are quite prolific when

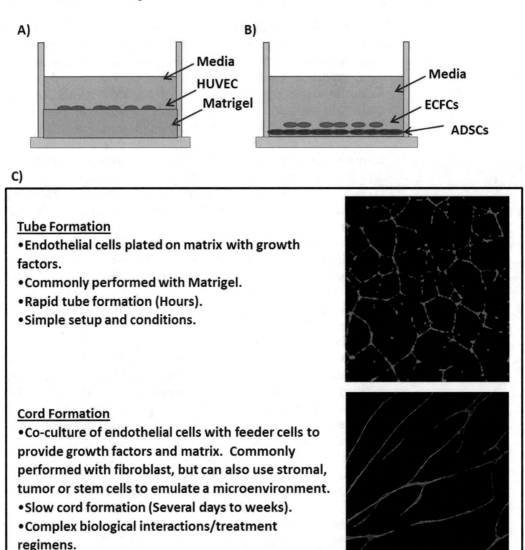

Fig. 1 The cartoon above shows the layout of two different forms of cord formation assay system. (**a**) The traditional matrigel assay has the endothelial cells seeded upon a layer of matrigel and form cords without addition of exogenous growth factors. (**b**) The basal cord formation assay has the endothelial cells seeded upon a layer of adipose derived stem cells and the cords form in response to the exogenously added growth factors. (**c**) Key features of the Matrigel Tube formation and Co-culture cord formation assays

in the same media with ADSCs present. Working in this "basal" media setting allows for a low level of cord formation where both inductions and inhibitions can be readily observed. For screening of anti-angiogenic agents, growth factor can be supplemented into the media to bias the assay toward a particular pathway and effectively increase the dynamic range to better observe inhibitory responses. Likewise, more complex growth factor combinations or even conditioned cell media may be used to skew the assay toward a particular biology.

We have optimized the ECFC/ADSC cord formation assay to be quantitative, multiparametric, scalable, and automatable and thereby ideal for drug discovery purposes. As an image-based high content assay, standard fixatives and cell labeling reagents are used to assess dozens of features, such as total cord area, Angio Index, branchpoints, and smooth muscle actin area (Fig. 2). A built-in toxicity assessment can also be provided by performing nuclear area counts (assessing both ECFC and ADSC nuclei), which dramatically decreases in the presence of acute cytotoxic agents. While the standard format for this assay (as described

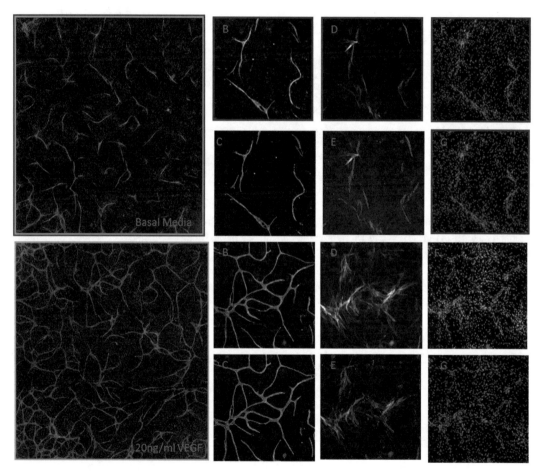

Fig. 2 Representative images from control wells (*top*) and VEGF treated wells (*bottom*). (**a**) Composite image montage of all nine fields taken by the Arrayscan of the three channels used for analysis. (**b**) Grayscale image of the detection of the cords by CD31, and (**c**) the mask overlay the algorithm puts onto the image to calculate the Total Tube Area of connected (*green*) and unconnected (*blue*) cords. (**d**) Grayscale image of smooth muscle actin positive cells, and (**e**) the mask overlay the algorithm puts onto the image to calculate the SMA Total Intensity. (**f**) Grayscale image of the detection of nuclei by Hoechst, and (**g**) the mask overlay to calculate the Nuclei Count

A)

B)

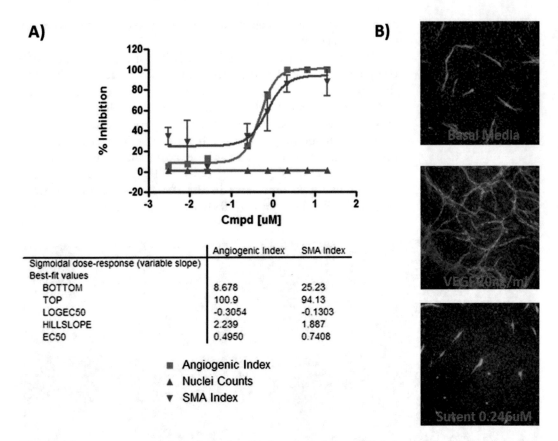

	Angiogenic Index	SMA Index
Sigmoidal dose-response (variable slope)		
Best-fit values		
BOTTOM	8.678	25.23
TOP	100.9	94.13
LOGEC50	-0.3054	-0.1303
HILLSLOPE	2.239	1.887
EC50	0.4950	0.7408

■ Angiogenic Index
▲ Nuclei Counts
▼ SMA Index

Fig. 3 (**a**) Dose response of Sutent shows the effect on Angio Index, SMA Index and Nuclei Count on cords induced by VEGF in the cord assay. (**b**) Representative images of the Basal Media and VEGF cords compared to the cords left after Sutent treatment

here) is 96-well, we have also scaled versions of this assay to accommodate automated plating and compound delivery in 384-well plates to enable medium-throughput screening (MTS). Variations of this assay are quite robust and reproducible and have been mainstays at Lilly to support the profiling of anti-angiogenic activity for many inhibitors (Fig. 3; [20–24]), as well as the primary screening assay for the Angiogenesis Module of the Lilly Phenotypic Drug Discovery (PD²) effort [31].

2   Materials

1. HCS cell imaging equipment: the method described herein was developed and optimized on the Thermo Scientific (Cellomics) Arrayscan VTI automated imaging platform with a 5× objective (2 × 2 binning) or the Cell Incyte imaging platform with a 4× or 10× objective using the Tube Formation Bioapplication. We anticipate that it could be adapted to other automated HCS platforms with similar capabilities and image analysis algorithms.

2. Adipose-Derived Stem Cells (ADSC) from Lonza, Endothelial Colony Forming Cells (ECFC) from Lonza.

3. Human Recombinant VEGF (20 ng/mL).

4. 96-well black, clear-bottom plates.

5. Anti-human CD31 antibody from R&D Systems, (1:250 dilution).

6. Anti-smooth muscle actin antibody directly conjugated with Cy3, from Sigma, (1:250 dilution).

7. Alexa Fluor-488 (AF488) donkey anti-sheep IgG secondary, from Invitrogen, (1:400 dilution).

8. Hoechst (1:1000 dilution) (*see* **Note 1**).

9. PBS (Ca^{++}/Mg^{++} free).

10. FBS.

11. EBM2MV medium + bullet kit from Cambrex (*see* **Note 2**).

12. Orange cap ($225cm^2$) flasks from Corning (*see* **Note 3**).

13. Collagen I coated ($150 \ cm^2$) flasks.

14. TrypLE Express—recombinant cell-dissociation enzymes.

2.1 Basal Media

1. MCDB-131 medium.

2. Insulin (10 µg/mL).

3. Dexamethasone in 95% ethanol (10^{-6} M).

4. L-Ascorbic Acid (30 µg/mL).

5. Tobramycin (50 ng/mL).

6. Cell Primer r-Transferrin AF(10 µg/mL).

2.2 Culturing the Adipose Derived Stem Cells

1. Thaw stock vial of ADSC 10E7 (*see* **Notes 4–6**).

2. Add one vial per T225 orange cap flask in 40 mL of EGM2MV medium.

3. Allow cells to settle for 3–4 h.

4. Remove medium and feed with 40 mL of fresh medium (*see* **Note 7**).

5. Cells will be confluent and ready for the assay in 3 days.

2.3 Culturing the Endothelial Colony Forming Cells

1. Thaw stock vial of ECFC 10E6.

2. Add one vial per T150 Collagen coated flasks with 40 mL of EGM2MV medium (*see* **Note 5**).

3. Allow cells to settle for 3–4 h.

4. Remove the medium and feed with 40 mL of fresh medium (*see* **Note 8**).

5. Once the cells recover and begin to grow from the thaw they can be split twice a week for 2 weeks (*see* **Note 9**).

3 Methods

Described below is a general setup for running the ECFC/ADSC coculture cord formation assay in "neoangiogenic" mode stimulated with VEGF over 3 days. Several variations to this assay can be made, such as changing stimulators to bias the system to a specific signaling pathway or waiting until day 3 or 4 when cords are already developed before adding inhibitors ("established cord" mode; [20]). Additionally, a kinetic, rather than fixed-endpoint, version of this assay can be run to obtain more discrete, temporal data [20], but this requires other labeling techniques or the use of engineered cells for the assay and is beyond the scope of this discussion.

3.1 Harvesting and Seeding the ADSCs

1. Take the flask of ADSCs and aspirate the medium.
2. Wash one time with 10 mL of PBS and aspirate the liquid.
3. Add 5mL of TrypLE, tilt flask to cover completely (*see* **Note 10**).
4. Aspirate the TrypLE and set in the hood for a minute.
5. Tap the side of the flask and see whether the cells are lifting from the flask.
6. Add 5 mL of Basal medium to harvest cells (*see* **Note 11**).
7. Take out an aliquot to count cell numbers.
8. Seed 40,000 ADSCs in 100 μL of Basal medium to the inner 60 wells of the 96-well plate (*see* **Note 12**).
9. Place in the 37 °C incubator with 5% CO_2 overnight.

3.2 Harvesting and Seeding the ECFCs on the Next Day

1. Take the flask of ECFCs and aspirate the medium.
2. Wash one time with 10 mL of PBS and aspirate the liquid.
3. Add 5 mL of TrypLE, tilt flask to cover completely (*see* **Note 10**).
4. Aspirate the TrypLE and set in the hood for a minute.
5. Tap the side of the flask and see whether the cells are lifting from the flask.
6. Add 5 mL of Basal medium to harvest cells (*see* **Note 11**).
7. Take out an aliquot to count cell concentration.
8. Carefully aspirate the medium from the wells of the plate (*see* **Note 13**).
9. Seed 4000 ECFCs in 50–100 μL of Basal medium to the wells containing previously seeded ADSCs of the 96-well plate (*see* **Notes 14** and **15**).
10. Place in the 37 °C incubator with 5% CO^2 for 2–3 h.

3.3 Reagent Addition to the Plates and Incubation of the Plates

1. Design a plate map with Basal media only wells and VEGF stimulated only control wells (*see* **Note 16**).

2. Dilute VEGF to 20 ng/mL based on the number of wells requiring each reagent.

3. Add the growth factors to the plates at a 2–4× solution so the final concentration is 20 ng/mL in the 200 μL.

4. Total volume for the assay wells on the plate should be 200 μL.

5. Place in the 37 °C incubator with 5% CO_2 for 3 days.

3.4 Assay Takedown and Staining

1. After incubation for 72 h the plates are removed from the incubator.

2. Aspirate the medium and add 100 μL of room temperature 80% Ethanol (*see* **Note 17**).

3. Incubate on the bench top for 20 min only (*see* **Note 18**).

4. Carefully remove the ethanol with gentle aspiration and wash 3× with 100 μL of PBS (*see* **Note 19**).

5. Dilute the anti-human CD31 (1:250) and the anti-smooth muscle actin (1:250) antibodies in PBS+ 2.5% FBS at 100 uL per well.

6. Incubate at 37 °C for 2 h.

7. Remove and wash 3× with PBS.

8. Dilute the Alexa Fluor-488 donkey anti-sheep IgG secondary (1:400) and the Hoechst (1:1000) in PBS + 2.5% FBS at 100 μL per well, incubate at room temperature for 30 min.

9. Remove and wash 3× with PBS.

10. Add 100 μL PBS.

11. Seal the plate with a black vinyl film and read on a suitable automated HCS imaging platform.

3.5 Image Capture and Data Analysis

1. Fixed and stained plates are read on the Thermo Cellomics Arrayscan VTi or Cell Incyte using the Tube Formation Bioapplication. Images are acquired in three channels using the XF93 (Omega) filter set with the first channel (AF488, excitation 488 nm/emission 520 nm) used to image CD31-positive endothelial cords, the second channel (CY3, excitation 555 nm/emission 565 nm) to image SMA-expressing cells, and the third channel (DAPI, excitation 350 nm/emission 450 nm) to image cell nuclei (Fig. 2).

2. Plates are imaged using 4×, 5×, or 10× objective lens and 9 fields/well are captured to cover the maximal area within the well.

3. Cords are masked as the primary objects (using a value of Fixed Threshold = 100 with Background Correction "On"), while

SMA is contoured as a maximally dilated sub-mask of the primary object (using a value of Fixed Threshold = 150 with Background Correction "On").

4. Using this algorithm, several quantitative parameters can be extracted to express the level and quality of endothelial cord formation including total cord area, # of branchpoints, Length-to-width ratio (LWR), total tube length and Angio Index (index of area covered by connected cord networks) and SMA Index (ratio of SMA area to cord area) (Fig. 3).

5. To assess potential toxicity effects, images are reanalyzed using a one-channel Target Activation Bioapplication using the DAPI channel to assess total area of Hoechst-stained objects (*see* **Note 20**).

4   Notes

1. Hoechst and DAPI may be used interchangeably for the visualization of cell nuclei. This protocol has been optimized using Hoechst and imaging on the DAPI channel, as they share similar emission and excitation spectra.

2. This medium is specific for both the ECFC and ADSC cell lines for normal growth.

3. A number of other flasks have been tested to grow the ADSC cell line. Unfortunately, the orange cap flasks from Corning are the only ones that allow the cells to seed and grow properly. They will not adhere well to other tissue culture flasks causing cell differentiation.

4. A new vial of cells is brought out of the liquid nitrogen freezer fresh for each experiment.

5. The cells should not be passed from the frozen stock. If they are split again, it will cause the cells to differentiate into fat cells during the assay.

6. Carefully thaw in a water bath at 37 °C, swirl gently while suspended in the water.

7. The media does not need to be replaced again. They will grow and divide in the flask for the next 3 days.

8. These cells are brought out of the liquid nitrogen freeze from the frozen stock and passed for only 2 weeks. Try to limit the splitting to two times a week to maintain the ability for the cells to form cords.

9. Pass them at 1:5 during the week and then pass them at $0.5 \times 10E6$ cells per flask for the weekend. This will insure they do not over grow. Do not let the flasks become confluent, because this will change the ability of the cells to form cords.

10. Watch under the microscope to see the cells begin to round up. This should be under a minute.

11. Tap the flask to cause the cells to release from the bottom of the flask. Now add an additional 5 mL of Basal Media, rinse the bottom to collect all the cells, and remove the cell suspension to a 50 mL conical tube.

12. Due to edge effects that have been noted, only the inside 60 wells are used for this assay. To each of the edge wells 200 μL of PBS is added to help control evaporation during the assay.

13. Carefully aspirate the media from the wells without disturbing the ADSC layer present in each well. Residual media could add "conditioned" ADSC media to the assay and some factors from this media can induce cord formation.

14. Carefully add the cell suspension to each well using a multi-channel pipet. Try not to disturb the ADSC layer in each well.

15. The volume to seed the ECFCs depends on the additional reagent that will be added to the assay.

16. In order to determine if the assay works correctly you need to have the control wells on the assay plate, and each condition should be tested in triplicate to ensure reproducibility.

17. Carefully pipette the media from the assay wells, trying not to disturb the cells on the bottom of the wells.

18. Only permeabilize cells with ethanol for 20 min. Longer exposures will result in cell tearing and disassociation.

19. It is important to carefully remove and add the liquids in the following steps. The cell layer is very delicate and can tear and lift off if not handled with care.

20. The Tube Formation Bioapplication does not currently allow for the quantitation of all cell nuclei (including those within tubes and outside of tubes) in the imaged fields. A virtual scan reanalysis using the Target Activation Bioapplication is used to analyze the DAPI channel images acquired from the Tube Formation Bioapplication.

References

1. Folkman J (1971) Tumor angiogenesis: therapeutic implications. N Engl J Med 285:1182–1186

2. Hanahan D, Weinberg RA (2000) The hallmarks of cancer. Cell 100:57–70

3. Hanahan D, Weinberg RA (2011) Hallmarks of cancer: the next generation. Cell 144:646–674

4. Leung DW, Cachianes G, Kuang WJ et al (1989) Vascular endothelial growth factor is a secreted angiogenic mitogen. Science 246:1306–1309

5. Senger DR, Galli SJ, Dvorak AM et al (1983) Tumor cells secrete a vascular permeability factor that promotes accumulation of ascites fluid. Science 219:983–985

6. de Vries C, Escobedo JA, Ueno H et al (1992) The fms-like tyrosine kinase, a receptor for vascular endothelial growth factor. Science 255:989–991

7. Terman BI, Dougher-Vermazen M, Carrion ME et al (1992) Identification of the KDR tyrosine kinase as a receptor for vascular endothelial cell growth factor. Biochem Biophys Res Commun 187:1579–1586

8. Jayson GC, Hicklin DJ, Ellis LM (2012) Anti-angiogenic therapy–evolving view based on clinical trial results. Nat Rev Clin Oncol 9:297–303

9. Ebos JM, Kerbel RS (2011) Antiangiogenic therapy: impact on invasion, disease progression, and metastasis. Nat Rev Clin Oncol 8:210–221

10. Bergers G, Hanahan D (2008) Modes of resistance to anti-angiogenic therapy. Nat Rev Cancer 8:592–603

11. Gabrielli MG, Accili D (2010) The chick chorioallantoic membrane: a model of molecular, structural, and functional adaptation to transepithelial ion transport and barrier function during embryonic development. J Biomed Biotechnol 2010:940741

12. Smith LE, Wesolowski E, McLellan A et al (1994) Oxygen-induced retinopathy in the mouse. Invest Ophthalmol Vis Sci 35:101–111

13. Stahl A, Connor KM, Sapieha P et al (2010) The mouse retina as an angiogenesis model. Invest Ophthalmol Vis Sci 51:2813–2826

14. Vidavalur R, Swarnakar S, Thirunavukkarasu M et al (2008) Ex vivo and in vivo approaches to study mechanisms of cardioprotection targeting ischemia/reperfusion (i/r) injury: useful techniques for cardiovascular drug discovery. Curr Drug Discov Technol 5:269–278

15. Jensen LD, Rouhi P, Cao Z et al (2011) Zebrafish models to study hypoxia-induced pathological angiogenesis in malignant and nonmalignant diseases. Birth Defects Res C Embryo Today 93:182–193

16. Nagy JA, Shih SC, Wong WH et al (2008) Chapter 3. The adenoviral vector angiogenesis/lymphangiogenesis assay. Methods Enzymol 444:43–64

17. Eklund L, Bry M, Alitalo K (2013) Mouse models for studying angiogenesis and lymphangiogenesis in cancer. Mol Oncol 7:259–282

18. Nyberg P, Salo T, Kalluri R (2008) Tumor microenvironment and angiogenesis. Front Biosci 13:6537–6553

19. Goodwin AM (2007) In vitro assays of angiogenesis for assessment of angiogenic and anti-angiogenic agents. Microvasc Res 74:172–183

20. Falcon BL, O'Clair B, McClure D et al (2013) Development and characterization of a high-throughput in vitro cord formation model insensitive to VEGF inhibition. J Hematol Oncol 6:31

21. Burkholder TP, Clayton JR, Rempala ME et al (2012) Discovery of LY2457546: a multi-targeted anti-angiogenic kinase inhibitor with a novel spectrum of activity and exquisite potency in the acute myelogenous leukemia-Flt-3-internal tandem duplication mutant human tumor xenograft model. Investig New Drugs 30:936–949

22. Lee JA, Uhlik MT, Moxham CM et al (2012) Modern phenotypic drug discovery is a viable, neoclassic pharma strategy. J Med Chem 55:4527–4538

23. Meier T, Uhlik M, Chintharlapalli S et al (2011) Tasisulam sodium, an antitumor agent that inhibits mitotic progression and induces vascular normalization. Mol Cancer Ther 10:2168–2178

24. Tate CM, Blosser W, Wyss L et al (2013) LY2228820 dimesylate, a selective inhibitor of p38 mitogen-activated protein kinase, reduces angiogenic endothelial cord formation in vitro and in vivo. J Biol Chem 288:6743–6753

25. Evensen L, Link W, Lorens JB (2013) Image-based high-throughput screening for inhibitors of angiogenesis. Methods Mol Biol 931:139–151

26. Traktuev DO, Prater DN, Merfeld-Clauss S et al (2009) Robust functional vascular network formation in vivo by cooperation of adipose progenitor and endothelial cells. Circ Res 104:1410–1420

27. Butler JM, Kobayashi H, Rafii S (2010) Instructive role of the vascular niche in promoting tumour growth and tissue repair by angiocrine factors. Nat Rev Cancer 10:138–146

28. Merfeld-Clauss S, Gollahalli N, March KL et al (2010) Adipose tissue progenitor cells directly interact with endothelial cells to induce vascular network formation. Tissue Eng Part A 16:2953–2966

29. Arnaoutova I, George J, Kleinman HK et al (2009) The endothelial cell tube formation assay on basement membrane turns 20: state of the science and the art. Angiogenesis 12:267–274

30. Arnaoutova I, Kleinman HK (2010) In vitro angiogenesis: endothelial cell tube formation on gelled basement membrane extract. Nat Protoc 5:628–635

31. Lee JA, Chu S, Willard FS et al (2011) Open innovation for phenotypic drug discovery: the PD2 assay panel. J Biomol Screen 16:588–602

Chapter 22

High-Throughput Automated Chemical Screens in Zebrafish

Manush Saydmohammed and Michael Tsang

Abstract

Zebrafish are increasingly used to perform phenotypic screens to identify agents that can alter physiology in a whole organismal context. Here, we describe an automated high-content chemical screen using transgenic zebrafish embryos to identify small molecules that modulate Fibroblast Growth Factor Signaling. High content multi-well screening was further refined with a particular emphasis on automated imaging and quantification that increases sensitivity and throughput of whole organism chemical screens.

Key words Chemical genetics, Small molecule screens, Drug discovery, Transgenic reporter, FGF hyper-activators

1 Introduction

Classical genetics have played a central role in elucidating the relationship between genes and phenotypes in animal models over the past centuries [1–3]. Forward genetics identifies gene mutations responsible for observed phenotypes induced by mutagenizing agents. However, in the past 30 years, the use of reverse genetic approaches to create knock-out animals has been at the forefront for understanding aberrant biological pathways [4–6]. Here, a target gene of interest is disrupted first to reveal the phenotype. Alternatively, biologists study the perturbation of protein function using small molecules, which is often referred to as chemical genetics. Initially, small molecules are identified from a diverse chemical library that induces a specific phenotypic effect in animals, followed by the identification of the target protein that is modulated by the compound. One of the central goals of chemical genetics is to identify small molecules that can modulate a biological pathway or a specific protein function. There are two approaches for carrying out small molecule screening. Positive hits are selected based on either (1) their ability to reveal an abnormal phenotype when treated in wild-type embryos (often referred to as chemical

Paul A. Johnston and Oscar J. Trask (eds.), *High Content Screening: A Powerful Approach to Systems Cell Biology and Phenotypic Drug Discovery*, Methods in Molecular Biology, vol. 1683, DOI 10.1007/978-1-4939-7357-6_22,
© Springer Science+Business Media LLC 2018

screening) or (2) their ability to reverse an abnormal mutant phenotype to wild-type embryo (often called therapeutic screening).

For identifying small molecules with biological activity, whole
organisms offer key advantages over cell lines, by providing information on tissue specificity, toxicity, and accounting for bioavail-
ability. Indeed, a number of animal models have been successfully
used in chemical screens that include *Mus musculus* [7], *Caenor-
habditis elegans* [8–10], *Drosophila melanogaster* [11, 12], *Xenopus*
[13], and zebrafish *Danio rerio* [14–19]. Although there are significant genetic, physiological, and anatomical similarities to humans,
mouse models are not commonly used for chemical genetic screens
due to significant cost and ethical concerns. On the other hand,
although simple invertebrate animal model systems such as *C.
elegans* and *D. melanogaster* have a high degree of functional con-
servation in cellular processes with mammals, there are however
some diseases that cannot be modeled because the animal does not
have the corresponding genes or organs. Also, *D. melanogaster*
cannot be grown in a liquid medium, which limits their use for
chemical screening with respect to delivery of chemicals to the
organism. On the contrary, zebrafish is an ideal model for high-
throughput chemical screening because they are closer to mammals
in the evolutionary tree than either *C. elegans* or *D. melanogaster*.
Also, attractive features including cost-effective maintenance, high
fecundity, and rapid development highlight the suitability of this
model organism for chemical screens. Above all, embryo optical
transparency allows in vivo monitoring of morphological defects in
real time as well as screening for compound effects on larval organ
development.

Our laboratory has generated a number of transgenic zebrafish
lines that use *Dual Specificity Phosphatase 6* promoter (*dusp6*, a gene
that is regulated by active FGF signaling) to drive Green Fluorescent
Protein (GFP) expression in an FGF-sensitive manner [20, 21].
From a screen of 5000 compounds, a novel FGF hyper-activator,
(E)-2-benzylidene-3-(cyclohexylamino)-2,3-dihydro-1H-inden-1-
one (BCI), was identified using the *Tg(dusp6:d2EGFP)^{pt6}* line [16].
Increase in GFP expression could be quantified through automated
image analysis using Definiens Cognition Network Technology
(CNT), which is an object-based analysis method uniquely suited
to the analysis of specimens that contain spectral features across
multiple scales, such as tissues or organs. CNT mimics the human
cognitive process in a computerized environment that allows the user
to translate visual observations into object features and relations
[22]. With a specific ruleset (algorithm), we were able to correctly
detect and quantify GFP expression in transgenic embryos both at
baseline and after chemical exposure [23]. We recently implemented
the CNT analysis in 96-well microplates on high content imagers
(ImageXpress Ultra High Content Reader) and identified two novel,
but mild FGF hyper-activators (Oxyquinoline and Pyrithione Zinc)

from a well-characterized USFDA approved drug library [18]. We would not have identified such mild FGF hyper-activators using manual visual inspection if we had not applied the CNT high content imaging and analysis protocol. Here, we describe our protocol for chemical genetic screening for FGF hyper-activators using $Tg(dusp6: d2EGFP)^{pt6}$ embryos.

2   Materials

1. $Tg(dusp6EGFP)^{pt6}$ embryos at 24 h post fertilization (hpf): These fish models were generated and described previously [20].

2. Round bottom 96-well plates: All embryos are in the chorion at 24hpf, so remain in the center of the wells.

3. Petri-plates: 100×20 mm, for collecting embryos.

4. Transfer pipettes: For loading embryos into round-bottom 96-well plates.

5. Multi-channel pipettes (50–300 μl, 1–10 μl range): Use 50–300 μl pipettes for remove excess E3 (see below) from wells after embryo loading and replenishing with fresh E3 (200 μl). 1–10 μl pipettes can be used for dispensing drugs into treatment plates.

6. Micropipettes (2–20 μl range): For removing excess E3 from wells after embryo loading.

7. Nitrile gloves (*see* **Note 1**).

8. Fluorescent stereomicroscope: For observing 24hpf embryos for GFP expression. Also, embryos after the treatment can be visually screened for positive hits.

9. Aluminum foil (*see* **Note 2**).

10. Incubator (28.5°C): For incubating the embryos during drug treatment.

11. Tricane Methanesulfonate (MS-222). Prepare a stock solution of 4 mg/ml and adjust the pH to 7–7.5 (*see* **Note 3**). Use 4.2 ml of stock solution and make up to 100 ml using E3. Store frozen (−20°C) until use.

12. E3 solution: 5 mM NaCl, 0.17 mM KCl, 0.33 mM $CaCl_2$. $2H_2O$, 0.33 mM $MgSO_4.7H_2O$: Prepare a $60\times$ E3 stock solution by adding 172 g NaCl, 7.6 g KCl, 29 g $CaCl_2$. $2H_2O$, 49 g $MgSO_4.7H_2O$ and add distilled water and make up to 10 l. For working stock, add 160 ml of $60\times$ stock and make up to 10 l distilled water.

13. DMSO: 100%.

14. (E)-2-benzylidene-3-(cyclohexylamino)-2,3-dihydro-1H-inden-1-one (BCI): Prepare a 20–50 mM stock solution in DMSO and solubilize it completely. Dilute it to 1 mM working stock and use 2 μl stock solution into 200 μl E3 containing embryos that serve as the positive control. Tap the plate to mix it.

15. Drug plates: In a typical drug screening library, individual wells of a 96 well from lanes 2–11 are filled with a stock solution (typically 50–100 mM). Before performing actual screening, we need to decide the desired dose of compound. Zebrafish chemical screens use doses between 10 and 60 μM as reviewed by Wheeler and Brandli [24]. In our USFDA-approved drug library screening, we used 20 μM for treating embryos [18].

16. ImageXpress Ultra high content reader (Molecular Devices, Sunnyvale, CA).

3   Methods

An outline of the screening protocol is shown in Fig. 1.

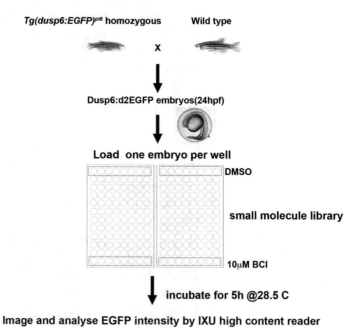

Fig. 1 An outline illustrating semiautomated chemical genetic screening for identifying FGF hyper-activators using *Tg (Dusp6:d2EGFP)* embryos. Load 24 hpf embryos onto 96-well plates in 200 μl E3 and treat with drug library in duplicates at 20 μM concentration (in lanes 02 to 11). Load Lane 01 with vehicle DMSO as negative control and lane 12 with 10 μM BCI. After treating for 5 h, scan for GFP intensity in IXU high content reader using cognition network technology

3.1 Setting Up of Fish

1. Outcross transgenic parents with wild-type zebrafish in a breeding tank. Transfer homozygous transgenic zebrafish of male or female into individual mating tanks. Use dividers to keep fish in separate sections of the tanks prior to mating. Next, add wild-type AB or Tü strain of the opposite sex to each tank and keep overnight at 28.5 °C.

2. On the next morning, replenish tanks with fresh fish water and remove the separators. Typically, within an hour, the fish will begin to mate. Collect the embryos using a plastic sieve and transfer them to 100 mm dishes with E3 solution. Incubate embryos at 28.5 °C for 6–8 h at such point examine the embryos under a stereomicroscope. Remove any unfertilized or malformed embryos using a transfer pipette. Replenish with fresh E3 solution and incubate overnight at 28.5 °C.

3. The next day, place embryos under a fluorescent stereomicroscope to sort the transgenic embryos for similar stage and uniform GFP expression. 24 hpf embryos usually show a GFP within the mid-hind brain boundary, trigeminal ganglia, and dorsal retina (*see* **Note 4** and **Fig. 2**).

4. Using a plastic pasteur pipette, transfer a single 24 hpf embryo into the individual wells of a round-bottom 96-well plate (*see* **Note 5**).

5. Using a micropipette, remove any excess E3 solution from the wells. Next, using a multi-channel pipettor, add 200 μl of E3 solution into each well (*see* **Note 6**).

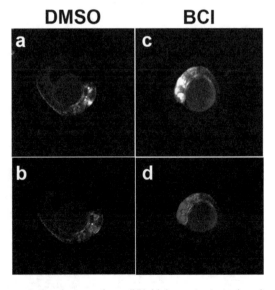

Fig. 2 Representative images from IXU high content reader showing visual difference in EGFP intensity (**A** and **C**). Images are processed and quantified for EGFP intensity by the specific algorithm designed for this experiment (**B** and **D**)

3.2 Treatment of Embryos

6. Add 2 μl of 1% DMSO to all the wells in lane 1. Then add 2 μl BCI (prepared in 1% DMSO) to all wells in lane 12 (*see* **Note 7**). The remaining wells are filled with 2 μl of screening compounds (*see* **Note 8**). Then, cover the plates using aluminum foil and incubate for 5 h at 28.5 °C.

7. To limit embryos from moving during imaging, Tricane (MS-222) is added to anesthetize the embryos after drug treatment. Add 10 μl of Tricaine (final concentration) to each well using a multi-channel pipettor.

3.3 Imaging of Chemical-Treated Embryos

8. Place the microplate into an ImageXpress Ultra (IXU) high content reader and acquire images using Open MetaXpress software. We use a 4× objective at fully open aperture and a fixed offset from the plate bottom for image capture. Capture images of single embryos per well using an Argon laser at excitation/emission wavelengths of 488/525 nm to detect GFP.

9. Select one of the wells containing a DMSO-treated embryo and choose the option to "find sample" on the image acquisition and control panel. The instrument will then perform an autofocus scan and acquire a single image of zebrafish embryo showing the regions with fluorescence.

10. Select a few vehicle-treated wells and wells treated with positive control to ascertain that head structures are in focus and that the fluorescence intensity in the positive controls is within the range of the 16-bit camera. If necessary, laser power and photomultiplier tube (PMT) gain can be adjusted such that brightest regions in the positive control images do not saturate the detector.

11. To start a plate scan, "acquire plate" is chosen in the plate acquisition and control window. The instrument will automatically acquire each image of the entire 96-well microplate and archive them as TIFF files in a SQL server database.

12. Once the plate scan is finished, further processing of the images is carried out in Definiens Developer software. For this, upload the images from the microplate reader into Developer using the Cellenger module and create a new workspace. Multiple plates can be loaded into the same workspace.

13. Load a custom-created rule set to detect and quantify GFP-expressing regions. The ruleset for this transgene is designed to automatically identify the whole embryo, yolk, head, and GFP-expressing regions within the head (eye, retina, mid-hind brain boundary, rhombomers, and trigeminal ganglia) [23].

14. Apply the ruleset on a few positive and negative controls, adjusting thresholds for yolk and head detection until both the compartments are properly identified and detection of GFP-positive head structures matches visual observation [23].

15. Apply the ruleset on entire plates, using the integrated job scheduler, which distributes the analysis to four parallel processors ("engines"). During the scan, a heatmap is generated that is continuously updated. Both the data and images can be accessed during the scan for assessing ruleset performance.

16. The ruleset can be further customized to calculate and export any numbers of features; the most informative parameter is the total GFP intensity in GFP-expressing head structures. Numerical data and processed images are automatically saved. Once all of the images have been collected and processed, the raw data can be exported to Excel, GraphPad Prism, or other statistical analysis software for further analysis. Here is a representative image that is captured from Definien's software (Fig. 2a, c) and after applying CNT (Fig. 2b, d) in DMSO or after treating with BCI.

17. The success of high content screening depends on how well assay is responding to the negative and positive controls. One of the strategies to improve the assay performance is by limiting inherent variability of GFP in embryos by selecting uniform stage and all heterozygous embryos are used for screening (*see* **Note 9**). Also, we use the strictly standardized mean difference (SSMD) value to ensure quality control performance of our zebrafish studies [18] (*see* **Note 10** and Fig. 3a).

4   Notes

1. Most of the drugs used for screening are dissolved in DMSO in concentrated stocks. Nitrile gloves are impermeable to DMSO and therefore will prevent chemical hazard to the personnel performing the screening protocol.

2. Small molecules can be sensitive to light. Therefore, aluminum foil is used for covering the treated plate while incubating the embryos with drugs in the dark.

3. While preparing stock solution of Tricane, it is important to adjust the pH to 7–7.5. Because if not properly buffered, Tricane is toxic to fish.

4. Embryos at 28–36 hpf often present in a lateral view allowing direct imaging of the GFP expression in the MHB and dorsal retina, two domains that have the highest GFP levels.

5. In the case of zebrafish screens using $Tg(Dusp6:EGFP)^{pt6}$ the supply of 24hpf staged embryos is a critical bottleneck that defines throughput of the screen. For each screen, we set up on average 20–30 pairs (as described in Subheadings 3.1–3.3) to obtain enough healthy transgenic embryos to load at least 3 chemical library plates in duplicates (total 6 plates). We select

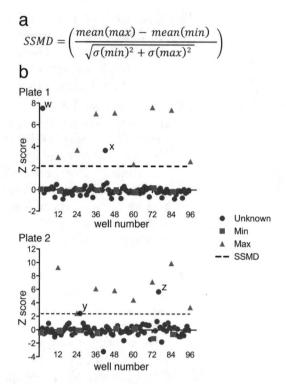

a

$$SSMD = \left(\frac{mean(max) - mean(min)}{\sqrt{\sigma(min)^2 + \sigma(max)^2}} \right)$$

b

Fig. 3 Quality control parameters used in determining positive hits in a chemical screening assay. (**a**) SSMD equation. (**b**) Example of two plates from the USFDA screen showing Z scores. The SSMD calculated for each plate was 2.17 and 2.38 for Plates 1 and 2, respectively. In Plate 1 compounds "w" and "x" (oxyquinoline) were deemed to be "hits" and in Plate 2 only compound "z" (pyrithione zinc) satisfied the criteria of having a Z score above the Plate SSMD. Plate 2 compound "y" was deemed marginal

only those embryos, which are developing synchronously in the right stage for treatment and expressing uniform GFP at the time for loading.

6. Up to 5 embryos can be loaded into each well of a 96-well plate in a typical chemical screen. With respect to the methodology we have developed for automated imaging and analysis, only single embryos are loaded per well. While loading the embryos and removing excess E3, care must be taken not to disrupt the chorions, as that can damage the embryos and lead to poor survivability.

7. We carefully count embryos from a single mating pair to ensure that there are enough to load one plate and avoid mixing embryos from different mating pairs. This ensures uniformity in the response to both negative (DMSO) and positive (BCI) controls, which yields better plate performance. The whole process of loading embryos and compounds into 96-well plates

takes approximately 30–40 min for each duplicate set. After 5 h treatment imaging of the plates in IXU high content reader as described in Subheading 3.3, **step 8–17** takes approximately 15–20 min per plate. Working with these timelines with one person, we were able to screen two to three chemical screening plates in duplicates per day.

8. Most of the screening libraries are usually diluted in DMSO. Zebrafish embryos and larvae tolerate up to 2.0–2.5% DMSO [25]. In our screening experiments, we have used 1% DMSO as the vehicle [18]. Other solvents include polyethylene glycol (PEG400), propylene glycol, ethanol and methanol are also found to be well tolerated (1–2.5%) by zebrafish embryos [25].

9. In order to circumvent the variability in fluorescence in transgenic embryos, $Tg(Dusp6:EGFP)^{pt6}$ homozygous lines were generated that are expected to follow Mendelian inheritance [20]. When these are outcrossed with wild-type AB or Tü, 100% of embryos are expected to exhibit equal GFP intensity. In maintaining homozygous $Tg(Dusp6:EGFP)^{pt6}$ lines there is a possibility of epigenetic silencing of reporter gene expression. To avoid this, we maintain the homozygous lines by alternatively outcrossing with AB or Tü, strains followed hemizygous inbreeding.

10. Plates are accepted if the SSMD between positive and negative controls exceed a value of 2, which is considered a "good" assay [26]. An added advantage of the SSMD is that, in contrast to Z-factor, the SSMD can be used to set a hit selection criterion [27]. To incorporate SSMD in our hit selection paradigm, z-scores, which denote how many standard deviations a sample is away from the plate mean were calculated for all wells, and compounds were classified as hits if their z-score exceeded that of the plate SSMD (Fig. 3b) [28]. With these criteria, the screen identified two FGF hyper-activators, oxyquinoline and pyrithione zinc [18]. Both compounds confirmed in dose-response and in orthogonal follow-up assays for FGF target gene activation, validating both the utility of the transgenic zebrafish model and the reliability of automated high content imaging and analysis for identifying novel FGF hyper-activators [18].

Acknowledgments

Our chemical screening work is supported by NIH grant (Grant numbers: NIH 1R01HD053287 and NIH 1R01HL088016). This project used the UPCI Chemical Biology Facility supported in part by award NIH P30CA047904. Our sincere thanks go to Andreas Vogt, UPDDI, University of Pittsburgh for providing the Images in Fig. 2 from IXU high content reader, CNT data for Fig. 3 and critically reading the manuscript.

References

1. Mendel G (1866) Versuche uber Pflanzen-Hybriden Verhandlungen des naturforschenden Vereines in Brunn 4:3–47
2. Morgan TH (1910) Sex limited inheritance in *Drosophila*. Science 32(812):120–122. doi:10.1126/science.32.812.120
3. Muller HJ (1927) Artificial transmutation of the Gene. Science 66(1699):84–87. doi:10.1126/science.66.1699.84
4. Capecchi MR (1989) The new mouse genetics: altering the genome by gene targeting. Trends Genet 5(3):70–76
5. Lawson ND, Wolfe SA (2011) Forward and reverse genetic approaches for the analysis of vertebrate development in the zebrafish. Dev Cell 21(1):48–64. doi:10.1016/j.devcel.2011.06.007
6. Struhl K (1983) The new yeast genetics. Nature 305(5933):391–397
7. West DB, Iakougova O, Olsson C, Ross D, Ohmen J, Chatterjee A (2000) Mouse genetics/genomics: an effective approach for drug target discovery and validation. Med Res Rev 20(3):216–230
8. Gosai SJ, Kwak JH, Luke CJ, Long OS, King DE, Kovatch KJ, Johnston PA, Shun TY, Lazo JS, Perlmutter DH, Silverman GA, Pak SC (2010) Automated high-content live animal drug screening using *C. elegans* expressing the aggregation prone serpin alpha1-antitrypsin Z. PloS One 5(11):e15460. doi:10.1371/journal.pone.0015460
9. Lendahl U, Orrenius S (2002) Sydney Brenner, Robert Horvitz and John Sulston. Winners of the 2002 Nobel prize in medicine or physiology. Genetic regulation of organ development and programmed cell death. Lakartidningen 99(41):4026–4032
10. Putcha GV, Johnson EM Jr (2004) Men are but worms: neuronal cell death in C elegans and vertebrates. Cell Death Differ 11(1):38–48. doi:10.1038/sj.cdd.4401352
11. Arias AM (2008) *Drosophila melanogaster* and the development of biology in the 20th century. Methods Mol Biol 420:1–25. doi:10.1007/978-1-59745-583-1_1
12. Manev H, Dimitrijevic N (2004) *Drosophila* model for in vivo pharmacological analgesia research. Eur J Pharmacol 491(2-3):207–208. doi:10.1016/j.ejphar.2004.03.030
13. Tomlinson MLHA, Wheeler GN (2012) Chemical genetics and drug discovery in Xenopus, Methods Mol Biol, vol 917, pp 155–166
14. Burns CG, Milan DJ, Grande EJ, Rottbauer W, MacRae CA, Fishman MC (2005) High-throughput assay for small molecules that modulate zebrafish embryonic heart rate. Nat Chem Biol 1(5):263–264. doi:10.1038/Nchembio732
15. de Groh ED, Swanhart LM, Cosentino CC, Jackson RL, Dai WX, Kitchens CA, Day BW, Smithgall TE, Hukriede NA (2010) Inhibition of histone Deacetylase expands the renal progenitor cell population. J Am Soc Nephrol 21(5):794–802. doi:10.1681/Asn.2009080851
16. Molina G, Vogt A, Bakan A, Dai W, Queiroz de Oliveira P, Znosko W, Smithgall TE, Bahar I, Lazo JS, Day BW, Tsang M (2009) Zebrafish chemical screening reveals an inhibitor of Dusp6 that expands cardiac cell lineages. Nat Chem Biol 5(9):680–687. doi:10.1038/nchembio.190
17. Peterson RT, Link BA, Dowling JE, Schreiber SL (2000) Small molecule developmental screens reveal the logic and timing of vertebrate development. Proc Natl Acad Sci U S A 97(24):12965–12969. doi:10.1073/pnas.97.24.12965
18. Saydmohammed M, Vollmer LL, Onuoha EO, Vogt A, Tsang M (2011) A high-content screening assay in transgenic Zebrafish identifies two novel activators of FGF signaling. Birth Defects Res C-Embryo Today 93(3):281–287. doi:10.1002/Bdrc.20216
19. Yui PB, Hong CC, Sachidanandan C, Babitt JL, Deng DY, Hoyng SA, Lin HY, Bloch KD, Peterson RT (2008) Dorsomorphin inhibits BMP signals required for embryogenesis and iron metabolism. Nat Chem Biol 4(1):33–41. doi:10.1038/Nchembio.2007.54
20. Molina GA, Watkins SC, Tsang M (2007) Generation of FGF reporter transgenic zebrafish and their utility in chemical screens. BMC Dev Biol 7:62. doi:10.1186/1471-213X-7-62
21. Wang G, Cadwallader AB, Jang DS, Tsang M, Yost HJ, Amack JD (2011) The rho kinase Rock2b establishes anteroposterior asymmetry of the ciliated Kupffer's vesicle in zebrafish. Development 138(1):45–54. doi:10.1242/dev.052985
22. Vogt A, Cholewinski A, Shen X, Nelson SG, Lazo JS, Tsang M, Hukriede NA (2009) Automated image-based phenotypic analysis in zebrafish embryos. Dev Dyn 238(3):656–663. doi:10.1002/dvdy.21892
23. Vogt A, Codore H, Day BW, Hukriede NA, Tsang M (2010) Development of automated imaging and analysis for zebrafish chemical screens. J Vis Exp 40. doi:10.3791/1900

24. Wheeler GN, Brandli AW (2009) Simple vertebrate models for chemical genetics and drug discovery screens: lessons from zebrafish and Xenopus. Dev Dyn 238(6):1287–1308. doi:10.1002/dvdy.21967

25. Maes J, Verlooy L, Buenafe OE, de Witte PA, Esguerra CV, Crawford AD (2012) Evaluation of 14 organic solvents and carriers for screening applications in zebrafish embryos and larvae. PLoS One 7(10):e43850. doi:10.1371/journal.pone.0043850

26. Zhang XD (2008) Novel analytic criteria and effective plate designs for quality control in genome-scale RNAi screens. J Biomol Screen 13(5):363–377. doi:10.1177/1087057108317062

27. Zhang XD (2007) A pair of new statistical parameters for quality control in RNA interference high-throughput screening assays. Genomics 89(4):552–561. doi:10.1016/j.ygeno.2006.12.014

28. Malo N, Hanley JA, Cerquozzi S, Pelletier J, Nadon R (2006) Statistical practice in high-throughput screening data analysis. Nat Biotechnol 24(2):167–175. doi:10.1038/nbt1186

Erratum to: High Content Screening

Paul A. Johnston and Oscar J. Trask

Erratum to:
Paul A. Johnston and Oscar J. Trask (eds), *High Content Screening:
A Powerful Approach to Systems Cell Biology and Phenotypic Drug Discovery*,
Methods in Molecular Biology, vol. 1683
DOI 10.1007/978-1-4939-7357-6

The affiliation of one of the volume editors, Paul A. Johnston, was incorrect in the originally published volume. It has been revised in the current version of the book to reflect the actual affiliation:

Department of Pharmaceutical Sciences, School of Pharmacy, University of Pittsburgh, Pittsburgh, PA, USA

The volume frontmatter reflects this change, as well as the online metadata.

The updated original online version for this book can be found at
DOI 10.1007/978-1-4939-7357-6

Paul A. Johnston and Oscar J. Trask (eds.), *High Content Screening: A Powerful Approach to Systems Cell Biology and Phenotypic Drug Discovery*, Methods in Molecular Biology, vol. 1683, DOI 10.1007/978-1-4939-7357-6_23,
© Springer Science+Business Media LLC 2018

INDEX

Paul A. Johnston and Oscar J. Trask (eds.), *High Content Screening: A Powerful Approach to Systems Cell Biology and Phenotypic Drug Discovery*, Methods in Molecular Biology, vol. 1683, DOI 10.1007/978-1-4939-7357-6, © Springer Science+Business Media LLC 2018

Printed in the United States
By Bookmasters